GM Food Systems and Their Economic Impact

CABI Biotechnology Series

Biotechnology, in particular the use of transgenic organisms, has a wide range of applications including agriculture, forestry, food and health. There is evidence that it could make a major impact in producing plants and animals that are able to resist stresses and diseases, thereby increasing food security. There is also potential to produce pharmaceuticals in plants through biotechnology, and provide foods that are nutritionally enhanced. Genetically modified organisms can also be used in cleaning up pollution and contamination. However, the application of biotechnology has raised concerns about biosafety, and it is vital to ensure that genetically modified organisms do not pose new risks to the environment or health. To understand the full potential of biotechnology and the issues that relate to it, scientists need access to information that not only provides an overview of and background to the field, but also keeps them up to date with the latest research findings.

This series, which extends the scope of CABI's successful "Biotechnology in Agriculture" series, addresses all topics relating to biotechnology including transgenic organisms, molecular analysis techniques, molecular pharming, in vitro culture, public opinion, economics, development and biosafety. Aimed at researchers, upper-level students and policy makers, titles in the series provide international coverage of topics related to biotechnology, including both a synthesis of facts and discussions of future research perspectives and possible solutions.

Titles Available

1. Animal Nutrition with Transgenic Plants
   Edited by G. Flachowsky
2. Plant-derived Pharmaceuticals: Principles and Applications for Developing Countries
   Edited by K.L. Hefferon
3. Transgenic Insects: Techniques and Applications
   Edited by M.Q. Benedict
4. Bt Resistance: Characterization and Strategies for GM Crops Producing *Bacillus thuringiensis* Toxins
   Edited by Mario Soberón, Yulin Gao and Alejandra Bravo
5. Plant Gene Silencing: Mechanisms and Applications
   Edited by Tamas Dalmay
6. Ethical Tensions from New Technology: The Case of Agricultural Biotechnology
   Edited by Harvey James
7. GM Food Systems and their Economic Impact
   Tatjana Brankov and Koviljko Lovre

GM Food Systems and Their Economic Impact

Tatjana Brankov

University of Novi Sad, Serbia

and

Koviljko Lovre

University of Novi Sad, Serbia

CABI is a trading name of CAB International

CABI
Nosworthy Way
Wallingford
Oxfordshire OX10 8DE
UK

CABI
745 Atlantic Avenue
8th Floor
Boston, MA 02111
USA

Tel: +44 (0)1491 832111
Fax: +44 (0)1491 833508
E-mail: info@cabi.org
Website: www.cabi.org

Tel: +1 (617)682-9015
E-mail: cabi-nao@cabi.org

A catalogue record for this book is available from the British Library, London, UK.

Library of Congress Cataloging-in-Publication Data

Names: Brankov, Tatjana, 1975- author. | Lovre, Koviljko, author.
Title: GM food systems and their economic impact / Tatjana Brankov and
    Koviljko Lovre.
Other titles: CABI biotechnology series ; 7.
Description: Boston, MA : CABI, 2018. | Series: CABI biotechnology series ; 7
    | Includes bibliographical references and index.
Identifiers: LCCN 2018034921| ISBN 9781789240542 (hardback) | ISBN
    9781789240566 (epub)
Subjects: LCSH: Food--Biotechnology. | Food--Biotechnology--Economic aspects.
Classification: LCC TP248.65.F66 B73 2018 | DDC 664/.024--dc23 LC record available at
    https://lccn.loc.gov/2018034921

ISBN-13: 9781789240542 (hardback)
        9781789240559 (ePDF)
        9781789240566 (ePub)

Commissioning Editor: David Hemming
Editorial Assistant: Tabitha Jay
Production editor: Marta Patiño

Typeset by SPi, Pondicherry, India
Printed and bound in the UK by Severn, Gloucester

English Editors

Jonathan Boulting, MA English Literature, Trinity College, Cambridge.
Michael Wise, BSc, University of Portsmouth.

Contents

Contributors

Brankov, Tatjana, Department of Agricultural Economics and Agribusiness, Faculty of Economics, University of Novi Sad, Serbia. E-mail: tbrankov@ef.uns.ac.rs

Lovre, Koviljko, Department of Agricultural Economics and Agribusiness, Faculty of Economics, University of Novi Sad, Serbia. E-mail: klovre@ef.uns.ac.rs

Preface

We are at a unique moment in history, as intensive transgenic agriculture and its products are being widely distributed, despite the lack of a broad consensus among the public and in the scientific community. The Gene Revolution has expanded the possibilities opened up by the Green Revolution, and led to further industrialization of agriculture. Unlike the Green Revolution, which was designed to help small farmers in developing countries but was also suitable for large-scale farms, it seems that the Gene Revolution led by multinational companies will never reach the small farmers. Thus, in contrast to the Green Revolution that spread enhancing technologies without meeting any organized resistance, transgenic technology has provoked the organization of social movements, and non-adopter farmer countries have become symbols of freedom in the modern food system. However, the motives that determine whether a country will become an adopter or non-adopter of transgenic technology, and the consequences of adoption at the national, regional and global levels, as well as the consequent challenges and changes to the modern food system, have not been fully studied. This monograph attempts to fill this major knowledge gap. Until now, the majority of the discussion over the spread of transgenic crops has been conducted by either extreme proponents or extreme opponents of the technology. This book instead seeks a middle ground. Working from the large body of economics literature, we have demonstrated that: Genetically Modified Organisms (GMOs) have broken almost all the barriers in the food system; national GMO politics should be analysed as an integral part of overall agricultural politics; and food regime concerns in relation to GMOs take precedence over health and safety concerns. With our 'middle-ground' aim in mind, we discuss transgenic technology on both a macroeconomic (global and national) and a microeconomic level. In microeconomic terms, this means looking at individual crops and their cost structure. A unique feature of the book is that we have synthesized the most recent knowledge about all the GMOs approved for direct use in food or as additives. This provides the readers with a window into the various possibilities of exposure to transgenic ingredients in different countries. In addition, we explore the impact of transgenic technology on food production and its prices as well as its agrochemical use on a global scale, establishing a clear distinction between two periods – the period before and the period after the commercialization of GMOs.

This book consists of six chapters. Chapter 1 presents a description of the historical evolution of biotechnology, and gives definitions of key terms. Chapter 2 moves into the key elements of the book: how transgenic technology issues can be understood in the light of food regime concepts and lead to different experience with GMOs in different countries. It analyses GMO neoregulation in the USA, the EU (with particular reference to Spain), Brazil, Russia, China, India, South Africa, and Serbia within a broader context, as an integral part of general agricultural policy. This chapter therefore provides a *brief description of these countries' agricultural performances*, agricultural supports and trade relations. Each selected country has its own specific policy regarding GMOs. From the very beginning, US government policy has strongly favoured the use of Genetically Modified (GM) varieties. The USA leads the list of countries in the production of GM crops, and rules GM foods substantially equivalent to conventionally produced foods. On the other side, EU countries apply the precautionary principle as a guiding approach for trans-border movement of GMOs. GMO cultivation in the EU is very limited because of concerns expressed by stakeholders about adverse effects on the environment, farmland and biodiversity. The central hub of transgenic technology in the EU is located in Spain, which produces almost 95% of the EU's transgenic crops. Brazil is the second largest producer of transgenic crops in the world, planting transgenic soybean, maize and cotton. The country's position on GMO issues is quite similar to that of the USA, but it does not fully adhere to the concept of substantial equivalence, since products containing more than 1% of transgenic ingredients need to be labelled. Under the current GMO legislative framework Russia *de jure* prohibited transgenic crop cultivation. Although this country has one of the most restrictive laws in the world, it also needs to import transgenic feeds for its growing livestock sector. India produces only one GM crop – cotton – in significant quantities. Unlike many other countries, it does not depend on soybean feedstuffs, but it is heavily dependent on soybean oil import used as food. India is an example of resistance to international pressure in order to protect indigenous peoples. To ensure equitable benefit sharing with farmers, it has developed a sui generis system for protection of plant varieties. Despite being a producer, China is considered to be a country whose policies slow down further diffusion of transgenic crops. China is unique because it produces only transgenic crops obtained from its own research, meaning that no foreign crops have been approved for commercialization. On the other hand, the country allows importation of transgenic crops to be used as feed and for processing. South Africa is the first and the largest producer of transgenic crops in Africa, and has a GMO policy similar to the USA. South Africa has failed, however, to significantly improve its transgenically produced food exports and has remained dependent on imports. This example shows that a strong neoliberal stance on GMOs is not a guarantee for success. Finally, Serbia has unique GMO policies compared with most countries in the world. In Serbia, the production and commercialization as well as importation of transgenic crops and products is strictly forbidden by law. Moreover, it has no urgent need for GMO soybeans to feed livestock. Chapter 3 considers in detail an important question – 'Does transgenic food production affect world food prices?' Long-term trends in international prices, the effects of increased Asian demand on food prices, and US ethanol and maize prices constitute some of its subsections. Food security at the global and regional levels, and the linkages between GMOs and world hunger, are explored in Chapter 4. The question, 'Is transgenic technology an environmentally friendly technology?' is the subject of the discussion in Chapter 5. Pesticide and fertilizer usage and efficiency over a period spanning five decades is analyzed, with a clear distinction between the two periods, i.e. before and after the adoption of transgenic technology. In addition, a comparison of pesticide consumption in GMO and non-GMO producing countries is also made. Finally, Chapter 6 demonstrates all the possible ways for transgenic ingredients to get into the food chain, and draws up a list of finished products that potentially contain transgenic ingredients.

This monograph is an attempt to address the issue of transgenic technology diffusion in a sophisticated but neutral way. It synthesizes current knowledge about the GM food systems and provides many illustrative examples to better understand the factors affecting them. The book is intended for a wide range of professionals and researchers whose interests relate to all aspects of the global food system, including policy makers, policy advisers and analysts, NGOs, students, and other interest groups. At the end, a *glossary* of all the uncommon and specialized terms will enable the book to be comprehensible to anyone interested in a better understanding of GMOs.

Tatjana Brankov
Belgrade, Serbia

Koviljko Lovre
Novi Sad, Serbia

List of Figures

List of Tables

List of Acronyms

_____

2,4-D	2,4-Dichlorophenoxyacetic acid
ABPI	Brazilian Intellectual Property Association
ACS	American Chemical Society
AGY	Altered growth/yield
AMS	Aggregate measure of support
APHIS	Animal and Plant Health Inspection Service
AST	Abiotic stress tolerance
BC	Before Christ
BRICS	Brazil, Russia, India, China and South Africa
Bt	*Bacillus thuringiensis*
cal BP	Calendar years before the present
CAGR	Compound annual growth rate
CAP	Common Agriculture Policy
CBD	Convention on Biological Diversity
CCC	Commodity Credit Corporation
CFA	Committee on Food Aid Policies and Program
CFS	Committee on Food Security
CIS	Commonwealth of Independent States
CMV	*Cucumber mosaic virus*
CPI	Consumer Price Index
CPVO	Community Plant Variety Office
DAP	Diammonium phosphate
DNA	Deoxyribonucleic acid
DR	Disease resistance
EAEU	Eurasian Economic Union
EAP	Economically active population
EC	European Commission
EEC	European Economic Community
EFSA	European Food Safety Authority
EPA	Environmental Protection Agency
EPC	European Patent Convention
ERS	Economic Research Service

EU	European Union
F1	The first generation
F2	The second generation
FAO	Food and Agriculture Organization
FDA	Food and Drug Administration
FFDCA	Federal Food, Drug and Cosmetic Act
FPA	Food for Peace Act
G6	France, Germany, Italy, Japan, the UK, the USA
GAIN	Global Agricultural Information Network
GATT	General Agreement on Tariffs and Trade
GDP	Gross domestic product
GE	Genetically engineered
GFR	Gross farm receipts
GHI	Global Hunger Index
GIEWS	Global International Early Warning System
GM	Genetically modified
GMOs	Genetically modified organisms
GSSE	General services support estimate
HFCS	High-fructose corn (maize) syrup
HIV	*Human immunodeficiency virus*
HPP	Hydrolysed plant protein
HSP	Hydrolysed soy protein
HT	Herbicide tolerant
HVFs	High-value foods
HVP	Hydrolysed vegetable protein
IDEC	Brazilian Institute for Consumer Defense
IEFR	International Emergency Food Reserve
IFAD	International Funds for Agricultural Development
IFPRI	International Food Policy Research Institute
IMF	International Monetary Fund
IPR	Intellectual Property Rights
ISAAA	International Service for the Acquisition of Agri-Biotech Applications
ITO	International Trade Organization
LDCs	Least-developed countries
LMOs	Living modified organisms
MDG	Millennium development goal
MPQ	Modified product quality
MPS	Market price support
mRNA	Messenger RNA
MSG	Monosodium glutamate
NACs	New agricultural countries
NATO	North Atlantic Treaty Organization
NCDs	Non-communicable diseases
NGOs	Non-governmental organizations
NICs	Newly industrialized countries
NPK	Nitrogen-phosphorus-potassium (fertilizer)
OECD	Organization for Economic Cooperation and Development
OPEC	Organization of the Petroleum Exporting Countries
PCR	Polymerase chain reaction
PCS	Pollination control system
PGR	Plant growth regulator

PPP	Purchasing power parity
PPV	Plum pox virus
PRSV	Papaya ringspot virus
PSE	Producer support estimate
PVPA	Plant Varieties Protection Act
RASFF	Rapid Alert System for Food and Feed
R&D	Research and development
RFS	Renewable Fuel Standard
RNA	Ribonucleic acid
RR	Roundup ready
SPS	Sanitary and Phytosanitary Measures Agreement
SRISTI	Society for Research and Initiatives for Sustainable Technologies and Institutions
TRIPS	Agreement on Trade-Related Aspects of Intellectual Property Rights
TSE	Total support estimate
TSP	Triple superphosphate
UN	United Nations
UPOV	International Union for the Protection of Plant Varieties
US	United States (of America)
USA	United States of America
USDA	United States Department of Agriculture
WFC	World Food Council
WHO	World Health Organization
WTO	World Trade Organization

1 Ancient, Classical and Modern Biotechnology

The process of mankind's development and food production has proceeded *pari passu* through the millennia. As Sauer pointed out, 'man evolved with his food plants, forming a biological complex', regardless of whether the man was a hunter-gatherer, a food-domesticator or a modern large-scale food manufacturer (Sauer, 1963, p. 155).

The first changes took place seamlessly, without increasing the food supply and without significant changes in the environment. An exception was the use and control of fire by early man, which caused the transformation of forest into grassland, permitted large migrations by enabling people to extend their ranges into habitats that were impossible to live in before, extended the period of activity independent of daylight, provided protection from predators and insects, and caused the evolution of man's digestive system due to adjustments to cooked food. Wrangham *et al.* (1999, p. 573) assumed that cooking 'doubled the energy value from carbohydrate in underground storage organs and increased it by 60% in seed.'

Animal and plant domestication had a special role in the further development of food production. Domestication can be defined as 'a complex evolutionary process in which human use of plant and animal species leads to morphological and physiological changes that distinguish domesticated *taxa* from their wild ancestors' (Purugganan and Fuller, 2009, p. 843). Domestication provided the impetus for humans to create a food surplus and build the world's first villages and cities near fields of domesticated plants. Consequently, 'this led to craft specializations, art, social hierarchies, writing, urbanization and the origin of the state' (Purugganan and Fuller, 2009, p. 843).

Many historians think domestication happened between 10,000 and 13,000 years ago. Numerous indications and evidence in the present suggest that the domestication of animals had to be preceded by the domestication of plants. Such examples can be found in the steppes of Iran and Afghanistan, or the Maasai ethnic group inhabiting southern Kenya and northern Tanzania who live in ways similar to man in the Neolithic period. There are numerous archaeological studies identifying the dynamics of domestication. The hunter-gatherer societies independently began food production in nine areas of the world: the Fertile Crescent (extending from the eastern Mediterranean upward through Anatolia and down into the valley of the Rivers Tigris and Euphrates) (Fig. 1.1), China, Mesopotamia, Andes/Amazonia, the American East, the Sahel, Tropical West Africa, Ethiopia and New Guinea (Fig. 1.2) (Diamond, 2002).

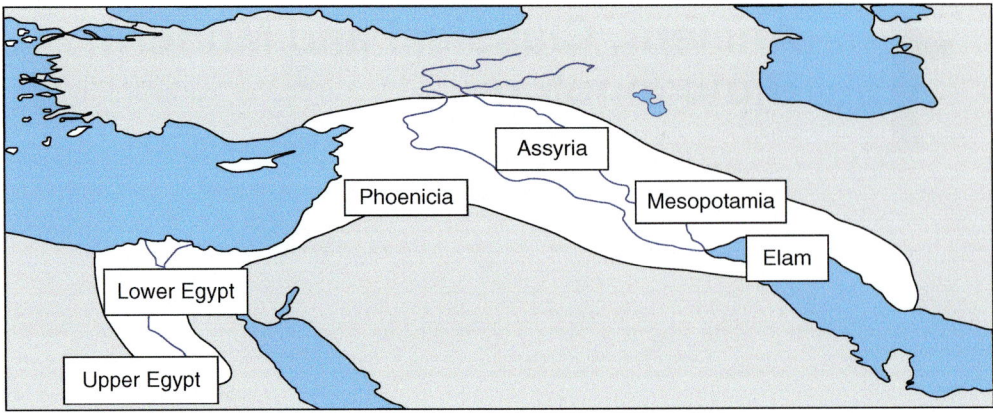

Fig. 1.1. Fertile Crescent.

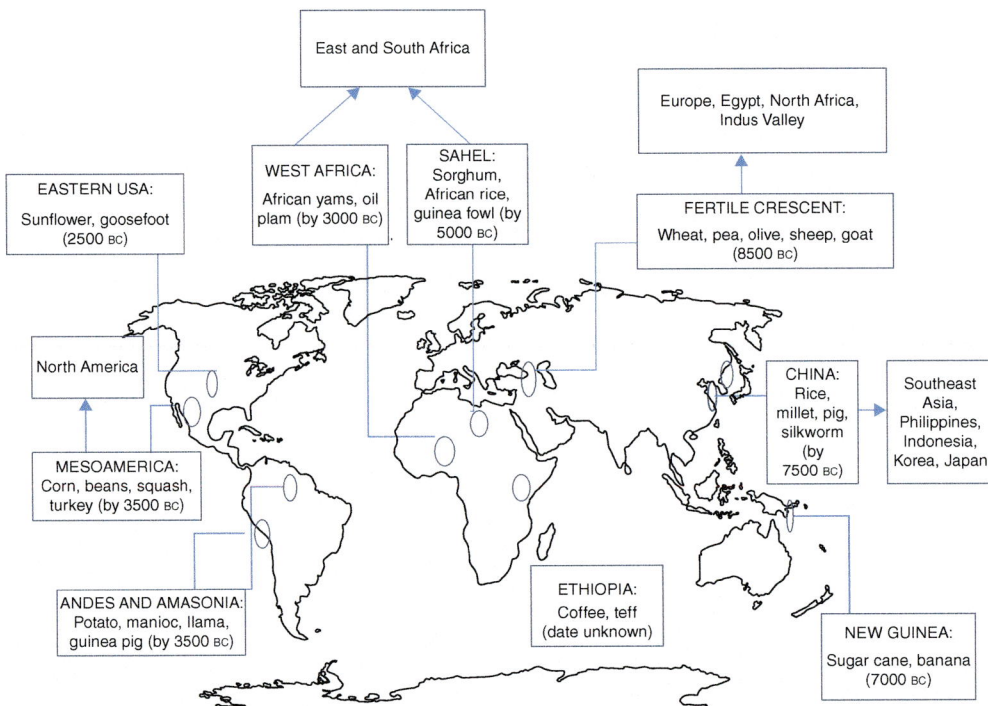

Fig. 1.2. Independent evolution of food production – the earliest known date of local domestication and the spread of food production. (Adapted from Diamond, 1997.)

Eight plants are considered to be the domesticated 'founder crops': three cereals (einkorn wheat *Triticum monococcum*, emmer wheat *Triticum turgidum* subsp. *dicoccum*, and barley *Hordeum vulgare*), four pulses (lentil *Lens culinaris*, pea *Pisum sativum*, chickpea *Cicer arietinum*, and bitter vetch *Vicia ervilia*), and a single oil and fibre crop (flax *Linum usitatissimum*) (Weiss and Zohary, 2011). The origins of our modern wheat developed by cultivating its wild ancestors *Triticum boeoticum* and *Triticum monococcum* in the Karacadag mountain region (southeastern Turkey). Emmer wheat was also domesticated in the same region in Turkey, while the earliest barley (wild relative *Hordeum spontaneum*) is recorded in Syria. Syria was also an early home for lentils developed out of *Lens c. orientalis* and chickpeas. The first appearance of the domestic pea is in the Near East, while the earliest flax seeds came from Jericho. Not counting the domestic dog, it is believed that animal domestication started with sheep (*Ovis aries*), the first 'meat' animals adapted from different wild subspecies of *Ovis gmelini* (wild mouflon) in the Fertile Crescent; then most probably came goats (*Capra hircus*), from *Capra aegargus* (bezoar ibex wild goat) in Turkey, Iran, Pakistan and China (Luikart et al., 2001). Wild cattle (*Bos primigenius*) most likely had three main loci of domestication: the Taurus Mountains (*Bos taurus*), Indus Valley (Pakistan) (*Bos indicus*), and Algeria (*Bos africanus*) (Decker et al., 2014).

Agriculture was launched in the Fertile Crescent, the home of the most valuable crops and livestock, such as wheat, barley, peas, sheep, goats, cows and pigs. As can be seen in Fig. 1.2, the Fertile Crescent has the earliest dates of animal and plant local domestication (8500 BC), followed by China, New Guinea and the Sahel.

The earliest date of domestication recorded in the American East is 6000 years after the Fertile Crescent (2500 BC). From these centres, food production spread around the globe at different speeds, firstly at locations with similar climate and habitats, but 'with the general axis oriented east–west for Eurasia and north–south for the Americas and Africa' (Diamond, 2002, p. 703). The above does not mean that one can clearly delineate the period between hunter-gathering and farming, because there was and actually still is some overlap between them. Along with the hunter-gathering indigenous cultures of the Pacific Northwest Coast living in a rich environment, the Apache also practised some farming. As Suttles (2009, p. 56) has suggested '...the Northwest Coast peoples seem to have attained the highest known levels of cultural complexity achieved on a food-gathering basis and among the highest known levels of population density. The Northwest Coast refutes many seemingly easy generalizations about people without horticulture or herds'. Knowledge about the existence of social inequality amongst the population that survived without animal

and plant domestication could be considered as one of the most important advances in anthropological research in the last few decades (Sassaman, 2004).

If we exclude complex hunter-gatherer social formations that already practised sedentary or semi-sedentary lifestyles, nomadic hunter-gatherers started their sedentary lifestyle by applying farming practices to their permanent agricultural areas instead of migrating to follow the seasonal shifts in wild food:

> *The sedentary lifestyle permitted shorter birth intervals. Nomadic hunter-gatherers had previously spaced out birth intervals at four years or more, because a mother shifting camp can carry only one infant or slow toddler... Food production also led to an explosion of technology, because sedentary living permitted the accumulation of heavy technology (such as forges and printing presses) that nomadic hunter-gatherers could not carry, and because the storable food surpluses resulting from agriculture could be used to feed full-time craftspeople and inventors.*
> (Diamond, 2002, p. 703)

The advent of agriculture and, with it, of food surpluses, increased the density of population; caused epidemics of infectious diseases; led to social stratification, political centralization and the formation of standing armies; and led to nutritional changes and adaptation to a diet quite different from that of the hunter-gatherer: 'a diet rich in simple carbohydrates, saturated fats and calories and salt, and lower in fibre, complex carbohydrates, calcium and unsaturated fats' (Diamond, 2002, p. 704).

Unconscious selection of plants for desirable traits (around 9000 BC) resulted in the elimination of dormancy and seed dispersal, and actually led to dependable germination and the predictable continuity of plants in the field. Conscious plant cultivation for desirable traits started, also BC. Conscious cultivation led to the diversification of crops and their local adaptation, higher seed yield combined with a higher degree of self-pollination, and many other traits related to consumer acceptance, such as culinary preferences and quality processing. Domestication continued after the birth of Christ and is still going on; the only changing factor is the technology for obtaining the desired properties. As Darwin observed in 1868: 'No doubt man selects varying individuals, sows their seeds, and again selects their varying offspring…Man therefore may be said to have been trying an experiment on a gigantic scale; and it is an experiment which nature during the long lapse of time has incessantly tried' (Darwin, 2010, p. 2).

1.1 Historical Evolution of Biotechnology

Among other things, domestication stimulated the magnification of food storage, a practice already followed in the pre-agricultural Near Eastern Early (14,500–12,800 cal BP) and Late Natufian periods (12,800–11,500 cal BP). Food storage coincided with the growth of microorganisms that caused the birth of the first biotech food applications – food fermentation. Fermentation can be defined as 'the transformation of simple raw materials into a range of value-added products by utilizing the phenomenon of growth of microorganisms and/or their activities on various substrates' (Prajapati and Nair, 2008, p. 2).

Fermentation is considered to be the world's oldest food preservation method apart from the drying of food in the hot sun, with roots dating back deep into the past of the Middle East (Nummer, 2002). The first fermented products were created from stored milk surplus. The earliest evidence dates back as far as 7000 BC in the Fertile Crescent, and refers to cheese making. Archaeological records confirm the dates and origins of other fermented products as follows: wine-making process – Western Iran (6000 BC); wheat bread making – Egypt (3,500 BC); preparation of meat sausages – Babylonia (1,500 BC); sour rye breads – Europe (800 BC); preservation of vegetables – China (300 BC). Thanks to their understanding of how to use yeasts separated from wine to prepare bread, the Romans opened 250 bakeries around 100 BC. The oldest known (before 3000 BC) man-made animal 'hybrid' product of mating between two different species is

the mule, a crossbreed between a male donkey and a female horse. A slightly less common 'hybrid', the hinny, was also bred in ancient times by the mating of a female donkey with a male horse. Highly valued in trade, the *perdum*-mule, a riding animal used mostly by kings, is found in Central Anatolia in the ancient town of Kaniš, and dates back to the 19th and 18th centuries BC (Michel, 2002).

Each of these 'discoveries' is accompanied by a legend. For example, it is believed that an Arabic trader accidentally discovered a way of making cheese. Preparing for a long journey through the desert, he put milk in a bag made of sheep stomach. Because of the rennet in the bag and the sun in the desert, the milk separated into curd and whey. Angels were said to have helped an ancient nomadic Turk prepare the first yoghurt, while the recipe for making kefir, which originates from the Caucasus Mountains, was kept secret for a long time because Mohammad strictly forbade transmitting it to people of faiths other than Islam. In the Old Testament, the mule replaced the donkey as the 'royal beast' and was ridden by King David and King Solomon at their coronations.

Leaving aside the legends and often accidental discoveries, it can be noticed that man travelled a long, gradual path from the hunter-gatherer to the modern producer dependent on technological innovation (Fig. 1.3).

The essential understanding of fermentation, which could not be performed without microorganisms, happened in the 17th century (around 1678), when Antony Van Leeuwenhoek developed his method of creating powerful lenses and applied them to the study of the microscopic world. The discovery of the microscope 'was the milestone for the development of classical biotechnology' (Pele and Cimpeanu, 2012, p. 6), and was a prerequisite for Louis Pasteur, in the middle of the 19th century, to do his great research into microbial fermentation, which revolutionized medicine, industry and agriculture.

The era of classical biotechnological evolution lasted until the 1970s. Although in 1665 the English scientist Robert Hooke had discovered cells while looking at a tiny slice of cork and van Leuwenhoek had discovered

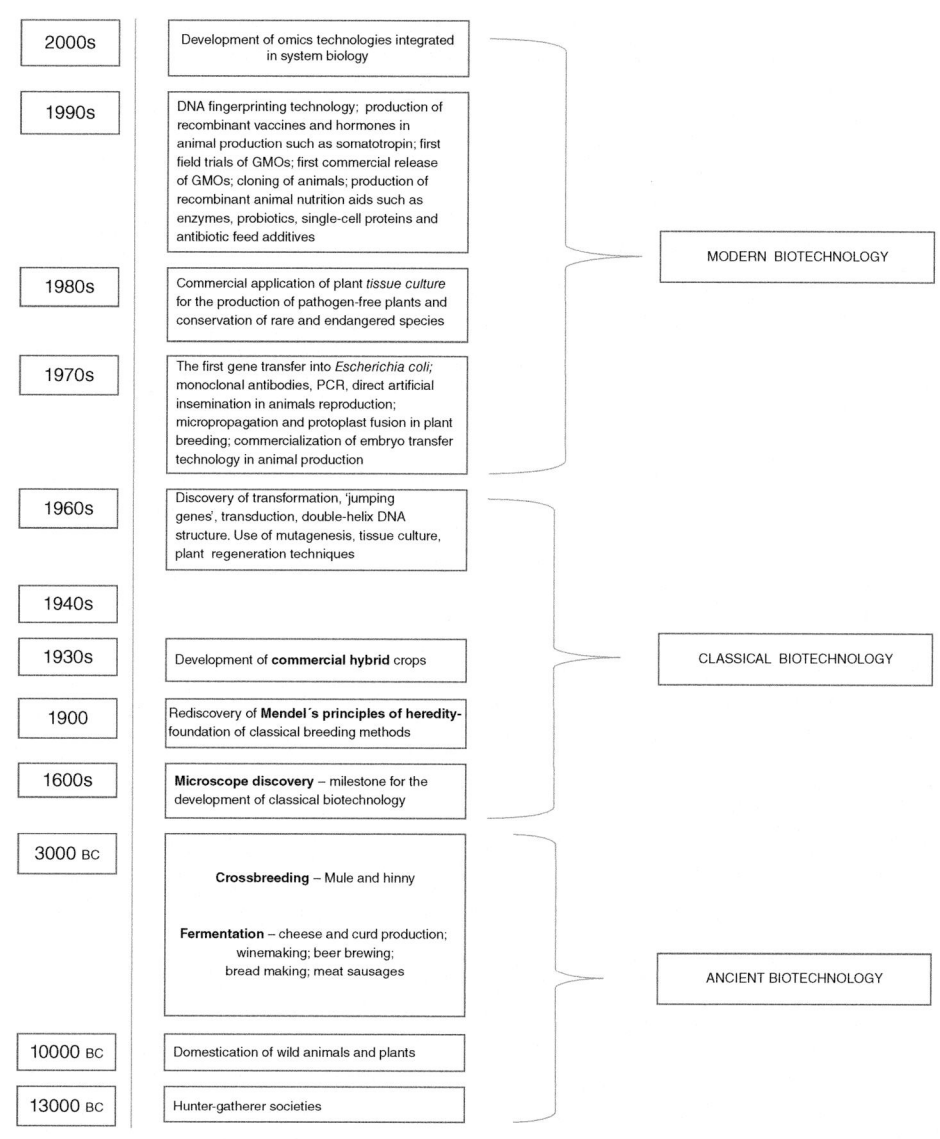

Fig. 1.3. The timeline of agriculture.

single-celled organisms in 1673 using a handmade microscope, further progress in cell investigation occurred almost 200 years later. Theodor Schwann (1810–1882) and Matthias Schleiden (1804–1881), the founders of cell theory (1838–1839), confirmed that all organisms are composed of one or more cells, and that cells are the basic units of life in all living things. The final major contribution to cell theory was made by the German pathologist Rudolph Virchow (1821–1902), who summed up his research as follows: 'Where a cell arises, there a cell must have previously existed (*Omnis cellula e cellula*), just as an animal can spring only from an animal, a plant only from plant' (Moore, 1999, p. 261).

Six years after Charles Darwin announced his theory of evolution to the world in 1859, in the book *On the Origin of Species by Means of Natural Selection*, Gregor Mendel presented his results on peas obtained after 12 years of systematic investigations in the famous paper *Versuche über Pflanzenhybriden* [Experiments in Plant Hybridization]. Mendel discovered the basic principles of heredity, and established the Law of Segregation (that there are *dominant* and *recessive* traits passed on randomly from parents to offspring) and the Law of Independent Assortment, later known as Mendel's Laws on Inheritance (that traits were passed on independently of other traits from parent to offspring). He also proposed that this heredity followed basic statistical laws (Box 1.1). The scientific community failed to recognize the huge importance of this research for 34 years, until the three botanists Carl Erich Correns, Erich Tschermak and Hugo de Vries in their research independently came to the same

conclusion about inheritance as Mendel. In this way, Mendel's work was rediscovered in 1900 and Mendel himself posthumously became 'the father of genetics'.

During the era of classical biotechnology the development of vaccines and immunization started. Edward Jenner, a country doctor living in Berkeley, England, created the world's first vaccination for smallpox in 1796 (Baxby, 1981). His assertion 'that the cow-pox protects the human constitution from the infection of smallpox', published in the work *Inquiry into the Causes and Effects of the Variolae Vaccine*, laid the foundation for modern vaccinology (Benenson *et al.*, 1952). Louis Pasteur (1822–1895), a French chemist and biologist, successfully tested the first human vaccine created in a laboratory, made of an extract gathered from the spinal cords of rabies-infected rabbits in 1885. Another

---

Box 1.1. Mendel's experiment on *Pisum sativum*. (Adapted from Mendel, 1865, pp. 6–11).

The pea plant that was the subject of Mendel's experiment can either self-pollinate or cross-pollinate, because it has both male and female reproductive organs. Mendel successfully cross-pollinated pure bred plants with particular traits in order to observe the offspring over many generations. For the purpose of the experiment different characteristics of peas were selected: difference in the form of the ripe seeds, difference in the colour of the seed albumen, difference in the colour of the seed-coat, difference in the form of the ripe pods, difference in the colour of the unripe pods, difference in the position of the flowers, difference in the length of the stem. Each pair with differentiating characteristics were united by cross-fertilization.

1st experiment 60 fertilizations on 15 plants.
2nd experiment 58 fertilizations on 10 plants.
3rd experiment 35 fertilizations on 10 plants.
4th experiment 40 fertilizations on 10 plants.
5th experiment 23 fertilizations on 5 plants.
6th experiment 34 fertilizations on 10 plants.
7th experiment 37 fertilizations on 10 plants.

Mendel found that the hybrid plants obtained looked like only one parental strain, but produced progeny that resembled both parental strains. Mendel referred to the trait that was expressed in the F_1 plants as *dominant* and to the alternative form, which was not expressed in the F_1 plants, as *recessive*. In the first generation from the homozygous parent plants for all of the crosses, the progeny resembled both parental strains in the ratio of three dominants to one recessive. The plants in the F_1 generation were all heterozygous. The same ratio apparently occurs in the later F_2 generations as well, but Mendel distinguishes that 'it is actually 1:2:1, the ratio of true-breeding dominant to non-true-breeding dominant to true-breeding recessive'. The following figures show the example of yellow and green pea seeds (Y–yellow allele, G–green allele):

	YY GG		Parental generation
	YG YG YG YG		F_1 generation
	YY YG YG GG		F_2 generation
YY YY YY YY	YY YG YG GG	YY YG YG GG	GG GG GG GG

of his achievements was the development of a vaccine for anthrax, a breakthrough important for agriculture (Pasteur and Chamberland, 2002).

A lot of genetics-related discoveries were made in the 19th century: in 1831 a botanist, Robert Brown, discovered the cell nucleus and described it in the paper *On the Organs and Mode of Fecundation in Orchideae* and *Asclepiadeae*; and in 1868 a Swiss doctor, Frederich Miescher, performed laboratory experiments that led to deoxyribonucleic acid (DNA) discoveries (because he had isolated it from the cells' nuclei, he named it nuclein). However, an understanding of the importance of Miescher's discovery, which can be seen as the birth of molecular genetics, did not occur until 75 years later. In 1881 at the International Medical Congress in London, Robert Koch, a German doctor, demonstrated a new solid medium technique which could be used both to isolate pure cultures of bacteria and to sub-culture them. One year later Fannie Hesse suggested replacing gelatin with agar, and after that agar became the most commonly used solid medium because of its remarkable physical properties. Clearer than gelatine, it resists digestion by bacterial enzymes, melts when heated to around 85°C, and yet when cooled, does not gel until it reaches 34–42°C. And in 1888, the famous German anatomist Heinrich Wilhelm Gottfried von Waldeyer-Hartz coined the word 'chromosome', in his paper *Über Karyokinese und ihre Beziehungen zu den Befruchtungsvorgängen* [Karyokinesis and its Relation to the Process of Fertilization].

After 1900, 'systematics, evolution, physiology, cytology, embryology and practical breeding were in close contact at the birth of genetics' (Roll-Hansen, 2014, p. 2432). In 1902, a German physiologist, the father of tissue culture Gottlieb Haberlandt, developed the concept of *in vitro* cell culture (whereby 'cultured plant cells could grow, divide and develop into embryo and then to whole plant', or, in one word, 'totipotency', as coined by Steward in 1968) (Rai, 2007). In 1909, the Danish biologist Wilhelm Johannsen coined the terms 'gene' (an abbreviation of Darwin's and De Vries' *pangene*, from Greek *gennao*, to breed), 'genotype'

(from *gennao*, to breed, and *typos*, an imprint), and 'phenotype' (Greek *phain-omai*, to appear and *typos*, an imprint). He used the word gene to refer to the discrete determiners of inherited characteristics: 'The word gene is fully free from any hypothesis; it only expresses the securely ascertained fact that at least many properties of the organism are conditional on individual, separable and thus independent "states", "bases", and "dispositions" found in the gametes – briefly, just what we want to call genes.' (Johannsen, 1909) [English translation (by N.R.)].

In the 3rd edition of his book, Johannsen wrote a sentence similar to what appeared the 2nd edition from 1913: 'On one side there is the genotype as the constitution of the organism; on the other side, there is the environment – and the often rather complicated cooperation between the genotype and the environment conditions the realized personal character of any organism, its phenotype.' (Johannsen, 1926).

In 1926, Thomas Hunt Morgan, an American embryologist and Nobel Prize winner, used the name 'gene' in his book *The Theory of the Gene*, where he described the five principles of the gene: segregation, independent assortment, crossing over, linear order and linkage groups. In the same year, Fritz Went discovered the first plant growth regulator (PGR), indoleacetic acid, important for further improvement in tissue culture (Rai, 2007). A new era in genetics development was opened by the Nobel Prize winner Hermann Joseph Muller, an American geneticist known as the founder of the field of radiation genetics. Muller's 1927 paper, *Artificial Transmutation of the Gene*, demonstrates that X-rays can induce mutations in the fruit fly *Drosophila melanogaster* at a much higher frequency than occurs in nature: within a few weeks, more than 100 mutations were discovered in the resulting progeny, about half the number of all mutations discovered in *Drosophila* over the previous 15 years. H. J. Muller was also the first scientist to propose that bacteriophages might be related to genes; and he made a prophetic statement in 1922: 'Perhaps we may be able to grind genes in a mortar and cook them in a beaker after all' (Sapp, 2003, p. 141).

The first decades of the 20th century brought big changes in agricultural production, with the invention and commercialization of high-yield hybrid crops. Twenty years after Harrison Shull and Edward Murray East, working independently in 1908, rediscovered and deepened Charles Darwin's research into inbreeding depression and hybrid vigour in maize ('heterosis' was the term Shull coined for this process), and 10 years after Donald F. Jones, in 1918, made an announcement of the double cross method, US farmers started to produce hybrid crops on a large-scale. Diffusion of hybrid crops in the 1930s had an important socioeconomic impact, because farmers began to abandon the practice of keeping their own seeds and started buying hybrid maize seed on an annual basis (Duvick, 2001).

The first half of the 20th century brought discoveries that saved millions of lives. In 1928, Sir Alexander Fleming discovered penicillin and described the results of the experiments in a paper for the *British Journal of Experimental Pathology* (Fleming, 1929). Twelve years later, in 1940, Howard Florey and Ernst Chain at Oxford University performed a rat and mouse test and published their results in the *Lancet*. The test was of huge importance, because human testing of penicillin began soon after that. The discovery of penicillin led to the exploration of many other antibiotics and metabolites and was a huge step technically towards the first scaled-up microbial mass culture under sterile conditions (Fiechter, 2000). This saw the start of 'the golden age of industrial microbiology', and a new phase in the development of a large number of commercially important primary and secondary metabolites (Demain and Fang, 2000).

The mid-20th century was characterized by findings which paved the way for the development of genetic engineering. In 1944, Oswald T. Avery, Colin M. MacLeod and Maclyn McCarty reported that DNA is the substance that causes bacterial transformation (*Streptococcus pneumoniae*); in 1950 Barbara McClintock, a pioneer in the field of cytogenetics, discovered transposable elements known as 'jumping genes' using maize as a model organism, and proved that 'the genome is not a stationary entity, but rather is subject to alteration and rearrangement' (Pray and Zhaurova, 2008, p. 169). Two years later, Joshua Lederberg and Norton Zinder discovered transduction, the process by which a virus transfers genetic material from one bacterium to another. A groundbreaking moment in the development of genetics and the cornerstone of the development of molecular genetics, which would lead to the creation of genetically modified organisms (GMOs) a few decades later, was the discovery of the structure of DNA. In 1953 Francis Crick, James Watson and Maurice Wilkins discovered that DNA consists of two chains twisted around each other – double helixes, but in the opposite direction. The circular structure of *Escherichia coli* DNA and enzyme polymerase was discovered in 1957. Interbacterial gene transfer was first described in Japan by Ochiai and colleagues in a publication in 1959 that demonstrated the transfer of antibiotic resistance between different species of bacteria *Shigella* and *Escherichia* with a plasmid, extrachromosomal circular DNA (Kasuya, 1964). This transfer has proved very important for the development of transgenic techniques. The 1960s brought discoveries about the regulation of gene expression and protein synthesis. A DNA segment (lac region) was moved from *Escherichia coli* to another microorganism. Thus it was shown that genes can be transferred and chromosomes can be redesigned (Beckwith and Signer, 1966).

In the 1960s other fields of science also evolved, each at their own pace. The field of biochemical engineering began to develop after Hixson and Gaden's article about oxygen transfer was published in 1950 (Hixson and Gaden, 1950). In the same period, 'the field of chemical engineering was maturing' (Katzen and Tsao, 2000, p. 79) as well.

During the 1970s with the advent of restriction enzymes, tissue culture headed towards a new research area (Rai, 2007). It was demonstrated that DNA could be cut into pieces by restriction enzymes, which were isolated for the first time by Smith and Nathans from *Haemophillus influenzae*, and that the clipped DNA part can be transferred to the clipped plasmid part. Recombinant DNA created in this way may be/is biologically active, and can replicate in the host bacterial cell (Dimitrijević and Petrović, 2004). Regarding the technology of recombinant DNA, the most important discoveries were: (i) the tumour-inducing principle of *Agrobacterium tumefaciens* (Ti plasmid) in 1973 by Zaenen *et al.* (Rai, 2007), the first gene transfer into *Escherichia coli* (Fiechter, 2000), which is generally accepted as the first true success of gene technology; and (ii) the creation, by Hargobind Khorana in 1976, of an artificial gene that functions in a bacterial cell. A new era of technology related to diagnostics and therapeutics started after the first production of monoclonal antibodies in 1975 (Liu, 2014). Regarding agricultural production, the 1970s brought, among other things, production of the first somatic hybrid of *Nicotiana* by protoplast fusion (Rai, 2007), cryopreservation development, and the commercialization of the embryo transfer technology, stimulated by worldwide acceptance of artificial insemination technology (Foote, 2002).

In the 1980s, great progress was made in agriculture production, medicine and genetics. This period saw the beginning of the commercial application of plant tissue culture for the production of pathogen-free plants and of the conservation of rare and endangered species (Tsay, 2002); Karl Mullis invented the polymerase chain reaction (PCR), a process that has multiple applications in medicine, genetics, biotechnology, forensics and paleobiology; insulin, the first pharmaceutical made by genetically engineered bacteria, was approved for use in the UK and the USA (1982); and the first transgenic animals (mice) were produced.

In the early 1990s DNA fingerprinting methods found extensive use in the areas of forensics and establishing paternity, but also in agriculture for the characterization and detection of genetic diversity, origin authentication, variety identification, parentage testing and breeding of animals, and detection of GMOs. The production of recombinant vaccines and hormones used in animal production also started in that period. In addition, the 1990s saw the largest life science project ever conducted, the Human Genome Project, aimed at 'reading' the whole genome; Dolly the sheep was the first animal to be cloned from an adult cell; and the first field trials of genetically engineered plant varieties were followed by the first commercial release, as a result of which, GMOs entered the food chain.

The 2000s were the years of the development of the 'omics' technologies 'aimed primarily at the universal detection of genes (genomics), [messenger ribonucleic acid] mRNA (transcriptomics), proteins (proteomics) and metabolites (metabolomics) in a specific biological sample in a non-targeted and non-biased manner', integrated into systems biology, which represents 'biological research focusing on the systematic study of complex interactions in biological systems using integration models' (Horgan and Kenny, 2011, p. 189).

1.1.1 Biotechnology as a reflection of human history

The brief review of the development of biotechnology presented above is also a reflection of human history. From the very beginning, basic human needs have remained the same. Ensuring a reliable food supply has certainly been an equally vital concern of all states, whether ancient or modern. In other words, from the first proper cities in southern Mesopotamia (beginning with the city of Uruk) during the Uruk (*c.* 3400–3200 BC) and Jamdet Nasr (*c.* 3200–3000 BC) periods, until the present time, food technology and rise of the state have been interconnected and have continued to develop in parallel.

When we consider the number of plant and animal species used for human consumption today, we can see that not much has changed when compared with ancient times. Our survival still depends on the 100 higher plants that were domesticated out of 200,000 accessible species, and the 14 domesticated large terrestrial mammalian herbivores and omnivores weighing 45 kg or more (out of 148 possible species) (Diamond, 2002). Wealthy Romans enjoyed a wider range of animals than is commonly eaten these days: pork, mutton, goat, lamb, horse, wild boar, deer, hare, and beef, but also ass, fox, hedgehog, and dog (including puppy). They also routinely consumed oat and millet, beans and pulses, peas, chickpeas, lentils, vetches, linseed, sage, lupin seeds, hedge-mustard, cucumber seeds, sesame, safflower, eggs, cheese, wine, vinegar and honey. Their diet also included a wide range of fruits and vegetables such as apples, pears, mulberries, quinces, figs, grapes, almonds, lettuce, cabbage, anise, onion, garlic, asparagus, mint, spinach, etc. (Wilkins and Hill, 2006). The world oldest recipes, the 'Yale Culinary Tablets', dating from 1700 BC, describe a varied and sophisticated cuisine in the ancient Near East (Curtis, 2001), with dishes prepared by cooking on an open fire – using different kinds of meat, such as lamb, mutton and stag, spiced with onion, garlic, leek, mustard, and cumin and coriander – and a large selection of stews and broths.

The Romans experienced food shortages, with meat and fish often being high-status foods. Homer's heroes feasted on beef. We are still fighting against hunger; and meat demand is still associated with higher income and, compared with other commodities, meat is characterized by high production costs and high output prices.

What has changed? Man has walked away from nature. Firstly, he largely destroyed it; and then he has tried to re-establish the former equilibrium. The most obvious reflection of this can be seen in the high valuation of organic, pure food in the modern diet, which is almost the same as the food consumed by the vast majority of ancient Romans, particularly poor Romans, who survived on vegetables, chickpeas, beans, apples, figs and birds as the most common meat. Secondly we have seen the severe environmental collapse of the cradle of civilization, the Fertile Crescent, particularly regarding soil erosion, deforestation, salinization and climate change threat. Thirdly, we have seen changes in nutritional diseases. In ancient times, cases of metabolic, diet-related diseases were rare, although an early African *Homo erectus,* classified as KNM-ER 1808, could have had hypervitaminosis A (Ungar, 2007). Even in the BC era, the importance of proper nutrition and appetite control was

recognized, including by those of the highest rank. A good example of this was Alexander the Great (356 BC–323 BC), a moderate eater who successfully resisted pressure from his mother Olympia to eat more cakes and bakery products. Despite this early understanding of nutrition, in the 21st century we are facing a global obesity pandemic as the leading cause of soaring rates of different metabolic diseases.

In addition, the centres of power in food production have changed over time. As Diamond points out, there is almost no overlap between the most productive areas for farming in the past and today. In ancient times, the most productive areas were: 'The Fertile Crescent, China, Mesoamerica, Andes/Amazonia, Eastern North America, Sahel, Tropical West Africa, Ethiopia and New Guinea', areas rich in native plant and animal species. Today, the most suitable areas for farming include 'California, North America's Great Plains, the pampas of Argentina, the South African Cape, the Indian subcontinent, Java and Australia's wheat belt' (Diamond, 2002, p. 702). The reasons for this change are very complex and touch upon economic, political and cultural issues.

Advances in science and huge technological changes have certainly improved the quality of life, but have also increased man's expectations and led to more sophisticated methods of conflict with rivals, compared with ancient warfare. In opposition to the efforts of many scientists who put their knowledge at the service of mankind and who have been working towards the common good in order to save human lives, anti-humanist movements have also developed. One example is the eugenics movement. Eugenic ideology is based on the belief that genetic traits determine social stratification. Francis Galton, a European theorist who coined the term 'eugenics' in 1883, in his book *Inquiries into Human Faculty and Its Development* stated, 'Eugenics is the study of all agencies under social control which can improve or impair the racial quality of future generations' (Signil, 2012, p. 114). Some members of the white American elite have accepted a belief that lower classes and minorities (black Americans,

Mexicans, non-assimilated European immigrants, poor white Americans) are genetically inferior and therefore only capable of reproducing genetically inferior offspring. Under the leadership of Charles Davenport, a zoologist, and with the financial help of the Carnegie Institute, the Rockefeller Foundation and wealthy individuals, political leaders and other supporters of the movement started to promote the superiority of the Nordic race. This politics led to the sterilization of more than 80,000 people in the USA, as the idea of eugenics began to spread around the globe. A particularly fertile ground was found in Nazi Germany (Daniels, 2005). Under the influence of the eugenic ideology, the USA adopted the Immigration Act in 1924, the federal law that set immigration quotas for individual countries in order to exclude Asians and people from Southern and Eastern Europe (primarily Jews and Slavs), while allowing significant immigration from northern and Western Europe.

Similarly, new transgenic technology, apart from having unsuspected beneficial applications, can also be used for destructive purposes such as bioterrorism; it can lead to the deepening of social inequalities, establishing the monopolistic position of multinational corporations, the leadership of certain countries and the subjection of others. Unlike maize hybrids, which were accepted without public outcry in the 1930s (Duvick, 2001), GMOs have caused a lot of dissatisfaction, fear and resistance around the globe, and opened a broad debate that is still going on. Two decades after the first transgenic crops were commercialized, we are still 'not in heaven or on earth' when it comes to definitive answers from the scientific community. There is still no consensus about whether or not genetically modified foods are safe for human health and the environment. Different countries follow different policies regarding GMOs; trade wars cease, and start again; multinational corporations propagate their policies, often using all possible means; and non-governmental organizations (NGOs) opposing transgenic technology continue with their activities, too.

Nevertheless, genetically modified food has entered the food chain, and there is no doubt about that.

1.2 Definitions of Biotechnology

The word biotechnology is a cross between the Greek words 'bios' (everything to do with life) and 'technikos' (involving human knowledge and skills). The term 'biotechnology' was coined by a Hungarian expert, Karl Ereky, in his book *Biotechnologie der Fleisch-, Fett- und Milcherzeugung im landwirtschaftlichen Grossbetriebe* [The Biotechnology of Meat, Fat, and Milk Production in the Agricultural Plant] published in 1919. The term was used to denote production processes by which products (bread, cheese, wine) were derived from raw materials with the help of living organisms.

There is no single definition of biotechnology. Different countries and different organizations define it differently. Most often the European Commission (EC) and the Food and Agriculture Organization (FAO) use the definition stated by the United Nations Convention on Biological Diversity (UN CBD): 'any technological application that uses biological systems, living organisms, or derivatives thereof, to make or modify products or processes for specific use' (EC, 2010, p. 3).

The European Federation of Biotechnology, a non-profit organization aimed at promoting biotechnology, defines biotechnology as 'the integrated use of biochemistry, microbiology, and engineering sciences in order to achieve technological (industrial) application of the capabilities of micro-organisms, cultured tissue cells, and parts thereof' (Bull *et al.*, 1982, p. 60).

The Organization for Economic Cooperation and Development (OECD) has used a single broad definition, which says that biotechnology consists of 'the application of science and technology to living organisms, as well as parts, products and models thereof, to alter living or non-living materials for the production of knowledge, goods and services' (OECD, 2016a).

The American Chemical Society (ACS) considers that 'Biotechnology (biotech) involves the study and use of living organisms or cell processes to make useful products' (ACS, 2016).

There are many other definitions of biotechnology, such as:

'Biotechnology is the use of living organisms or other biological systems in the manufacture of drugs or other products or for environmental management, as in waste recycling: includes the use of bioreactors in manufacturing, microorganisms to degrade oil slicks or organic waste, genetically engineered bacteria to produce human hormones, and monoclonal antibodies to identify antigens' (Academic Dictionaries and Encyclopedias, 2016).

'Biotechnology is the controlled use of biological agents such as microorganisms or cellular components for beneficial use', *US National Science Federation* (Biotechnology 4u, 2016).

Biotechnology is 'the application of biochemistry, biology, microbiology and chemical engineering to industrial processes and products and on the environment', *International Union of Pure and Applied Chemistry* (Biotechnology 4u, 2016).

'Biotechnology is an activity to modify living organisms or processing biological material according to anterior design, which is based on modern life science, and combined with advanced engineering technology and other basic science theory, to provide products or services', *China, Shanghai Science and Technology Commission* (OECD, 2016b).

'The technology of living systems: (i) a discipline that studies the possibility of using living organisms, their systems or their metabolic products to solve technological problems, as well as the possibility of creating living organisms with the necessary properties by genetic engineering; and (ii) use of biological structures for production of food and industrial products and for targeted transformations. Biological structures in this case are the microorganisms, plant and animal cells, cell components, such as membrane cells, ribosomes, mitochondria, chloroplasts, as well as biological macromolecules (DNA, RNA, proteins – mostly enzymes)', *Russian Ministry of Economic Development* (GAIN, 2012).

From the foregoing definitions, although each is different from the others, it can be concluded that biotechnology encompasses a wide segment of activities; and that it covers traditional, borderline and modern technologies used in industry, medicine and agriculture, thanks to its ability to apply to all living entities and organisms: viruses, bacteria, plants and animals. Thus, it is more appropriate to narrow the definition, and in fact to define traditional biotechnology and modern biotechnology separately, as has been done in the Cartagena Protocol on Biosafety. The Protocol defines modern biotechnology as the application of: (i) *in vitro* nucleic acid techniques, including recombinant DNA and direct injection of nucleic acid into cells or organelles; or (ii) fusion of cells beyond the taxonomic family, that overcome natural physiological reproductive or recombination barriers and that are not techniques used in traditional breeding and selection.

In addition to the widely accepted definitions of the CBD, the FAO have used a narrow definition which states that biotechnology is 'a range of different molecular technologies such as gene manipulation and gene transfer, DNA typing and cloning of plants and animals' (FAO, 2004, p. 8).

When one considers agriculture alone, there are many biotechnologies that differ significantly, 'from biotechnologies that are relatively "low-tech" (such as biofertilizers, biopesticides or tissue culture in crops/trees; artificial insemination in livestock; fermentation and use of bioreactors in food processing), to those that are more "high-tech" (such as use of PCR-based methodologies for disease diagnosis, marker-assisted selection, genomics or *in vitro* fertilization in livestock)' (FAO, 2016). Unlike recombinant DNA technology, these technologies 'do not normally require any specific regulatory approval, meaning that they can be quickly adopted by farmers and that the costs of release are low' (FAO, 2016).

1.2.1 Biotechnology classification by colour

Biotechnology classification by colour in relation to application, as proposed by Martinez

(Martinez, 2010), is quite accepted nowadays. According to this classification, there are five types of biotechnology: red, white, grey, green, and blue. Red biotechnology is used to define various types of biotechnologies applicable in medical science, such as developing of new drugs and therapies, production of vaccines, antibodies and antibiotics, cell and gene therapy, and regenerative medicine. White biotechnology or industrial biotechnology brings together all biotechnologies used in industrial processes. Relevant examples of white biotechnology are industrial fermentation (use of microorganisms for this purpose), production of new materials (plastic, paper, detergents, and textiles), or production of biofuels. Grey biotechnology refers to applications used in regards to the environment. There are two main issues in the focus of grey biotechnology: biodiversity maintenance and contaminant removal. For this purpose, different techniques (e.g. cloning) and tools (e.g. microbes) of molecular biology are used for species analysis or for isolation and disposal of harmful substances. Green biotechnology includes all activities, procedures and approaches connected with agriculture, regardless of whether they are traditional (e.g. selection and crossbreeding) or modern (e.g. transgenic technology), with the tendency to improve resistance to pests and crop diseases and nutritional value, and to develop bio-factory plant varieties. Finally, blue biotechnology is focused on marine and aquatic resources – in fact, on their exploitation in order to obtain industrially important products.

1.2.2 Definitions of genetic engineering, GMOs and GM food

For a single definition of genetic engineering, also known as recombinant DNA technology, genetic modification, gene technology or transgenic technology, the following formulation might be useful: 'The formation of new combinations of heritable material by the insertion of nucleic acid molecules into any virus, bacterial plasmids or other vector system, so as to allow their incorporation into a host organism in which they do not naturally occur, but in which they are capable of continued propagation' (EC, 1998).

Using recombinant DNA technology, one or more genes (called transgenes) from one organism can be introduced into the genetic material of another organism. This gives birth to a new organism without sexual reproduction, which is called a GMO. The World Health Organization (WHO) describes GMOs as: 'organisms (i.e. plants, animals or microorganisms) in which the genetic material (DNA) has been altered in a way that does not occur naturally by mating and/or natural recombination,' while 'foods produced from or using GMOs are often referred as to genetically modified foods' (WHO, 2016).

The diversity of the use of biotechnology for scientific, medical, agricultural, and industrial purposes, as well as the diversity of techniques used, regardless of whether they are traditional or modern, clearly indicate a need to separate the term biotechnology from the term genetic engineering. That is why we cannot agree with the authors who suggest the term biotechnology can be used as a synonym for (a single biotechnology-) transgenic technology, i.e. genetic engineering.

Considering that the term agriculture refers to crops, livestock, fish, and forestry products, the phrase genetic engineering in agriculture covers the use of transgenic technology in any of these sectors. Since there is no commercially important application of recombinant DNA technology in livestock, fish, and forestry production, the phrase genetic engineering in agriculture usually refers to crops only.

References

Academic Dictionaries and Encyclopedias (2016) Biotechnology. Available at: http://universalium. academic.ru/82074/biotechnology (accessed 8 June 2018).

ACS (2016) Biotechnology. Available at: https://www.acs.org/content/acs/en/careers/college-to-career/chemistry-careers/biotechnology.html (accessed 8 June 2018).

Baxby, D. (1981) Jenner's smallpox vaccine: The riddle of vaccinia virus and its origin. *Trends in Immunology* 3, 26.

Beckwith, J.R. and Signer, E.R. (1966) Transposition of the lac region of Escherichia coli: I. Inversion of the lac operon and transduction of lac by Φ80. *Journal of Molecular Biology* 19, 254–265.

Benenson, A.S., Kempe, H.C. and Wheeler, R.E. (1952) Problems in Maintaining Immunity to Smallpox. *American Journal of Public Health and the Nations Health* 42, 535–541.

Biotechnology 4u (2016) Introduction. Available at: http://www.biotechnology4u.com/introduction.html (accessed 8 June 2018).

Bull, A.T., Holt, G. and Lilly, M.D. (1982*) Biotechnology International Trends and Perspectives*. OECD, Paris, France.

Curtis, R.I. (2001) *Ancient Food Technology*. Brill Academic Publishers, Leiden, Netherlands.

Daniels, R.J. (2005) War against the weak: Eugenics and America's campaign to create a master race, by Edwin Black. *Perspectives in History* xx, 77–80.

Darwin, C. (2010) *The Variation of Animals and Plants under Domestication*. Cambridge University Press, New York, New York.

Decker, J.E., McKay, S.D., Rolf, M.M., Kim, J., Alcalá, A.M., et al. (2014) Worldwide patterns of ancestry, divergence, and admixture in domesticated cattle. *PLoS Genetics* 10, e1004254, 1–14.

Demain, A.L. and Fang, A. (2000) The natural functions of secondary metabolites. In: Scheper, T. and Fiechter, A. (eds) *History of Modern Biotechnology I*. Springer, Berlin, Germany, pp. 1–39.

Diamond, J. (1997) *Guns, Germs, and Steel: The Fates of Human Societies*. WW Norton & Company, New York, New York.

Diamond, J. (2002) Evolution, consequences and future of plant and animal domestication. *Nature* 418, 700–707.

Dimitrijević, M. and Petrović, S. (2004) *Genetically Modified Organisms – Questions and Dilemmas*. Green Network of Vojvodina, Novi Sad, Serbia.

Duvick, D.N. (2001) Biotechnology in the 1930s: The development of hybrid maize. *Nature Reviews Genetics* 2, 69–74.

EC (2010) Biotechnology Report. Available at: http://ec.europa.eu/public_opinion/archives/ebs/ebs_341_en.pdf (accessed 9 June 2018).

FAO (2004) *The State of Food and Agriculture 2003–2004. Agriculture Biotechnology: Meeting the needs of the poor?* FAO, Rome, Italy.

FAO (2016) Frequently Asked Questions about FAO and Agricultural Biotechnology. Available at: http://www.fao.org/fileadmin/user_upload/biotech/docs/faqsen.pdf (accessed 10 June 2018).

Fiechter, A. (2000) *Preface to History of Modern Biotechnology I*. Springer, Berlin, Germany, pp. ix–x.

Fleming, A. (1929) On the antibacterial action of cultures of a penicillium, with special reference to their use in the isolation of *B. influenza*. *British Journal of Experimental Pathology* 10, 226–236.

Foote, R.H. (2002) The history of artificial insemination: Selected notes and notables. *Journal of Animal Science* 80, 1–10.

GAIN (2012) Program on Development of Biotechnology in Russia through 2020. Available at: http://economy.gov.ru/minec/activity/sections/innovations/development/doc20120427_06 (accessed 25 December 2016).

Hixson, A.W. and Gaden, E.L. (1950) Oxygen Transfer in Submerged Fermentation. *Industrial and Engineering Chemistry* 42, 1792–1801.

Horgan, R.P. and Kenny, L.C. (2011) Omic technologies: genomics, transcriptomics, proteomics and metabolomics. *The Obstetrician & Gynaecologist* 13, 189–195.

Johannsen, W. (1909) Elemente der exakten Erblichke itslehre. In: Roll-Hansen, N. (2014) The holist tradition in twentieth century genetics. Wilhelm Johannsen's genotype concept. *The Journal of Physiology* 592, 2431–2438.

Johannsen, W. (1926) Elemente der exakten Erblichke itslehre. In: Wanshner, J.H. (1975) An analysis of Wilhelm Johannsen's genetical term genotype 1909–26. *Hereditas* 79, 1–4.

Kasuya, M. (1964) Transfer of drug resistance between enteric bacteria induced in the mouse intestine. *Journal of Bacteriology* 88, 322–328.

Katzen, R. and Tsao, G.T. (2000) A view of the history of biochemical engineering. In: Sheper, T. and Fiechter A. (eds) *History of Modern Biotechnology II*. Springer, Berlin, Germany, pp. 77–91.

Liu, J.K.H. (2014) The history of monoclonal antibody development – Progress, remaining challenges and future innovations. *Annals of Medicine and Surgery* 3, 113–116.

Luikart, G., Ludovic, G., Excoffier, L., Vigne, J.D., Bouvet, J. and Taberlet, P. (2001) Multiple maternal origins and weak phylogeographic structure in domestic goats. *Proceedings of the National Academy of Sciences* 98, 5927–5932.

Martinez, V.D. (2010) The Colors of Biotechnology. Available at: https://biotechspain.com/en/article.cfm?iid=colores_biotecnologia (accessed 10 June 2017).

Mendel, G. (1865) Experiments in plant hybridization (1865). Available at: http://old.esp.org/foundations/genetics/classical/gm-65-a.pdf (accessed 10 June 2018).

Michel, C. (2002) The perdum-mule, a mount for distinguished persons in Mesopotamia during the first half of the second millennium BC. In: Frizeli, B.S. (ed.) *PECUS. Man and Animal in Antiquity. Proceedings of the Conference at the Swedish Institute in Rome.* Swedish Institute, Rome, Italy, pp. 190–200.

Moore, J.A. (1999) *Science as a Way of Knowing: The Foundations of Modern Biology.* Harvard University Press, Cambridge, Massachusetts.

Nummer, B. (2002) Historical Origins of Food Preservation. Available at: http://www.uga.edu/nchfp/publications/nchfp/factsheets/food_pres_hist.html (accessed 8 June 2018).

OECD (2016a) Statistical Definition of Biotechnology. Available at: http://www.oecd.org/sti/biotech/statisticaldefinitionofbiotechnology.htm (accessed 10 June 2017).

OECD (2016b) Biotechnology Statistics – China. Available at: http://www.oecd.org/sti/biotech/biotechnologystatistics-china.htm (accessed 10 June 2018).

EC (1998) Council Directive 98/81/EC of 26 October 1998 amending Directive 90/219/EEC on the contained use of genetically modified microorganisms. *Official Journal of the European Communities* L 330/13. Available at: https://ec.europa.eu/health/sites/health/files/files/eudralex/vol-1/dir_1998_81/dir_1998_81_en.pdf (accessed 10 June 2018).

Pasteur, L. and Chamberland, R. (2002) Summary report of the experiments conducted at Pouilly-le-Fort, near Melun, on the anthrax vaccination, 1881. *The Yale Journal of Biology and Medicine* 75, 59–62. [Originally published in *Comptes Rendus de l'Academie des Science* 92, 1378–1383, June 13, 1881.]

Pele, M. and Cimpeanu, C. (2012) *Biotechnology: An Introduction.* WIT Press, Southampton, England.

Prajapati, J.B. and Nair, B.B. (2008) The history of fermented foods. In: Farnwort, E.R. (ed.) *Handbook of Fermented Functional Foods.* CRC Press, Boca Raton, Florida, pp. 1–25.

Pray, L. and Zhaurova, K. (2008) Barbara McClintock and the discovery of jumping genes (transposons). *Nature Education* 1, 169.

Purugganan, M.D. and Fuller, D.Q. (2009) The nature of selection during plant domestication. *Nature* 457, 843–848.

Rai, R. (2007) *Introduction to Plant Biotechnology.* Available at: http://nsdl.niscair.res.in/jspui/bitstream/123456789/668/1/revised%20introduction%20to%20plant%20biotechnology.pdf (accessed 18 December 2016).

Roll-Hansen, N. (2014) The holist tradition in twentieth century genetics. Wilhelm Johannsen's genotype concept. *The Journal of Physiology* 592, 2431–2438.

Sapp, J. (2003) *Genesis: The Evolution of Biology.* Oxford University Press, New York, New York.

Sassaman, K.E. (2004) Complex hunter–gatherers in evolution and history: A North American perspective. *Journal of Archaeological Research* 12, 227–280.

Sauer, O.C. (1963) *Land and Life: A Selection From the Writing of Carl Ortwin Sauer.* University of California Press, Oakland, California.

Signil, C. (2012) Race, Faith, and Politics: 7 Questions Every African American Christian Must Answer. Creation House, Lake Mary, Florida.

Suttles, W. (2009) Coping with Abundance: Subsistence on the Northwest Coast. In: Lee, R.B. and Devore, I. (eds) *Man the Hunter.* Aldine Transaction, New Brunswick, New Jersey, pp. 56–68.

Tsay, H.S (2002) *Use of tissue culture for the mass propagation of pathogen-free plants.* Available at: http://www.fftc.agnet.org/htmlarea_file/library/20110805115956/tb158.pdf (accessed 10 June 2018).

Ungar, P.S. (2007) *Evolution of the Human Diet: The Known, the Unknown, and the Unknowable.* Oxford University Press, New York, New York.

Weiss, E. and Zohary, D. (2011) The Neolithic southwest Asian founder crops. *Current Anthropology* 52, 237–254.

WHO (2016) Frequently asked questions on genetically modified food. Available at: http://www.who.int/foodsafety/areas_work/food-technology/faq-genetically-modified-food/en/ (accessed 10 June 2018).

Wilkins, J.M. and Hill, S. (2006) *Food in the Ancient World.* Blackwell Publishing, Malden, Massachusetts.

Wrangham, R.W., Jones, J.H., Laden, G., Pilbeam, D. and Conklin-Brittain, N. (1999) The raw and the stolen. *Current Anthropology* 40, 567–594.

2 Genetically Modified Foods in the Light of Food Regimes

Understanding of the world food system in a broader context, its crises, food prices, environmentally hazardous agro-industrialization, and food sovereignty movements, as well as the diffusion of transgenic foods, is almost impossible without analyses of food regimes. Buttel (2001, pp. 21–24) wrote: 'Beginning in the late 1980s, the sociology and political economy of agriculture began to take a dramatic turn. The extent of the shift in the literature was not entirely apparent at the time, because at a superficial level the concepts and vocabulary of late 1980s and early 1990s agrarian studies did not depart sharply from those of the new rural sociology. The lexicon continued to be primarily that of Marxist/class categories. But only 5 years after the seminal piece – Friedmann and McMichael's 1989 *Sociologia Ruralis* paper on food regimes – was published, the sociology of agriculture had undergone a dramatic transformation... this article on food regimes was arguably the seminal piece of scholarship in the abrupt shift away from the new rural sociology, and "regime-type" work has proven to be one of the most durable perspectives in agrarian studies since the late 1980s, in large part because it is synthetic and nuanced.'

Thus, this chapter first presents the concept of the food regime and some major events affecting the formation of the first (1870–1914), the second (1950s–1970s) and the third food regime (late 1980s–). We then describe the supranational regulatory context relevant to transgenic technology: the role of the World Trade Organization (WTO) and its agreements – the Agreement on Trade-Related Aspects of Intellectual Property Rights (TRIPS) and the Sanitary and Phytosanitary Measures Agreement (SPS) – as well as their opposition, the biodiversity-related CBD agreement. Based on the supranational trade agreements, many national initiatives have been taken place around the globe. However, the principal drivers of worldwide regulatory activity on GMOs are two superpowers, the USA and the EU. GMOs have become big business in the USA, while the EU created serious obstacles to the export of the transgenic products from the USA. Countries that rely on the EU or the USA have aligned their regulations accordingly. Other countries, less dependent on EU or US markets, such as China, Brazil, India and Russia 'have adopted regulation whose stringency lies somewhere in between the EU and the US model' (Bernauer and Aerni, 2008, p. 7). Unlike all the countries mentioned so far, and despite pressure from the USA, EU and WTO, Serbia has adopted a unique GMO policy that forbids all production, commercialization and importation of transgenic crops and products.

National GMO regulation cannot be properly understood without knowing the

© CAB International 2018. *GM Food Systems and their Economic Impact*
(T. Brankov and K. Lovre)

role of agriculture in each country. Therefore in this chapter, we describe the interests and positions of the different countries (the USA, the EU, BRICS, and Serbia) regarding the GMO issue through basic agricultural performance, trade relations and public attitudes toward GMOs. On the whole, the chapter suggests that national GMO politics should not be analysed separately, but as an integral part of agricultural policy. Furthermore, analyses suggest that the countries whose main goal is to achieve self-sufficiency have adopted strong regulatory powers. By contrast, the countries whose main goal is to expand exports have approved a weak regulatory approach to GMOs. This chapter shows that transgenic technology should be interpreted as a logical continuation of the intensification and industrialization of agricultural production that started during the second food regime. The biggest argument for this claim is soy. Since 1930s soy has been an excellent source of foodstuffs and feedstuffs, and was at the centre of the post-war transformation of agriculture and international division of labour. Today, transgenic soy has broken all the barriers of national regulations and found its way into all parts of the world. Regardless of whether countries adopted a *de jure* prohibition of cultivation, or accepted substantial equivalence approaches, all have accepted transgenic soybeans. Even Serbia has imported certain quantities of transgenic soybean meal in years when there have been food shortages. Thus, it is hardly to be expected that any kind of effort can fully eliminate products of transgenic technology from the food system. Most probably, the benefit of resistance to GMOs will be to create some GMO-free niches.

2.1 Food Regime Theory

The concept of the food regime relies on a combination of two theories: regulationism and world-systems theory (Magnan, 2012). According to the regulationist perspective, there is a specific type of time interval within capitalism called a regime of accumulation, which is 'a systematic organization of production, income distribution, exchange of social product,

and consumption' characterized by relatively stable economic development. Such moments are typified by extensive accumulation, as in the 19th century, 'in which the investment of constant capital, including investments in iron and steel, railway construction, and shipbuilding itself, validated the growth of department 1;[i] or intensive accumulation, particularly after World War II, 'in which the conditions of existence of the wage-earning class were transformed through the articulation of mass production and mass consumption' (Dunford, 1990, pp. 305–306). These regimes have been separated from each other by crises of capitalism: the cumulative collapse in 1929, causing a crash of the extensive accumulation, and the crisis of the 1970s, causing a deterioration of the intensive accumulation and opening the door for the establishment of a new regime. At the same time, 'world-systems theory conceives of social change as occurring from the dynamic interplay of global capital accumulation and a hierarchical system of states, over successive periods of hegemony (economic, social, and political leadership vested in a dominant state) and transition' (Magnan, 2012, pp. 463–464).

By combining these two theories, the food regimes concept has been defined, referring to 'a relatively bounded historical period in which complementary expectations govern the behavior of all social actors, such as farmers, firms, and workers engaged in all aspects of food growing, manufacturing, services, distribution, and sales, as well as government agencies, citizens and consumers' (Friedmann, 2009, p. 125). 'Key elements identified, and that bear on the determinants and drivers, "shape" and consequences, and struggles of different food regimes, are: the international state system; international divisions of labour and patterns of trade; the "rules" and discursive (ideological) legitimations of different food regimes; relations between agriculture and industry, including technical and environmental change in farming; dominant forms of capital and their modalities of accumulation; social forces (other than capitals and states); the tensions and contradictions of specific food regimes, and transitions between food regimes' (Bernstein, 2015).

Friedmann and McMichael in their famous paper of 1989 recognized three food regimes: the first (1870–1914), the second (1950s–1970s), and the third possibly emergent food regime (late 1980s–). Subsequently McMichael dated the first food regime from the 1870s to the 1930s (McMichael, 2009). The same author in his individual work made use of some new notions. Instead of 'the food regime', there was 'the food regime project'; instead of 'the first food regime', he used 'the colonial project'; 'the second food regime' he called 'the development project'; while 'the third food regime' he designated as 'the globalization project' (Bernstein, 2015). From the other side, Friedmann in her work designated the first regime as 'the settler-colonial food regime', the second as 'the mercantile-industrial food regime', and the third as 'the corporate-environmental' (Friedmann, 2005, 2009). Between these two authors, as well as among other food regime scholars, there is a longstanding debate about whether a food regime is currently in place. According to Friedmann we are now in a period of transition to a potentially new food regime characterized by 'the proliferation of private food standards, supermarket power, and consumer-led food politics' (Magnan, 2012, p. 474) which we have not completely established yet. On the other hand, McMichael holds the position that the third food regime has emerged, 'organized around a politically constructed division of agricultural labor between Northern staple grains traded for Southern high-value products (meats, fruits and vegetables)' (McMichael, 2009, p. 148), and driven by WTO rules. Apart from disagreements about the focus of the third regime, the authors basically agree on the concepts of the first two food regimes. The first food regime took place under British hegemony and was 'centered on the Atlantic trade between England and the Americas, which was consolidated after 1870, but also included settler regions that have since declined, such as the Danube Basin and Punjab in British India; the second expanded after World War II to include all the former colonies of Europe, but excluded the Soviet bloc until its collapse, and hinged on the USA as a rule-maker and

consequently a dominant exporter' (Friedmann, 2009, p. 125).

2.1.1 The first food regime

During the first food regime, a new division of labour among the three main actors – European powers, settler-states (North America and Australasia), and 'occupied' colonies – was established. Europe, already at an impressive stage of economic growth thanks to associated profits from colonialism and slavery in the post-1500 period, and the USA, a new settler-state, colonized the rest of Africa and Asia, and subjected them as well as the other colonies to direct metropolitan rule. Colonies were obliged to supply the major powers with 'tropical and sub-tropical crops including sugar, coffee, tea, tobacco, and cocoa, many of which had assumed a new importance in working class European diets' (Magnan, 2012, p. 468) as well as with 'raw material for industry, cotton, rubber, indigo, jute, copper, and tin' (Friedmann and McMichael, 1989, pp. 97–98). A momentous innovation, with the first SS *Frigorifique* meat transport from Argentina in 1877, the ship refrigerator opened a new era in trade and international relations (Hyatt, 1997) because it enabled the extension of products offered and increased transportation distances. In parallel with the process of colonialism's culmination, during this extensive stage of capitalism settler-states developed independent national economies which become competitive with the imperial power. The new international division of labour, in which Europe imported meat and wheat from settler-states in exchange for export of capital, labour and processed products, represented the core of the first food regime and paved the way for the trading of competitive products. 'The most important fact to note is that in the settler-colonial food regime, power and wealth resided in the importing countries, which exported capital and labour to *improve* (or, as we would now say, *develop*) lands taken by force from indigenous peoples. The settler-colonial regime also laid the basis for the later industrialization of agriculture, paradoxically through the invention of the modern *family farm*' (Friedmann 2009, p. 125), where there is a new class of farmers from settling European emigrants who are dependent on the export market. 'A final legacy of the settler-colonial regime was the globalization and simplification of a wheat-beef diet'. In general, the quality of consumed food declined, because inexpensive wheat and meat, and canned and sugared food, replaced coarse grains and a variety of local foods rich in nutrients, while 'opium-based medicines blunted hunger and its effects' (Friedmann, 2009, p. 127).

As the demand for wheat products grew, millions more acres were ploughed and planted for monoculture. Uncontrolled exploitation of North American virgin land led to the Dust Bowl of the 1930s, while low prices led to the Great Depression. All these things, associated with the consequences of World War I and the inability of the gold standard to pass a test of war, marked the breakdown of the first food regime. Simultaneously, this signalled the end of the first food regime framework, as expressed in relatively free trade in goods, labour and capital. 'The collapse of the settler-colonial food regime left four enduring legacies which helped to shape new relations of power, property, and trade in the subsequent regime: a labor shortage in agriculture; deeply commodified farms; efficiency measures based on land-extensive monocultures; and the globalization and *democratization* of a diet based on wheat and beef' (Friedmann, 2009, p. 127).

2.1.2 The second food regime

After the first food regime collapsed, a new era of comprehensive attempts at administrative intervention in agricultural policy started. The New Deal Agriculture Adjustment Act was passed by the US Congress in the spring of 1933. The New Deal instituted price support for basic agricultural commodities in order to achieve a more equitable distribution of national wealth between agriculture and industry (Skocpol and Finegold, 1982). Since new technologies and farming methods increased production efficiency, the ultimate effects of the production control

based on the restriction of a planted area were seriously challenged. In order to directly subsidize agricultural exports and to make loans to farmers for buying and selling commodities, the USA introduced two new institutions: the Grain Stabilization Board and the Commodity Credit Corporation (CCC). 'The CCC became a vehicle for managing agricultural surpluses and the basis for the first structured US food aid programs drawing on the mounting food surpluses. The scope of the CCC was widened by an amendment to the Agricultural Adjustment Act in 1935, which authorized its use of customs revenues to subsidize agricultural exports and encourage domestic production' (Shaw, 2007, p. 13). 'US farm policy was designed not to be transparent (what were effectively subsidies were called "loans") and to raise agricultural prices rather than directly subsidizing farm incomes. An elaborate system of government purchases removed enough wheat or other commodities from the market to achieve target prices set by Congress. The result was surpluses held by government agencies. These put downward pressure on prices and therefore became self-perpetuating' (Friedmann, 2005, p. 239). Completely opposite to the US policy was Britain's system of deficiency payments. This was an example of a transparent policy without trade-distortive effects, and consistent 'with the World Food Board proposal and with liberal international trade... The government set target incomes for farmers and paid the difference between actual and target incomes out of general revenues. No surpluses accumulated anywhere, and prices to consumers were not affected' (Friedmann, 2005, p. 239).

As supply largely exceeded demand, several international control schemes mainly aimed at defending prices appeared after the economic catastrophe in 1929. Apart from agreements associated with tin, rubber and copper in the 1930s, a control scheme was also established for sugar, wheat and tea. An Inter-American Coffee Agreement setting the export quotas to the USA was signed between the USA and many Latin American countries in 1940. But the outbreak of World War II interrupted the implementation of this scheme as well as negotiation about the Wheat Agreement.

In contrast to the pre-war period when supply was higher than demand, in the post-war period the world faced a shortage of food which lasted until 1948 (Raffaelli, 2009). Under those circumstances, the measures of state intervention changed, because there was an obvious need for pricing strategy, cutting depression and stabilizing primary commodities production. In order to increase food production the USA increased price support for crops and introduced support for animal products. In addition to changes in food security, the entire political picture of the world changed after World War II. Germany was totally defeated, Japan suffered the consequences of extensive bombing, France had to recover from the Nazi occupation, the UK had lost the status of imperial power, while Russia (Soviet Union) was devastated by land warfare, but with a possibility to become a superpower thanks to its great resources and population. The US economy was the only one greatly stimulated by the war, so US hegemony began during the war, reaching its peak some 30 years later. The USA has used this new position, among other things, for the imposition of trade rules in order to protect and distribute its own surplus. The USA launched the idea of establishing the International Trade Organization (ITO). The International Conference on Trade and Employment was held in Havana from November 1947 to March 1948. The General Agreement on Tariffs and Trade (GATT), a multilateral agreement regulating international trade, was negotiated at the conference as an outcome of the failure of negotiating governments to create the ITO. The GATT agreement required removing trade barriers on internal trade, but at the same time it allowed the possibility of excluding agriculture from total liberalization. The Agreement 'allowed the US to retain domestic commodity programs, which were incompatible with the GATT since they depended on the use of both import controls and export subsidies' (Magnan, 2012, p. 470). Apart from this, the USA introduced new practices that shaped the second food regime. 'The most remarkable innovation of US hegemony was *aid* in the form of sales of US goods for *soft* currencies held by the US government as *counterpart funds*' (Friedmann, 2005, pp. 239–240).

Post-war price support in the USA and Canada led to a large food surplus that had to find its market. Following the Marshall Plan, the European Recovery Program, and taking advantage of the new monetary rules – a dollar-based international system instead of the gold standard, the largest ever transfer of bilateral aid was effected: 'Of the total aid package of $13.5 billion supplied between 1948 and 1953, about a quarter was committed in food, feed and fertilizer' (Shaw, 2007, p. 13). Marshall's transfer was very important because it 'simultaneously established the basis for Atlantic agro-food relations' and invented a foreign aid mechanism 'later adapted to the third world' (Friedmann, 1993, p. 35). In return for its openness to US exports of maize and soy used as feeds in 'livestock complex' development, the USA supported the high level of protection of European wheat and dairy products. US food aid also contributed to the ambitious development plans of the newly decolonized Third World. As Friedmann emphasized, the application of food aid to underdeveloped countries under US Public Law 480 in 1954 was the foundation of the second food regime (Friedmann, 2005, p. 241). 'This created the conditions of a potential scissors effect for many poor, primarily agricultural, countries: one blade being increasing food import dependence, the other the fluctuating but generally declining terms of trade for their historic export crops' (Bernstein, 2001, p. 35). Thus, food aid served the USA to achieve the defined objectives of its internal and external policies: to protect the income of American farmers from their own overproduction; to resolve fiscal problems of the state budget; to establish the dependence on inexpensive American wheat in new decolonized territories that hitherto had been largely self-sufficient in staple food production; and to encourage recipients and competitors to replicate the US model in production and consumption by adopting national regulation of agriculture and trade. In addition, food aid contributed to the achievement of anti-communist strategic goals during the Cold War.

The second food regime had a distinctly industrial character. The already difficult position of former colonies caused by dependence on cheap American wheat and price dumping worsened with the growth of industrially processed substitutes which eroded traditional Third World exports. Artificial sweeteners, high-fructose corn (maize) syrup and soy oil replaced tropical sugar and oils, two key tropical export products (Friedmann and McMichael, 1989). Family farms were encouraged to cooperate with corporations supplying these inputs. During the regime a Green Revolution campaign was launched to develop and apply high-yielding grains, primarily wheat and rice in the developing countries of the tropical and subtropical belts. The Green Revolution answered the problem of hunger in some parts of the world. Furthermore, it was used for solving some social and political problems, such as stopping the spread of communism (Fowler and Mooney, 1990). The Revolution also helped transform agriculture into agribusiness because the increased need for fertilizers, pesticides and machinery could be met only by the corporations (Paul and Steinbrecher, 2003). Technological solutions made farmers dependent on agro-input corporations, their production became more specialized, areas under monocropping arose, while the border between agriculture and industry was completely erased. Corporations increased their control of the food supply chain. The merging of agriculture and industry spawned a new manufactured diet based on 'fats and sweeteners, supplemented with starches, thickeners, proteins, and synthesized flavors' (Friedmann, 2009, p. 131). US consumers quickly accepted manufactured foods and changed their consumption pattern. As Huang observed for the period 1953–1983 with his statistical model for large-scale demand system estimation, 'if consumer spending increases consumption of certain processed food will increase significantly. These items are fruit juice, canned tomatoes, fruit cocktail, dried beans and peas, other processed fruit and vegetables, and cheese' (Huang, 1985, p. iii). Turning to a durable food complex resulted in a triple magnification of frozen foods consumption between 1950 and 1975 (Friedmann and McMichael, 1989).

Food manufacturing developed simultaneously with the spread of supermarkets

through which these products were placed on the market. Changes in America's consumption patterns, embodied in the acceptance of processed food, opened the space for the development of giants not only in manufacturing but also in retail. Gereffi estimated that during the 1960s and 1970s the US retail industry became oligopolistic with intensive price competitiveness (Gereffi, 1994). By that time buyers were becoming increasingly overwhelmed by an increasing number of processed products in supermarkets. The estimate stressed that the number of items on shelves increased from 800 in 1930 to 12,000 in 1980 (Friedmann and McMichael, 1989). The changes in America were reflected in the rest of the world, and first in Western Europe, because international trade caused convergence in food purchasing patterns (Connor, 1994). So the USA was a precursor of changes in consumption patterns in Western Europe. Certainly, Europe's previous replicating of the US farm model and development of the national livestock complex with American maize and soy as feedstuffs via their exemption from import control under the 1957 Common Agricultural Policy (CAP) were a fertile ground for these changes. Major changes in the food system caused discontent among certain groups, especially environmental groups, who organized themselves very well for the first time in the 1960s in North America, to request the rejection of heavily packaged foods and animal products from intensive agriculture (Connor, 1994).

The established system of the second food regime with its main features of non-free trade within the system of national agricultural regulation and protection framed in 1945 at Bretton Woods and 1947 under GATT; intensification of agriculture production; establishment of the power of the corporations in inputs, manufacturing and retail; mass production and mass consumption of standardized products; the wheat, livestock, and durable foods complex – all this experienced a serious crisis with the well-known Soviet–American Grain deals of 1972 and 1973. In 1972 the Soviet Union and the USA signed a trade agreement, historically the second attempt to establish economic cooperation.

After a trade break of 20 years, when the Commercial Relations Agreement signed in 1931 terminated in 1951 (Grzybowski, 1972), the two powers agreed on a valuable US$700 million transaction. The USA sent 440 million bushels of wheat or 30% of its average annual production to the Soviet Union. This transfer ended a long post-war period of food price stability: in 1 year wheat prices rose by 195%, soybean prices increased by 168%, broiler prices by 153%, and maize prices by 133%. In general, wholesale food prices rose by 29%, while farm commodity prices increased by 66% (Luttrell, 1973). The agreement was subsequently extended as of October 20, 1975, when the Soviet Union committed itself to purchasing at least 6 million tonnes of US wheat and grain annually for the next 6 years. The end of world food price stability and rising oil prices were closely associated with the 1972–1974 crises. The economic dependence of the Third World countries showed its negative side in the new circumstances, and scarcity increased, but unprecedented commercial demand for American grains significantly reduced the participation of food aid in the total US agriculture export: just 4% was intended for that purpose in 1974 (Kodras, 1993). Structural Adjustment policies or loans provided by the International Monetary Fund (IMF) and the World Bank on condition that borrowing countries agreed to implement certain policies had negative effects on the food security of indebted countries (Friedmann, 2005). Not only did the underdeveloped world face problems, but so did the USA. It encountered domestic and foreign problems and had to deal with the decline of its geopolitical hegemony, the dissatisfaction of citizens because of high prices, and the discontent of farmers, particularly as regards problems with the Japanese export market, strong European wheat and competition from Brazilian soy. Responding to the end of surpluses, the USA urged farmers to 'expand production, which they did enthusiastically, and by borrowing heavily. When surplus and price volatility returned later in the decade, heavily indebted farmers faced a financial crisis, laying the foundation for the farm income crisis of the last thirty years' (Magnan, 2012).

The second food regime mechanism failed to solve the problem of world hunger, so the FAO called the World Food Summit in 1974 to examine global food production and consumption, and the Universal Declaration on the Eradication of Hunger and Malnutrition was adopted, proclaiming that 'every man, woman and child has the inalienable right to be free from hunger and malnutrition in order to develop their physical and mental faculties'. The Declaration defined world food security system as follows: 'The well-being of the peoples of the world largely depends on the adequate production and distribution of food as well as the establishment of a world food security system which would ensure adequate availability of, and reasonable prices for, food at all times, irrespective of period fluctuations and vagaries of weather and free of political and economic pressures, and should thus facilitate, among other things, the development process of developing countries'; and it called on countries to 'cooperate in the establishment of an effective system of world food security' (Shaw, 2007, pp. 139–140). Among many other proposals, the Summit created International Funds for Agricultural Development (IFAD), the FAO Committee on Food Security (CFS), the Global International Early Warning System (GIEWS), the World Food Council (WFC), the Committee on Food Aid Policies and Program (CFA), and the International Emergency Food Reserve (IEFR). The USA changed its regulations on food aid, 'stipulating that at least 70 percent of Title 1 commodities be allocated to countries most seriously in need' (Kodras, 1993, p. 241), while the contribution from the Organization of the Petroleum Exporting Countries (OPEC) and European bilateral aid were expanded.

2.1.3 The third food regime

While the third food regime is still finding its final shape, some major events affecting its formation can be clearly distinguished. These are: the triumph of neoliberal policies, the creation of the WTO and signing of its Agreement on Agriculture, multifunctionality as the new frame of European agriculture, the positioning of New Agricultural Countries (NACs) as respected partners in the world market, the empowerment of corporations in all segments of agribusiness, internationalization of farm movements, the supermarket revolution in developing countries, land-grabbing, and the diffusion of genetically modified foods.

The formal inauguration of the era of neoliberal economics started in 1980 with the election of Ronald Reagan:

> Before 1980, economic policy was designed to achieve full employment, and the economy was characterized by a system in which wages grew with productivity... After 1980... globalization brought increased foreign competition from lower-wage economies and the prospect of off-shoring of employment. The new neoliberal model was built on financial booms and cheap imports... cheap imports ameliorate the impact of wage stagnation, thereby maintaining political support for the model. Additionally, rising wealth and income inequality makes high-end consumption a larger and more important component of economic activity, leading to the development of what Ajay Kapur, a former global strategist for Citigroup, termed a plutonomy.
>
> (Palley, 2010, pp. 16–17).

The WTO with the IMF became two of the pillars of the neoliberal model, as organizations, 'that expanded their jurisdictions and their respective capacities to intrude into national economic policies, and to incorporate countries into a global system of market-liberalizing economic rules' (Chorev and Babb, 2009, p. 460). The Agreement establishing the WTO, commonly known as the 'Marrakesh Agreement', was signed in Marrakesh, Morocco, on April 15, 1994, at the conclusion of the Uruguay Round of Multilateral Trade Negotiations. The round 'attempted to eliminate export subsidies on agricultural goods and textiles and dealt with non-tariff barriers, technical aspects of trade, and trade-related investment and intellectual property rights' (Hartwick and Peet, 2003, p. 191). The Uruguay Round was launched at a time when subsidized agricultural exports from the EEC (European Economic Community) went beyond

the export of these products from the USA. After long negotiations and what was almost an export subsidy war between the two agricultural superpowers, especially after the fall of the Eastern Socialist Bloc, the willingness of developed and less developed countries to make a radical change in agricultural policy was foreshadowed by the adoption of the WTO Agreement on Agriculture. A strong voice for agricultural reform was the Cairns Group, the second major agricultural exporting countries group, consisting at that time of 14 developed and developing countries led by Argentina, Canada and Australia. In order to provide better trading conditions in export markets for their products, the Cairns Group was seeking to eliminate high tariffs on agricultural imports and subsidies in the European Community, the USA and Japan. Subjecting agricultural commodities to multilateral trading rules marked the definitive end of the mercantile-industrial food regime, but 'the interests in the old regime died hard' (Friedmann, 2005, p. 246).

Europe in 1999, in its conclusions on the 'Millennium Round', chose a new paradigm for agriculture and rural development, i.e. multifunctional agriculture. The OECD defined multifunctionality as 'a characteristic of an activity which produces multiple and interconnected results and effects; these effects may be positive or negative, intentional or unintentional, synergetic or conflictive, and may have a value on the market or not' (Garzon, 2005, pp. 3–4). The EU with Japan, Korea, Norway and Switzerland set up an informal group, 'Friends of Multifunctionality.' Associating multifunctionality with 'blue box' (trade-distorting) and 'green box' (non-trade distorting) subsidies, caused new international tensions, because the USA and the Cairns Group used them as a pretext for maintaining protectionist agricultural policies (Majkovic et al., 2005). Nevertheless, the EU continues to use multifunctionality 'as a bargaining argument in trade negotiations to defend the right of countries to conduct domestic policies aiming at non-trade objectives', which has led to 'a modification of the trade policy and its discourse into a *pro-development* policy ... and ... an extension of the multifunctionality concept into specific concerns of developing countries' (Garzon, 2005, p. 17). The USA, pressed between the European and Japanese observations of food security in terms of self-sufficiency, increasing trade grain deficit and farmers seeking to raise grain prices, continued to act in the interest of agribusiness, and decided to increase 'farm subsidies to unprecedented levels' so 'domestic US grain prices continued to fall, benefiting increasingly concentrated livestock capitals' (Friedmann, 2005, pp. 246–247).

At the same time, some of the Third World countries intensified the agro-export of non-traditional products known as high value foods (HVFs). Fresh fruit and vegetables, cut flowers, shrimp, poultry, and dairy products were exports which replaced coffee, cacao, tea, and tobacco, and made those countries serious competitors with the EU and the USA. The aggregate export of meats and meat products, dairy products, fish and fish products, fresh and processed fruit, vegetables and nuts, feedstuffs, oilseeds, vegetable and animal oils, and spices exceeded US$500 million to 24 developing economies in 1989. Agricultural exports from the top six of those countries exceeded US$2000 million, as follows: Brazil 5852; Argentina 5017; China 4825; Thailand 4301; Malaysia 2463; Taiwan 2451 (Jaffee and Gordon, 1992). Four of them that account for 40% of the total HVF export from developing countries are the agro-industrial counterparts of the newly industrialized countries (NICs) referred to as NACs. 'Archetypal examples of these new agro-food systems are: Brazilian citrus, Mexican non-traditionals and exotics, Argentinean soy, Kenyan off-season vegetables and Chinese shrimp' (Rosset et al., 1999, p. 72). The reorganization of exports meant reshaping the dominant agro-food trade and encouraging fair trade, as well trade in organic products from the south to the north (Raynolds and Wilkinson, 2007).

The rise of supermarkets in developing countries has transformed the agro-food market and changed relations between capital and consumers. The main determinants of supermarket diffusion on the demand side were urbanization, entry of women into the workforce outside the home, and rapid growth in real mean-per-capita income, along with supply-side trade liberalization (Reardon and Berdegué, 2002). With the implementation of private quality standards for tracking and monitoring the supply-chain, multinational companies have become powerful controllers of the food network and guarantors of its quality and safety. By their involvement in the manufacturing sector through private-label products their power has been strengthened (Friedmann and McNair, 2008).

> Neoliberalism has significantly altered the dynamics of agrarian production and exchange relations within and between countries across the north–south divide. The simultaneous processes of globalization from above, partial decentralization from below and privatization from the side of the central state that used to play a key role in the maintenance or development of agrarian systems shook rural society to its core... providing even greater power to transnational and domestic capital to dictate the terms of agricultural production and exchange... Consequently, access to and control over land resources are being redefined and landed property rights restructured to favour private capital... While there are winners and losers in this global-local restructuring, working people and their livelihoods increasingly face ever more precarious conditions. Diversification of (rural and rural-urban, on-farm, off-farm or non-farm) livelihoods, forced or otherwise, has been widespread.
>
> (Borras et al., 2008, p. 1).

This reconstruction has affected farm movements in many ways, finally resulting in their internationalization during the 1990s, as typified in La Via Campesina. Representing about 200 million farmers from 73 countries in Africa, Asia, Europe, and the Americas that strongly oppose corporation-driven agriculture and free-trade agreements, the movement launched the concept of *food sovereignty* at the World Food Summit in 1996, i.e.:

> ...the right of peoples to healthy and culturally appropriate food produced through sustainable methods and their right to define their own food and agriculture systems... Food sovereignty prioritizes local food production and consumption. It gives a

country the right to protect its local producers from cheap imports and to control production. It ensures that the rights to use and manage lands, territories, water, seeds, livestock and biodiversity are in the hands of those who produce food and not of the corporate sector.

(La Via Campesina, 2017.)

The basic ideas of this movement, now recognized as an important actor in the food and agricultural debates, are agrarian reform, fair trade, gender equality, appropriate technologies, preservation of biological and genetic resources, social justice, healthy foods, and environmentally friendly agriculture.

The crisis of neoliberalism, as expressed in the climate, energy, and food crises, launched 'the bioenergy economy, fusing global ecology and political economy, (which) depends on the enabling role of financialization in managing a transition in capital accumulation and its foundations towards a new extractive food/fuel regime enclosing the world's remaining land and water… Land-grabbing, large-scale land acquisition following the 2007/08 crises undertaken by combinations of development agencies (World Bank, FAO), investment banks (Goldman Sachs), funds (Carlyle Group) and philanthropists (Soros, Gates Foundation), and… sanctioned by the World Food Summit of 2008… [and] anticipates the rising value of the living biomass as the source of inputs into the new bioeconomy' (McMichael, 2011, pp. 1–5).

Unlike the stability of the first and second food regimes under British and American hegemony, the collapse of the Doha Round revealed all the instability of the third food regime and the 'incapacity of the WTO to act as an institution that brings into being a new food regime', and provided the '*fin de siècle* to attempts to use multilateral means to resolve global food inequities and inefficiencies' (Pritchard, 2009, pp. 304–306). Without an adequate helmsman, the ship continues to roll perilously. It has become quite clear that there are no stable rules of trade between countries at different stages of development. Both the global north and the global south continue to put pressure on the WTO; the transnational corporations persist in promoting their own interests;

alternative trade networks continue to develop, while environmental and social movements around the globe rise alongside consumer awareness; national democratic institutions are standing up for their own principles; new forms of regulation have not acquired a clear shape, and it is unclear to what extent independent countries are prepared to surrender economic sovereignty to supranational institutions. Good examples of these international tensions are provided by the issues surrounding transgenic foods.

2.2 Supranational Regulatory Context of Transgenic Technology

The most important supranational regulatory body relevant to transgenic technology is the WTO. The WTO replaced the GATT, which for almost half a century covered international trade in goods, but with limited success in liberalizing agricultural trade. With respect to transgenic technology, crucial agreements administrated by the WTO, TRIPS Agreement and SPS entered into force at the beginning of 1995. The TRIPS agreement prescribed minimum intellectual property rights (IPR) to all WTO members. Within 27 Article (1) the agreement provided protection for transgenic technology inventions, both processes and products capable of industrial application. Furthermore, Article 27 (3b) provided for some exclusions from patentability, as stated: '(3) Members may also exclude from patentability: (b) plants and animals other than micro-organisms, and essentially biological processes for the production of plants or animals other than non-biological and microbiological processes. However, Members shall provide for the protection of plant varieties either by patents or by an effective sui generis system or by any combination thereof.'

Article 27 (3) reflects the tension between some developed countries that wanted to provide strong patent protection for GMOs and developing countries that wanted to protect their traditional knowledge. Thus, with Article 27, WTO members are being offered dual policy options: either to provide plant and animal patent protection or to implement a sui generis system (a system unique to itself), which is in practice usually done by joining the Convention for the International Union for the Protection of Plant Varieties (UPOV), although UPOV membership is not an obligation under TRIPS. Of course, the possibility to implement both policies in parallel is the third option. Article 27 (3) in its last sentence states: 'The provisions of this subparagraph shall be reviewed four years after the date of entry into force of the WTO Agreement.' Thus, in the Doha Round, Paragraph 19 requests the expansion of the review to the relationship between the TRIPS and the CBD as well as the protection of traditional knowledge and folklore, as stated: '(19) We instruct the Council for TRIPS, in pursuing its work programme included under the review of Article 27.3(b), the review of the implementation of the TRIPS Agreement under Article 71.1 and the work foreseen pursuant to paragraph 12 of this declaration, to examine, inter alia, the relationship between the TRIPS Agreement and the Convention on Biological Diversity, the protection of traditional knowledge and folklore, and other relevant new developments raised by members pursuant to Article 71.1. In undertaking this work, the TRIPS Council shall be guided by the objectives and principles set out in Articles 7 and 8 of the TRIPS Agreement and shall take fully into account the development dimension.'

Accordingly, developed countries such as the USA and Japan use patent protection, while many developing countries have ratified UPOV, because it 'provides a generic solution, meaning that initially it is likely to be easier to administer, but in the long run could end up only protecting the interests of large-scale commercial breeders and biotechnology companies' (Robinson, 2007, p. 5). Anyway, TRIPS leaves the possibility for developing countries to build their own laws in order to protect traditional knowledge. Attracting worldwide attention – primarily through NGO campaigns, farmers' networks and organizations such as the Society for Research and Initiatives for Sustainable Technologies and Institutions (SRISTI), Third

World Network, BioThai, Navdanya – the governments of India, the Philippines and Taiwan are developing their own legal mechanisms (Robinson, 2007, p. 9).

The SPS agreement encouraged the harmonization of sanitary and phytosanitary measures in accordance with international standards and requested member states to use these measures only for the protection of human, animal or plant life or health, never as non-tariff barriers on international trade, as stated: Article 2(1) Members have the right to take sanitary and phytosanitary measures necessary for the protection of human, animal or plant life or health, provided that such measures are not inconsistent with the provisions of this Agreement. Article 2(2) Members shall ensure that any sanitary or phytosanitary measure is applied only to the extent necessary to protect human, animal or plant life or health, is based on scientific principles and is not maintained without sufficient scientific evidence, except as provided for in paragraph 7 of Article 5. Article 3(1) To harmonize sanitary and phytosanitary measures on as wide a basis as possible, Members shall base their sanitary or phytosanitary measures on international standards, guidelines or recommendations, where they exist, except as otherwise provided for in this Agreement, and in particular in paragraph 3.

The SPS agreement has been the focus of much attention. For example, its article 5(7) which includes a precautionary principle, has contributed to the deepening of the Transatlantic GMO conflict: Article 5(7) In cases where relevant scientific evidence is insufficient, a Member may provisionally adopt sanitary or phytosanitary measures on the basis of available pertinent information, including that from the relevant international organizations as well as from sanitary or phytosanitary measures applied by other Members. In such circumstances, Members shall seek to obtain the additional information necessary for a more objective assessment of risk and review the sanitary or phytosanitary measures accordingly within a reasonable period of time.

However, the SPS agreement has helped the proliferation of private standards and strengthened the role of the Codex Alimentarius

Commission in making decisions concerning GMOs, as clearly indicated in Article 3(4): Members shall play a full part, within the limits of their resources, in the relevant international organizations and their subsidiary bodies, in particular the Codex Alimentarius Commission, the International Office of Epizootics, and the international and regional organizations operating within the framework of the International Plant Protection Convention, to promote within these organizations the development and periodic review of standards, guidelines and recommendations with respect to all aspects of sanitary and phytosanitary measures.

In opposition to TRIPS there stands, with its conflicted objectives, the biodiversity related CBD, which entered into force in 1993, 2 years before TRIPS. The CBD and its supplementary 2003 Cartagena Protocol on Biosafety cover all ecosystems, species and genetic resources by dealing with 'conservation of biological diversity, the sustainable use of its components, and the fair and equitable sharing of the benefits arising from commercial and other utilization of genetic resources' (CBD, 2017). As of January 2014, the CBD had 193 nations that had ratified or acceded to it, while 163 nations had ratified or acceded to the Cartagena Protocol. Over 130 nations adhere to both treaties, TRIPS and CBD. The USA have signed, but not yet ratified the CBD, so as a non-party to it, cannot become party to the Cartagena Protocol. Developing countries are persistently 'lobbying for the incorporation of CBD values into the TRIPS', but the TRIPS agreement still dominates over issues related to transgenic technology (Pechlaner and Otero, 2008).

Based on these supranational trade agreements, many national neoliberal initiatives have been taken around the globe. National politics play the major role in acceptance or rejection of genetically modified food as an integral part of neoliberal policy promotion, so we argue in favour of Pechlaner and Otero, who have proposed using the term 'neoregulation' instead of 'deregulation', commonly used in literature for describing politics associated with GMOs (Pechlaner and Otero, 2008). The crown of neoliberal activities is IPR protection of transgenic technology, its

processes, methods and products. By eliminating the possibility of saving seeds as the last source of farmers' independence, the 'commodification of nature' was achieved, and seeds ironically became the engine of the capitalist development of agriculture (Mascarenhas and Busch, 2006). Taking into consideration that GMO supranational and national regulations, especially IPR, are integral to the third food regime and 'provide the means for biotechnology's ascendancy as a central technology for capitalist agriculture…while the technology itself provides a means for further corporate concentration and integration of the food regime' (Pechlaner and Otero, 2008, p. 352), we want later in the text to provide an answer as to what extent the transgenic technology tucked inside neoliberal supranational and national regulations has reorganized agricultural trade relations and been a point of trade stability. Furthermore, we want to assess the impact of contradictory forces in the form of local resistance to transgenic technology in a particular country as well as across the whole food regime.

2.3 Transgenic Technology Neoregulation

The principal drivers of worldwide regulatory activity on GMOs are the two superpowers, the USA and the EU. 'GMOs have become big business in the United States, where both government and industry have embraced the new technology of genetic engineering. The United States has surged ahead and remains the world leader in the development of GM foods. By contrast, the EU has taken a far more cautious approach to GMOs, dragging out the approval processes for new GM foods and insisting that such products be labeled as such for consumers. The EU's more cautious approach to GM foods has led many companies, such as Bayer AG and BASF AG, to move their biotechnology research facilities to the more biotech-friendly United States. More ominously, the EU's slow approval of new GM crops, coupled with its insistence on the labeling of such crops in the marketplace, have created serious obstacles to the export of agricultural products from the United States, and thus raised the prospect of a major transatlantic trade war' (Pollack and Shaffer, 2000, p. 41).

'Their policy-choices limit the options of other countries, particularly those that are economically dependent on the EU, the USA, or both. Switzerland, Norway, and Central and Eastern European countries have thus aligned with the EU, Canada with the USA. Other countries, which are less dependent on EU or US markets, e.g., China, Brazil, India, Japan, and Russia, have adopted regulations whose stringency lies somewhere in between the EU and the US model. GMO policy in these countries is very recent and very much in flux. Both the EU and the USA have been battling for influence on the regulatory policies of these countries' (Bernauer and Aerni, 2008, p. 7).

With respect to the above, we compare national GMO politics and local resistance in the two agriculture superpowers, the USA and the EU, with particular reference to Spain, because it shares a common European legal framework but has a completely opposite view on GMO issues. The BRICS – Brazil, the Russian Federation (hereafter Russia), India, the People's Republic of China (hereafter China) and South Africa – countries are selected for the analysis because of their huge influence on reshaping the existing food regime. The USA and EU, often competing with each other, have found a common interest in defending themselves against the BRICS bloc, which has the potential to become larger than the G6 (France, Germany, Italy, Japan, the UK, the USA), in US dollar terms, within 40 years (Wilson and Purushothaman, 2003). The BRICS countries have already changed the global picture of agricultural production and trade, as well as the picture of financial institutions, by setting up the New Development Bank, which will probably lead to a reduced dependence of developing countries on the IMF and the World Bank. Finally, Serbia, a small upper-middle income country from the Balkan region, is also included in the analysis, as providing an example of a strong social movement of GMO rejection, despite its minor role in global agriculture.

National GMO neoregulations cannot be properly understood without knowing the role of agriculture in each country. Thus, apart from the description of national GMO policies, we describe the basic agricultural performance, domestic support for agriculture and trade relations for each country. We also outline resistance to GMOs, and finally come to a conclusion about the transgenic technology impact on the evolution of the third food regime.

2.3.1   USA

It is well known that from the very beginning the USA has supported the transgenic industry, and that all administrations from Ronald W. Reagan's to Barack H. Obama's have pursued a policy that has considered transgenic and conventional foods substantially equivalent (Brankov and Lovre, 2012). Also, Donald J. Trump officially came out in favour of GMOs for the first time in January 2018 (Meyer, 2018).

The key US government document on biotechnology is *Coordinated Framework for the Regulation of Biotechnology,* issued as a formal policy in 1986. The Coordinated Framework is based upon existing laws, but with new regulations, policies and guidance for the application of these laws to transgenic technology-derived products (USDA, 2017). This framework specified the US Environmental Protection Agency (EPA), the USDA's Animal and Plant Health Inspection Service (USDA-APHIS) and the Department of Health and Human Services' Food and Drug Administration (FDA), as the three primary agencies responsible for oversight of the transgenic technology products. Under this framework, the EPA becomes responsible for pesticidal plants and genetically-engineered pesticides; the USDA for transgenic plants (import, interstate movement, release into the environment); and the FDA for the safety of transgenic products. 'According to a policy established in 1992, the FDA considers most GM crops as *substantially equivalent* to non-GM crops. In such cases, GM crops are designated as *Generally Recognized as Safe* under the Federal Food, Drug, and Cosmetic Act (FFDCA), and do not require pre-market approval' (FAS, 2018). Throughout the 1990s, the Agencies relaxed their GMO approval policies. APHIS introduced a notification instead of permit procedure. The FDA introduced voluntary consultation with GM crop developers to review the determination of 'substantial equivalence' before the crop is marketed, instead of conducting a comprehensive scientific review. 'In essence, this system allows biotechnology firms to decide for themselves whether a GM product is safe. The FDA is only consulted. This strong product-orientation is also reflected in policy on labeling. Labeling is only mandatory if a particular GM food product is no longer substantially equivalent to the corresponding conventional food in terms of composition, nutrition or safety. The label does not have to indicate that the food was produced with biotechnology. It must only state that a potential allergen has been added to the food' (Bernauer and Meins, 2003, pp. 663–664). Such labelling is very rare, and the vast majority of GM foods in the USA do not require labelling. However, there are some GMO products on the American market that are labelled. For example, transgenic canola oil with increased lauric acid content is labelled as 'laurate canola oil', soybean oil with a higher level of oleic acid as 'high oleic soybean oil', and stearidonic soybean oil (stearidonic acid is not found in conventional oil) as 'stearidonate soybean oil' (FDA, 2018).

Apart from a favourable regulatory climate, patent protection has had a profound role in the spread of transgenic technology in the USA. Going back in history, the Plant Variety Protection Act (PVPA) of 1970 allowed:

> seed companies to receive a 17-year protection on new varieties of the sexually produced seeds, other than hybrids, that they had developed... However, farmers were still allowed to save seed for replanting and to sell part of their seed to other farmers. In 1994, the PVPA was amended to increase incentives for private plant breeders. As a result, farmers were no longer permitted to sell seed without a license from the owner of the variety, and seed saving was only allowed for personal replanting. Attracted by the prospect of patent-like protection for new seed varieties, large chemical, oil and processing corporations [such as Dow, Dupont and Monsanto] acquired many of the independent seed companies and began to fund substantial research and development efforts... A second wave of mergers occurred congruent with, or in anticipation of, the extension of patents to seeds, plants and tissue cultures. Indeed, private investment in agricultural research tripled in real terms between 1960 and 1992.
> (Mascarenhas and Bush, 2006, p. 127).

The third stimulus to the rapid penetration of the transgenic food and feed into American markets was provided by the public. 'Public support for all applications of biotechnology has remained at rather high levels over time. General consumer acceptance of plant biotechnology has remained approximately constant and high, compared to Europe. Survey results from 1992, 1994 and 1998 show constant support from 70 percent of surveyed Americans' (Bernauer and Meins, 2003, p. 666). Similar trends continued in the following years. 'The majority of U.S. consumers expressed little to no concern about food and agricultural biotechnology, and were likely to buy food products produced from GM plants, although consumer awareness and knowledge about GM food was superficial' (Lucht, 2015). The root of the US public attitude towards GMOs can be seen in an individualistic culture and the weak influence of NGOs. In general, US policy was based primarily on industry interest, and NGOs were rarely involved in policy formulation or implementation' (Doh and Guay, 2006, p. 63). This contrasts strongly to the NGOs' powerful role in Europe, where broad social protest – the precursor to the NGOs – dates from the 18th century and takes on a modern appearance in the 1960s. In the communitarian European tradition, for a long time NGOs have received generous financial support and farmers are directly involved in governmental policy and corporate governance.

Thus, it is not surprising that from the very beginning the USA has been the absolute leader in production, accounting for 64% in the total area under transgenic crops in 1997; and although with time and the inclusion of other countries, its participation decreased to 39.4% in 2015 (Brankov et al., 2016), the USA is still the leader in hectarage planting. The country grew maize (35.5 mill ha), soybean (31.84 mill ha), cotton (3.70 million ha), alfalfa (1.23 million ha), canola (0.62 million ha), sugar beet (0.47 million ha), papaya (1000 ha) and squash (1000 ha) on 72.9 million ha in 2016. The soybean adoption rate is estimated at 94%, maize adoption at 92%, and cotton at 93% (James, 2016). Such a high adoption rate clearly indicates that GM crops dominate throughout the US food system.

Maize, wheat, soybeans, cotton, and hay account for 90% of harvested acreage in the USA (Table 2.1). Considering that transgenic alfalfa adoption rates have increased relatively slowly compared to other field crops, owing to the moratorium on new alfalfa seed in 2007 (suspended in 2010), and that it tends to be seeded (on average) once every 7 years (Fernandez-Cornejo et al., 2016), it can be seen that more than half of all the harvested area in the USA is transgenic.

Although the majority of farms in the USA are small farms, the bulk of production is concentrated in a small number of commercial farms that receive 'over half of the government's total commodity payments' (Table 2.2) (OECD, 2011). The number of farms has levelled off at about 2.2 million, with an average size of about 418 acres (169.1 ha) in 2016, as opposed to 155 acres (62.3 ha) in 1935. The average size of rural residence farms is about 67 ha, and of intermediate farms about 278 ha, while commercial farms average at around 710 ha. Although the number of farms has been a declining trend since World War II, the number of large farms grew rapidly in the period 1989–2007, with the most rapid growth of farms with sales of more than US$500,000. Concentration of production in agriculture is clearly indicated by the fact that 'in 1982, 431,634 farms produced 80% of the value of agricultural production, while in 2007 around half of this number produced 85%... The share of total sales accounted for by farms with sales of US$250,000 or more 'increased steadily, from 57% in 1982, to 85% in 2007' (OECD, 2011, p. 18). Also, farms in the USA have become increasingly specialized – about half produce just one single commodity. US agricultural production occurs in each of the 50 States, with California leading. Crop production that accounts for the largest share of the value of the country's production is concentrated in California, Iowa, Illinois, Minnesota, and Nebraska, while Texas, Iowa, California, Nebraska and Kansas lead the country in sales value of livestock and their products.

Highly concentrated interests groups gained the most benefits from the US agricultural policy. The Farm Bill, the permanent legal framework that governs a wide range of concerns related to agriculture, originally was designed to stabilize and boost farm income, but over time it has been amended to address additional objectives. By the end of the 1980s, it had become increasingly clear that in order to sustain prices and revenue, US farmers and agricultural firms relied heavily on export markets. Accordingly, a 1990 Farm Bill was particularly aimed to enhance exports. 'For the first time, this legislation included planning-flexibility provisions that promoted economic freedom by increasing farmers' ability to shift their resources in

Table 2.1. Harvested acreage in the USA, major crops, 2007. (From USDA ERS, 2013).

Commodity	Harvested area, 2007	
	Acres (mill)	%
Field crops	299.7	96.4
Barley	3.3	1.0
Corn (maize)	86.3	27.7
Cotton	10.5	3.4
Hay	58.1	18.7
Oats	1.5	0.5
Rice	2.8	0.9
Sorghum	6.7	2.1
Soybeans	63.9	20.6
Tobacco	0.4	0.1
Wheat	50.9	16.4
Other field crops	15.5	5.0
High-value crops	11.1	3.6
All crops	310.8	100.0

Table 2.2. Characteristics of US farms, 2007. (From OCED, 2011).

	Rural residence farms			Intermediate farms		Commercial farms			All farms
	Limited resource	Retirement	Residential/ Lifestyle	Farming occupation/ Lower sales	Farming occupation/ higher sales	Large family	Very large family	Non-family farms	
Number of farms (1000)	309	456	802	259	100	87	101	91	2205
Share of farms (%)	14	21	36	12	5	4	5	4	100
Land in farms (mill acres)	42	90	121	87	104	123	211	143	922
Average size (acres)	137	196	151	337	1040	1420	2085	1572	418
Total value of production (US$ bill)	3	7	11	6	17	31	157	66	297
Average per farm (1000 US$)	9	17	14	27	176	373	1577	732	138
Share of value of production (%)	1	2	4	2	6	10	53	22	100

response to world market conditions… After the Uruguay Round had concluded, various commodity organizations and agribusiness interests coalesced in the hope of eliciting a more market-oriented agricultural policy. Partly as a result of these groups' influence, target-price deficiency payments were replaced by fixed "production flexibility contract payments" that would be paid during the ensuing 7 years, and both the crop-basis and the acreage reduction programs were eliminated…' (Rausser and Zilberman, 2014, p. 519). Considering that 'each dollar received from agricultural exports stimulates another US\$1.64 in supporting activities to produce these exports' (OECD, 2011, p. 14) this political intervention continues in the following years. 'The trade title of the 2008 farm bill authorized and amended four kinds of export and food aid programs: direct export subsidies, export market development programs, export credit guarantees, and foreign food aid… During the period of 2002 through 2010, federal support of US agricultural exports—including the Food for Peace Act (FPA), credit guarantees, and generic and brand commodity promotion programs, averaged US\$5.5 billion annually' (Henneberry, 2013, p. 197–201). Such political intervention can be seen as a success because US agricultural exports is about 'a third of total US agricultural cash receipts' and the USA 'accounted for 23%, 52%, and 11% of world exports of wheat, corn, and rice, respectively' (Henneberry, 2013, pp. 197–201). Despite obvious benefits, the USDA's Foreign Agricultural Service partnership with non-profit trade associations representing commodity or regional interests often 'have been highly criticized as promoting corporate welfare' (Henneberry, 2013, p. 195). Such an example is transgenic technology promotion.

In acquiring its position, the USA has used a variety of aggressive means. As already mentioned, countries economically dependent on the USA have aligned their national legislation with the recommendations of the USA. European countries have imposed severe regulatory constraints on GMOs, with strong emphasis on the precautionary principle, and introduced a ban on US corn (maize) imports in the late 1990s. The moratorium caused economic losses for US companies estimated at around several hundred million US dollars. The US government was worried that other countries – important export destinations for US agricultural products – would follow the EU policy model for GMOs; with the support of Argentina and Canada, the USA initiated litigation in the WTO against the EU. The WTO's position was that the EU had created unlawful trade restrictions. The transatlantic GMO dispute 'has forced many developing countries to take sides and has crowded out systematic and pragmatic domestic debates in these countries about the types of biotech applications they may want and need' (Bernauer and Aerni, 2008, p. 7). For example, in 2002 US transgenic food aid intended for Malawi, Mozambique, Zambia, and Zimbabwe has nothing to do with ending hunger in the regions but much to do with promotion of transgenic crops in Africa and with expanding control of transnational corporations on that continent (Zerbe, 2004). Finding themselves between two superpowers, African governments have requested that US aid be milled before distribution. 'Unable to compete directly against American and European farmers, who are heavily subsidized and protected by their governments, African farmers were responding to European demands for non-GM agriculture through specialized production' (Zerbe, 2004, p. 607). Threatening to resort to the WTO has been a common threat for the USA; it was also applied against Croatia and Sri Lanka (Paul and Steinbrecher, 2003).

Contrary to the EU, where the first transgenic US soy shipment in 1996 caused intense protest, anti-GMO movements in the USA only intensified after high technology saturation. Demands for mandatory labelling are growing louder with time. 'In four states (California in 2012, Washington in 2013, and Colorado and Oregon in 2014) ballot initiatives for the mandatory labelling of GM food were put to public vote, and rejected by a narrow margin. Connecticut and Maine both passed legislation mandating GMO labels, but it is not clear when and whether these will enter into force, since they depend on the introduction of similar regulations in neighboring states. Legislation passed in Vermont (2014), that may make it the first U.S. state to enact the active requirement for GMO labels for food, is being contested in court' (Lucht, 2015). Between 2011 and 2013, the *Los Angeles Times, New York Times, Wall Street Journal*, and *Washington Post* published 207 favourable and 250 unfavourable articles about GMOs (Mintz, 2017). This late awakening of the American public is being confronted by great resistance from the industry, so it is unlikely that mandatory labelling at the federal level will be introduced. However, there has been a strong expansion of voluntary GMO labelling schemes, which may possibly open a profitable niche market for retailers (Lucht, 2015). A recent report showed that in the USA, 'out of the 7637 new food products introduced between February 12, 2010 and February 11, 2011, 2.6% advertise that they are GMO-free and 8% advertise that they are organic' (Goodwin *et al.*, 2016, p. 364). Logically, local food marketing channels have experienced growth, 'as of 2014, there were 8,268 farmers' markets in the United States, having grown by 180 percent since 2006' (Nestle, 2016, p. 2). Popular magazines have published articles with titles such as, 'Meet Your Farmers', 'Know Your Farmers', 'Every Family Need a Farmer'; even the USDA launched a campaign, 'Know Your Farmers, Know Your Food' (Myers and Sbicca, 2015). In that way, the USDA has involved itself in the promotion of alternative niche markets. Why? Two possible reasons can be suggested. The first and most important lies in the fact that, although the USA spends more on healthcare than any other country, 'the nation ranks lower than several other nations in life expectancy, infant mortality, and other healthy life indicators' (Benjamin, 2011). Healthcare spending in the USA started at the beginning of the 20th century at 0.25% of GDP and increased significantly with time, breaking through to 8% of GDP in 2016 (US Spending, 2018). According to the WHO list for 2015, the USA ranks as the 31st country in the world for life expectancy. Also, according to the CIA list for 2017 (CIA, 2017), the USA is ranked 170th out of 225 countries for infant mortality. That is why the National Prevention Strategy has identified seven

priority areas that require immediate focus to improve the health of the American people – among them, healthy eating (Benjamin, 2011).

The second possible reason is that the optimistic view of the transformative potential of organic farming is the result of the growing concern over the demand/supply gap, and of growing criticism about the lack of government support for organics. But, in its essence, this view is driven by corporate interests.

> The organic distribution system was transformed from one characterized largely by direct sales combined with natural foods retailers such as Whole Foods, to one fully incorporated into the conventional system, including mass retailers with their own private-label brands and the rapid growth of middlemen 'handlers' that coordinate the organic supply chain...As organics moved beyond its niche status in California, agribusiness entered the market to capture the monopoly rents associated with the price premium...As part of conventionalization, the organic label was co-opted by large firms, thereby blunting its transformative potential, as it was appropriated and subsumed by corporate actors.
> (Constance *et al.*, 2015, pp. 165–166).

Although the organic sector is growing, (occupies 5.3% of all agricultural land in the USA) and the potential exists for its concentration in some areas of the USA (Huffman, 2017), GM farming will remain as a dominant model and the US market will flood with new GMOs in the foreseeable future.

2.3.2 EU

The EU has possibly the most stringent GMO regulations in the world. It prescribes GMO safety, GMO thresholds, GMO labelling, GMO detection and coexistence. The building blocks of the GMO legislation are: Directive 2001/18/EC, Regulation (EC) 1829/2003, Directive (EU) 2015/412, Regulation (EC) 1830/2003, and Directive 2009/41/EC, supplemented by a number of rules and guidelines (EC, 2018).

The European Food Safety Authority (EFSA), in collaboration with its member states' scientific bodies, is responsible for GMO risk assessment. It implements Directive 2001/18/EC on deliberate release into the environment, and Regulation 1829/2003 on GM food and feed. On the basis of EFSA conclusions, the Commission drafts a proposal for supporting or rejecting the authorization for placing a GMO product on the market. Thanks to the newly adopted Directive 2015/412, member states may prohibit or restrict the cultivation of the GMO during the authorization procedure or after that. After the authorization, a member state may:

> prohibit or restrict the cultivation of the crop based on grounds related amongst others to environmental or agricultural policy objectives, or other compelling grounds such as town and country-planning, land use, socio-economic impacts, co-existence and public policy. Before the adoption of this Directive, Member States could provisionally prohibit or restrict the use of a GMO on their territory only if they had new evidence that the organism concerned constitutes a risk to human health or the environment or in the case of an emergency.
>
> (EC, 2015)

No one has ever been in a position before to provide such evidence as would activate the 'safeguard clause'. Therefore, these recent changes can be viewed as further tightening of the legislation. As stated in Regulations 1829/2003 and 1830/2003, the EU fix a threshold for the adventitious, or accidental, presence of GM material in non-GM food or feed sources of 0.9%, and this only applies to GMOs that have an EU authorization. Above this threshold, all foods should be labelled: *'This product contains genetically modified organisms [or the names of the organisms].'* In addition, the EU legislation does not forbid the use of 'GM-free' labels. Traceability (the ability to track GMOs at all stages of the production and distribution chain) regulation requests the sellers to 'inform trade buyers in writing that a product contains GMOs; communicate the unique identifiers assigned to each GMO under the regulation; identify each ingredient produced from GMOs. EU countries must carry out inspections, sample checks and tests, to ensure the rules on GMO labelling are followed' (EC, 2016). 'In the EU, field coexistence regulations are formulated by national governments, according to the principle of subsidiarity, while the EC simply issues guidelines. As a result there is great heterogeneity in coexistence regulations on buffer zones and isolation distances between GM and non-GM crops, which vary between 15 m (for silage maize in Sweden) and 800 m (for maize in Luxembourg), depending on the member state' (Davison, 2010, p. 96). Labelling, traceability, detection, and coexistence regulations as non food-safety issues, have the aim of providing consumers with the choice between GMO and non-GMO foods. All in all, the EU applies the precautionary principle as a guiding approach for GMOs. As the EUR-Lex glossary explains, this principle 'relates to an approach to risk management whereby, if there is the possibility that a given policy or action might cause harm to the public or the environment and if there is still no scientific consensus on the issue, the policy or action in question should not be pursued' EC (2016). This policy has been the subject of many criticisms by both opponents and proponents of GMOs. Often it is signed as too restrictive, a concealed trade barrier, political, and a manifestation of an 'anti-science attitude', leading 'to price rises in the meat and poultry industries' (Davison, 2010, p. 98). There are, on the other hand, opinions that such a legal regime is too weak and that the EU 'has accepted the inevitability of GMO-contamination to some level, and only seeks to control it' (Paul and Steinbrecher, 2003).

Anyway, GMO cultivation in the EU is very limited, 19 EU countries have 'opted out' of GMO crops within all or part of their territories (Eco Watch, 2015). In 2016, only four countries in the EU (out of 28) – Spain, Portugal, Slovakia and the Czech Republic – planted a single transgenic crop, *Bt* maize event MON 810, resistant to the European corn borer. Table 2.3 shows that the area planted with *Bt* maize in the EU reached its maximum in 2013. After a significant drop in 2014 and 2015, a recovery started in 2016 (James, 2016). In the Czech Republic and

Table 2.3. Transgenic maize area in the EU, by Member States (ha). (From James, 2016.)

Country	2011	2012	2013	2014	2015	2016
Spain	97,326	116,307	136,962	131,538	107,749	129,081
Portugal	7,724	9,278	8,171	8,542	8,017	7,069
Czechia	5,091	3,080	2,560	1,754	997	75
Romania	588	217	220	771	3	–
Slovakia	761	189	100	411	104	138
Poland	3,000	N/A	–	–	–	–
Total	114,490	129,071	148,013	143,016	116,870	136,363

Portugal, the area under *Bt* maize has gradually decreased, and Romania and Poland have stopped producing it, while Slovakia grew it on just 138 ha. All in all, Spain represents almost 95% of the total area in 2016.

Transgenic maize produced in the EU is used locally as animal feed and for biogas production. The EU does not export any transgenic crops or plants, but as a protein deficient area, imports large quantities of feeds. The share of transgenic products in total imports is estimated 'at around 90 percent for soybeans, less than 25 percent for corn [maize], and less than 20 percent for rapeseed' (GAIN, 2016a).

In relation to the attitude toward GMOs, there are great differences across EU countries. The GAIN report classified member states, depending on their acceptance of transgenic technology, into three categories: adopters, conflicted and opposed (GAIN, 2016a). 'Adopters' have 'pragmatic governments and industry generally open to the technology': Spain, Portugal, Slovakia, the Czech Republic, Denmark, Estonia, Finland, Flanders in Northern Belgium, the Netherlands, Romania, and England in the UK. 'Conflicted' states are those states where 'most scientists, farmers, and the feed/food industry are willing to adopt the technology, but consumers and governments, influenced by anti-biotech groups, reject it': France, Germany, Poland, Southern Belgium (Wallonia), Bulgaria, Ireland, Lithuania, Sweden, Northern Ireland, Scotland, and Wales. In the 'opposed' states, 'most stakeholders and policy makers reject the technology': Austria, Croatia, Cyprus, Greece, Hungary, Italy, Malta, Slovenia, Latvia and Luxembourg. European consumers generally have more negative perceptions about and lower intentions to purchase transgenic foods in comparison with consumers from North America. According to the Eurobarometer survey, Europeans 'do not see benefits of genetically modified food, consider genetically modified foods to be probably unsafe or even harmful and are not in favour of development of genetically modified food' (TNS Opinion & Social, 2010, p. 7). European NGOs, Green political parties and the organic movement have often organized 'sensationalistic campaigns multiplied by media articles' (Lucht, 2015).

In order to fully understand the EU position on GMOs, it is necessary to return to the past – to the emergence of the Common Agricultural Policy (CAP) in 1957. Article 39 of the Treaty of Rome specifies a set of objectives for the CAP, as follows: (i) to increase agricultural productivity by promoting technical progress and by ensuring the rational development of agricultural production and the optimum utilization of the factors of production, in particular labour; (ii) thus to ensure a fair standard of living for the agricultural community, in particular by increasing the individual earnings of persons engaged in agriculture; (iii) to stabilize markets; (iv) to ensure the availability of supplies; and (v) to ensure that supplies reach consumers at reasonable prices.

The CAP was determined by the agricultural policies of the six founding countries – Belgium, Luxembourg, the Netherlands, Germany, France, and Italy – in which agriculture played a major role in the economy (Table 2.4). Except for the Netherlands, all the other members of the original EEC Six (and then the UK, who later joined them) were a net importers of agricultural products

Table 2.4. Main agricultural indicators in the EEC Six and the UK, 1955–1960. (From Zobbe, 2002.)

	Agriculture share in GDP (%)		Agriculture share in total employment (%)		Agriculture share in total trade (%), average, 1955–59		Net foreign trade in agricultural products, 1960 US$, 1955–59, average
	1955	1960	1955	1960	Export	Import	
Belgium	7.9	7.3	9.3	7.6	5.4	17.2	−386.4
Luxembourg	9.3	7.6	19.4	16.4	–	–	–
Holland	11.4	10.5	13.2	11.5	33.6	19.6	+310.0
Germany	8.0	6.0	18.5	14.0	2.8	32.9	−2,124.6
France	11.4	9.7	26.9	22.4	14.9	29.2	−836.0
Italy	20.7	15.1	40.0	32.8	22.6	20.6	−114.2
EEC Six	11.5	9.0	21.2	17.5	15.9	23.9	–
UK	4.8	4.0	4.6	4.3	6.5	41.8	−4,013.6

(Zobbe, 2002). 'The main type of farm in Europe after the Second World War was a fairly small-scale family-owned farm which had structural problems to a greater or lesser extent. Increased production through increased productivity was seen as a solution to the farmers' income problem. In all six countries, a price policy combined with various structural policy measures had been chosen as the means to achieve this goal' (Zobbe, 2002, p. 13).

A high level of dependency on food imports was seen as political weakness – 'US farm exports to western Europe were worth over US$1 billion in 1961' (Ludlow, 2005, p. 365). Concern about the post-war provision of food security led the founders of the EEC to make certain decisions at the Stressa Conference in 1958. They formulated a policy of price support. Thanks to a highly expensive and protectionist system, the EEC and later the EU successfully increased agricultural self-sufficiency. 'In 2012, 96% of the total available meat in the EU was produced in the EU itself, while 92% of the available amount of meat was consumed. The comparable figures for fresh dairy products were both 99%. So, the self-sufficiency of the EU's consumer protein consumption based on animal-derived products is very high' (De Visser et al., 2014). However, 'the plant protein input for the EU's animal production industry shows an entirely different picture... (that of) an EU balance sheet for protein-rich feed materials showing that 69% is imported into the EU

(excluding fish meal)... the self-sufficiency of soya bean meal is only 3%, while this product supplies 64% of the protein-rich feed materials' (De Visser et al., 2014).

The current situation of the EU, with very little cultivation of transgenic plants but high imports, is the reflection of the desire of the community to maintain self-sufficiency in food production and the necessity of importing animal feed. The European view on GMOs can also be connected with the environmental policy. When the US leadership in environmental multilateral policy began to weaken in the 1980s, and when Japan failed in the 1990s to assume the position, 'Europe became the global leader' (Falkner, 2007, p. 7). It has adopted a progressive approach to sustainability and multifunctional agriculture (mutual supportiveness of economic, social and environmental objectives). The EU had a crucial role in the Cartagena Protocol negotiation and 'pushed for the adoption of the precautionary principle in risk assessment and sought to assist developing countries in their effort to strengthen domestic regulations, against the interests of the USA and other GMO exporters' (Falkner, 2007, p. 10). Europe has spent many years building a comprehensive system of regulation, and successfully exported its standards and model for precautionary regulation abroad. Thus, the EU will not easily give up its position on regulatory export, and global environmental matters will remain at the heart of its foreign policy. Consequently, it is not expected for the transatlantic GMO conflict to be resolved in the medium term.

Spain

The central hub of transgenic technology in the EU is located in Spain. In 2015 Spain was responsible for almost 95% of the EU's total corn MON810 production, on 107,748 ha with an adoption rate of 35% (GAIN, 2016b). Since MON810 is the only GMO event approved in the EU, the possibility for further diffusion of transgenic technology in Spain is limited to the size of the maize area. Spain follows EU monitoring and testing rules as well as the rules for labelling GMOs, thus establishing a threshold of 0.9% for accidental and technically unavoidable approved GMO ingredients, 0.1% for non-approved GMOs intended for feed use and 0% for non-approved GMOs that will enter the food chain. A threshold level for seeds has not been established yet, although the domestic seed industry insists on it in order to start trading with other partners (GAIN, 2016a). Regarding IPR, the Community Plant Variety Office (CPVO), an EU agency, manages a system of plant variety rights covering all 28 member countries, so breeders can protect their rights under CPVO or a national office. Plant varieties that are new, distinct, uniform and stable can be protected, but Article 53 (b) of the European Patent Convention (EPC) as of October 1973 excludes patents for plant varieties as stated: European patents shall not be granted in respect of: (i) inventions the commercial exploitation of which would be contrary to 'ordre public' or morality; such exploitation shall not be deemed to be so contrary merely because it is prohibited by law or regulation in some or all of the Contracting States; (ii) plant or animal varieties or essentially biological processes for the production of plants or animals; this provision shall not apply to microbiological processes or the products thereof.

There are some opinions that patent protection for plant varieties, including transgenic varieties, 'can be achieved indirectly by claiming a process to produce such a variety' (Antonow, 2010). However, at the moment transgenic varieties are not IPR issues in Spain. As a member of the EU, Spain has been a signatory of Cartagena since 2002. Although Spain shares a common European legal framework, it has a quite opposite view on GMO issues compared with other EU countries. The causes of the fairly easy acceptance of GMOs lies in the 'weak social mobilization and the shutting-out of NGOs and critical civil society voices... In 1998, the Spanish government authorized two varieties of *Bt* maize 176 for the first time, entrusting the biomonitoring process to the same companies that had created those varieties. The change of government in 2004, from right-wing to more center-oriented, made it possible for the protests coming from civil society to be heard, and a representative from the environmental sector was admitted to the National Commission on Biosafety' (Fernandez-Wulff, 2013). This is an example of how a country can suddenly change GMO policies because of elections, and change from a pro-GMO to an abstinent country over a few years, only to become a pro-GMO country again. Although the question of diffusion of transgenic technology is a highly politicized issue in all countries, in Spain it is associated with non-transparency. A good example is the refusal of the Spanish administration to publish a transgenic crops field register under the pretext that it could be misused (GAIN, 2016a). The reasons for Spain's adoption of GMOs have roots in agricultural transformation history. Starting from the 1950s under Franco's regime, Spain relatively quickly adopted a shift in agricultural policy and spread technological innovations, which implied an increase of capital stocks and a decrease of traditional inputs. For example, an annual average rate of capital use in the Spanish agricultural sector was 3.6% in the period 1950–2005, which was higher than in other Western Europe countries, and 'the use of capital continued to grow after 1985, when this growth was negative in the vast majority of the other European countries' (Clar *et al.*, 2016, pp. 5–6). 'While the agricultural modernization processes in the European democracies took place within a context of free participation by civil society (unions, cooperatives, entrepreneurs) ... and the consolidation of the welfare state, this was not the case in Spain, where modernization was facilitated by the authoritarian nature of the State, with very high social and environmental costs and without parallel measures that would increase the income of farmers' (Clar *et al.*, 2016, p. 13). The blatant elimination of small farms passed smoothly, with almost no resistance from farmers, who were forbidden to organize themselves in unions. The level of support was several times lower than in other European countries, even negative in some years. Spain also has faster growth in livestock production than Western Europe. Participation of livestock farming in agricultural production was 23% in 1961, 30.4% in 1970, 32.9% in 1985, 38% in 1995, and 36.8% in 2005. In the same period of 1961–2005, Europe increased participation of livestock farming in agricultural production by just 1.4%, from 46.3% to 47.7%, while the EU-9 decreased it by 0.5% from 52.4 to 51.9% (Clar *et al.*, 2016). Intensive livestock production contributed almost half of Spanish agriculture's productivity increase. It has been precisely the political decision to intensify livestock production that has opened the door for multinational companies, since Spain was not adequately supplied with feeds. As a solution, Spain started to import large quantities of soybeans and maize from the USA, and quickly established vertical integration with multinational suppliers of animal feeds, leaving farmers powerless. Spain

has not given up corporate agriculture favouritism at the expense of small farms, regardless of CAP policies.

Thanks to these measures, Spain is the world's fourth largest pork meat exporting country, just after the USA, Germany, and Denmark, and has managed to achieve a spectacular rise of pork meat exports, from 1544 tonnes in 1970 to 1.2 million tonnes in 2012 (Fig. 2.1).

The main destination for Spanish beef and pork, ahead of Japan, South Korea, and Hong Kong, is China, accounting for 48% of the exports in 2016 (Ecomercioagrario, 2017). Starting from 2015, Spain surpassed Germany in the number of pigs and became the EU's largest pig farming country, with 28.4 million pigs. However, its pork meat production of 3.9 million tonnes is still 1.7 million tonnes less than pork meat production in Germany (Pig Progress, 2016). Regarding the total meat production (Fig. 2.2), Spain is the ninth largest producer in the world after China, the USA, Brazil, Russia, Germany, India, Oceania, and Mexico, while in pork meat production Spain is ranked as the fourth largest country, behind only China, the USA, and Germany.

An increase in livestock production and dedication to a productivist approach makes Spain a large importer of transgenic feedstuffs, since domestic production is insufficient to meet the demand. In the period 1970–2014 Spain increased maize imports approximately

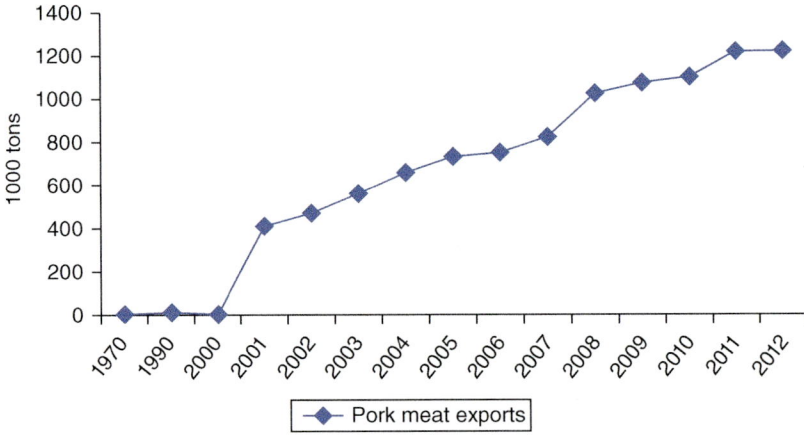

Fig. 2.1. Spain: Pork meat exports, by year. (Adapted from World Data Atlas (https://knoema.com/atlas).)

threefold (Fig. 2.3). In line with EU import approvals, and despite being a pro-GMO country, Spain imported a negligible share of transgenic corn (maize) from the USA, Argentina, and Brazil. Unlike with maize, for soybean imports (Fig. 2.3) there are no barriers in the EU, so most imported soybeans, if not all, are transgenic. Major suppliers are Brazil, USA, and Argentina.

To a large extent, the dependence on feedstuffs imports determines public perception of GMOs in Spain. The feed and livestock industries, as well as a large number of farmers' associations, are supporters of transgenic technology (GAIN, 2016a). Although some municipalities have declared themselves GMO-free, there is no strong public opposition to GMOs, as in some other EU countries. Almost all animal feed is transgenic and labelled in compliance with EU regulation. A somewhat different situation is found with final food products. Food with the label 'Contains GMOs' are rarely seen in Spain, because most manufacturers have eliminated GMO ingredients.

2.3.3 BRICS

The BRICS countries represent more than 40% of the world's population, while over a third inhabit just two of these countries:

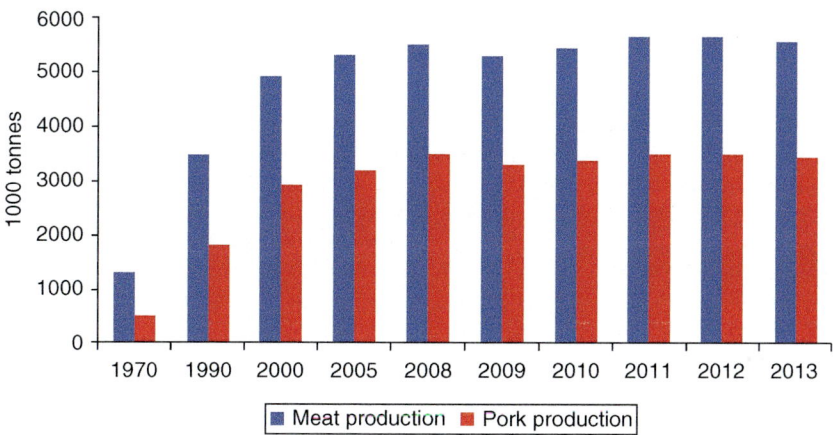

Fig. 2.2. Spain: Meat production, by year. (Adapted from FAOSTAT (http://www.fao.org/faostat/) and World Data Atlas (https://knoema.com/atlas).)

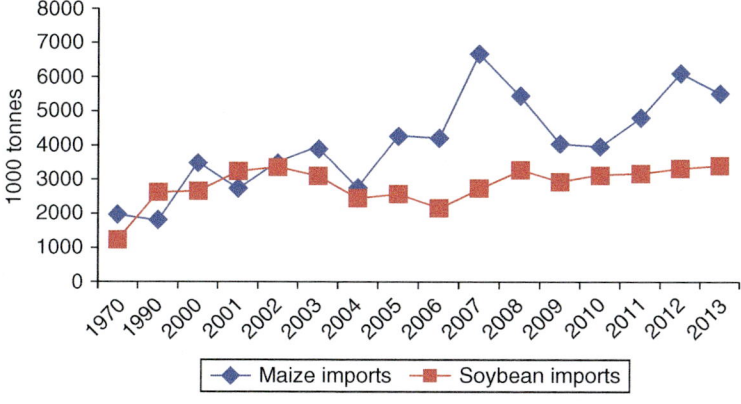

Fig. 2.3. Spain: Maize and soybean imports, by year. (Adapted from FAOSTAT (http://www.fao.org/faostat/).)

China and India (Table 2.5). BRICS account for 29.3% of the world territory, with Russia being the biggest country at 17,098 sq. km. BRICS contribute 21.5% of the world economic output, in which China excels, creating 13.3% of the world economy in 2014. Regarding gross domestic product (GDP) per capita, BRICS countries do not rank high, with only Brazil and Russia being somewhat above the world average of US$10,795, while India, China and South Africa are significantly below this. For example, GDP per capita in India in 2014 was just US$1,633, almost seven times lower than the world average.

Russia disposes of the greatest arable land per capita (0.86 ha), while China and India have the smallest surface area per capita, below the world average. BRICS contributed 38% to the total world cereal output and about 41% of world meat production in 2014 (Table 2.6). Just for comparison's sake, EU countries account for about 11% of the world cereal production and for 14% in world meat production, while the USA accounts for about 16% of the world cereal production and about 13% of meat production. BRICS produce more than 77% of the world's fruit and vegetables, led by China, which accounts for more than 58% on its own. In addition, China accounts for 61.6% of world aquaculture production (FAO, 2016).

The importance and role of the BRICS economies in the world food system can be seen from the above-mentioned indicators, particularly their impressive participation in food production. However, a considerably different picture of the BRICS agrarian sector can be obtained by analysing agriculture-generated income and capital stocks. As Table 2.7 shows, although BRICS agriculture employs a larger share of the economically active population (EAP), it generates less income per worker compared to the USA or Spain. The USA generates 81.7 times more income per agricultural worker than the worst ranked BRICS country, India, and 8.4 times more than the best ranked BRICS country, Russia. The ratio of agricultural GDP to agricultural EAP in Spain is higher than in Brazil, Russia, India, China, and South Africa, by factors of 3.7, 3.3, 32.5, 22, and 6.4, respectively. An even more unfavourable situation for the BRICS economy can be seen from agriculture capital stock value expressed per worker. For instance, the agriculture capital stock per worker is 159 times higher in the USA than in India.

As shown above, BRICS countries have different agricultural performances, depending on the arable land available and their different farm structures. Accordingly, the role that agriculture plays in each country is different. Consequently, each country adjusts domestic agricultural GMO policy in accordance with its own needs.

Russia

The Russian GMO legislative framework can be described as a wave between the *de facto* and the *de jure* ban of the cultivation thereof. On the eve of its WTO accession in 2010, Russia hinted that it would establish mechanisms for GMO cultivation. Due to the adoption of the 2013 Decree No.839 'On the

Table 2.5. BRICS population, territory and GDP. (From Rosstat (2015)).

	Population (million)	Share in the world population (%)	Area of territory (1000 sq. km)	Share in the world territory (%)	GDP (current prices, billion US$, 2014)	Share in the world GDP (%)[a]	GDP per capita/US$, 2014
Brazil	201.0	2.8	8,516	6.3	2,346	3.0	11,571
Russia	143.3	2	17,098	12.6	1,881	2.4	12,874
India	1,224	17.1	3,287	2.4	2,069	2.6	1,633
China	1,357	19	9,600	7.1	10,361	13.3	7,595
South Africa	53.2	0.7	1,221	0.9	350	0.4	6,483

[a]Authors' calculation.

Table 2.6. BRICS arable land, cereal, meat, fruit, and vegetable production (2014).

	Arable land (million ha)[a]	Arable land (ha per capita)[a]	Cereal production (million tonnes)[a]	Share in the world cereal production (%)[d]	Meat production (million tonnes)[b]	Share in the world meat production (%)[d]	Fruit and vegetable production (million tonnes)[c]	Share in the global fruit and vegetable production (%)[d]
Brazil	80.0	0.39	101.4	3.6	26.0	8.2	42 (2013)	2.4
Russia	123.1	0.86	103.1	3.7	9.1	2.9	19	1.1
India	156.4	0.12	294.0	10.5	6.6	2.0	256 (2013)	14.7
China	105.7	0.08	557.4	19.9	86.4	27.2	1,021	58.6
South Africa	12.5	0.23	17.3	0.6	3.2	1.0	10 (2011)	0.6

[a]World Bank (2017); [b]FAOSTAT (http://www.fao.org/faostat/); [c]Rosstat (2015); [d]Authors' calculation.

Table 2.7. Agricultural GDP, income, and capital stock per worker, selected countries.

	Agricultural GDP (billion US$)[a]	Agriculture in GDP (%, 2014)[a]	Agricultural EAP (thousand, 2010)[b]	Agricultural EAP (share of total)[b]	Ratio of agricultural GDP to agricultural EAP (US$/person)[c]	Agriculture capital stock, 2007 (million constant 2005 US$)[b]	Agriculture capital stock, per worker (constant 2005 US$)[b]
Brazil	130.1	5.8	11,049	11	11,775	206,250	17,328
Russia	82.3	4	6,251	8	13,166	161,586	24,280
India	366.6	17.9	269,740	54	1,359	355,253	1,363
China	1004.9	9.7	500,977	57	2,006	540,792	1,071
South Africa	8.2	2.4	1,188	6	6,902	42,668	33,178
USA	278.7	1.6	2,509	2	111,088	579,069	216,799
Spain	44.8	3.2	1,015	4	44,138	78,504	69,534

[a]CIA (2015); [b]FAO (2012); [c]Authors' calculation.

State Registration of Genetically Modified Organisms [GMOs] Intended for Release into the Environment, as well as Products Obtained with the use of such Organisms or Containing such Organisms', it was predicted that the initiation of GMO release into the environment would begin around 2023. However, the current GMO legislative framework presented in Federal Law 358-FZ 'On amendments to certain legislative acts of the Russian Federation concerning improvement of the state regulation in the sphere of genetic-engineering activities', dated July 3, 2016, *de jure* prohibited transgenic crop cultivation and animal breeding and strengthened state control of imported transgenic products.

Given transgenic technology, Russia has shown its ability to develop its own genetically modified plant varieties. Under the project 'The Development of Biotechnologies and Industrial Adaptation of High-Reproduction Agricultural Plant GM [genetically modified] Seed Production', two potato varieties were protected by patents and approved for marketing and food use in 2005 and 2006 (Korobko *et al.*, 2016). Although there is no commercial production of any transgenic crop, and transgenic seed imports are forbidden, importation of transgenic crops and processed products containing genetically modified ingredients is allowed if these are registered for food and feed use. Such registration has 'twelve corn [maize] lines, eight soybean lines, one rice line, and one sugar beet line' also registered in the Eurasian Economic Union (EAEU), and 'eleven corn[maize] lines and eight soybean lines' registered for feed use (GAIN, 2016c). According to Ministry of Health data in 2003, Russia registered 59 food products containing genetically modified ingredients (11 drinks and cocktails, 4 specialized products for athletes, 22 food additives, 3 kinds of ice cream, 3 types of vegetarian burgers, 16 other proteins) (Monostyrsky, 2004). Customs regulations request that products intended for sale must be labelled if they contain over 0.9% of genetically modified ingredients in registered lines or 0.5% in non-registered lines.

Russia chose to become a member of the UPOV in 1998 and it was a signatory to the 1991 version of the Act. A study shows that Russia actively pursues intellectual property protection, and that it established a transparent IPR court in 2013 (WIPR, 2016). This is a result of huge efforts to regulate the seed market, since in the Soviet Union 'all investments in breeding were from the State budget and the exclusive right for that property belonged to the State' (Malko, 2006, p. 175), which meant that exclusive private rights of plant breeders were not recognized by law. By the Breeder Achievement Law (1993) and Federal Seed Law (1997), rules for the seed industry business were established and patent holders gained the right to collect royalties. Since September 2013, pursuant to Article 103 of Federal Law No. 273-FZ and Article 5 of Federal Law No.127-FZ, the restriction on universities and scientific institutions participating in small innovation enterprises has been abolished. In addition, since January 2015, under Act 1406.1 GK RF, compensation in cases of patent infringement has been allowed. These changes will probably encourage Russian investment in universities as well as foreign investment in innovative technologies (Beier and Fiero, 2015). Russia ratified the CBD in 1995, but did not ratify its Cartagena supplement. Regardless of that, Russia is very active in the international arena related to nature conservation, and public pressure against GMOs is rather strong. The most important organizations and institutions active in the field of GMOs are the Commonwealth of Independent States (CIS) Biosafety Alliance, GMO-free-Caucasus, Greenpeace Russia, Irina Ermakova, and International Socio-Ecological Union. After the 2016 *de jure* prohibition of transgenic cultivation, their activities decreased. The US Department of Agriculture (USDA) suggested that this decrease in activity was due to a drop in consumer purchasing power and an increase in demand for cheaper products (GAIN, 2016c); but most probably the reason is much simpler – their anti-GMO goals have been incorporated into an official state policy, and currently there is no need for any collective action.

In order to understand Russian GMOs policies, it is necessary to look into basic agricultural indicators, agricultural policy, crop and livestock production as well as trade relations. First of all, out of all BRICS countries Russia disposes of the greatest arable land per capita (0.86 ha) and generates the biggest income per agricultural worker (13,166 US$/person) (Tables 2.6 and 2.7). Rural population in Russia was last measured at 26%, while the number of holdings was 23,224,000 (FAO, 2015). Both the huge potential of market demand and the export possibilities made the Russian 'transition of the agricultural and food sector from a centrally planned system to a market-oriented one uniquely important for the world food system' (Wehrheim *et al.*, 2000, pp. 1–2).

A land market in the Soviet era did not exist and farm size was decided by fiat (Lerman and Shagaida, 2007). Large-scale farms in the form of *kolhozi* (collective farms) and *sovkhozi* (state farms) dominated food production, while retaining for farmers involved in collectivization the right to handle their own land, usually less than 0.5 ha (Liefert and Liefert, 2012). As a result of the transition, three types of farms emerged: (i) enterprises (large and small), as corporate farms originating from the *kolhozi* and *sovkozi*; (ii) household plots; and (iii) new family farms. According to the 2006 Russian Agricultural Census, there are: 27,787 large enterprises with an average farm size of 3,864 ha; 20,392 small enterprises having 1,203 ha on average; 285,141 family farms, owning 85 ha on average; and 22,799,400 household plots, which dominate in number but have just 0.42 ha on average. Despite a huge difference between the holdings across the total agriculture area (71% corporate farms and 6% household plots), large enterprises and household plots contribute almost equally to the total agricultural output, with 45% and 47%, respectively, in 2009 (Liefert and Liefert, 2012). Household plots are specialized in the production of HVFs (fruits, vegetables and meat), while family farms and corporate farms are mainly oriented toward grain and sunflower production.

After the initial years of post-Soviet market liberalization, when farms were also privatized and reconstructed, and land reform was conducted, Russia – prompted by the 2007/08 food crisis – turned to the more protectionist policy as defined in the 2010 document 'Doctrine on Food Security' and specified in the State Program for Development of Agriculture for 2013–2020. It became one of the most important objectives of national agricultural policy to ensure self-sufficiency in the food supply at levels ranging from 80% to 95% as follows: for grains and potatoes at least 95%, for sugar and vegetable oil at least 80%, for salt, meat and meat products at a minimum 85%, for milk and dairy products at least 90%, and for fish and fish products 80% (Doctrine, 2010).

During the Soviet period, agriculture was subsidized through state budget subsidies and price policy. At their peak, state budget subsidies equalled about 10% of the state GDP, the prices of food products, especially beef and poultry, were above world prices, while input prices 'were set low relative to their production costs and to agricultural output prices' (Liefert and Liefert, 2012). After nearly two decades of negotiations, in August 2012 Russia became a member of the WTO as a developed country. Contrary to expectations and the usual reports, recent research has shown that accession will not lead to a decline of trade- and production-distortive measures. Instead, these types of measure will most probably become stronger (Sedik *et al.*, 2013). The present authors made their own projection for the aggregate measure of support (AMS) for agricultural production until 2020. Using the OECD/FAO commodity outlook projection until 2021 (OECD/FAO, 2015) to estimate market price support (MPS), and the State Program for Development of Agriculture for 2013–2020 for the budget support estimation, they concluded that the MPS will continue to be the main source of support and will more than double its participation in the producer support estimate (PSE) increasing from 59% in 2010 to 82% in 2020.

Livestock producers will continue to benefit from both budget subsidies and import tariffs. In addition, the government is encouraging domestic meat production with grain prices below world levels, and pig and poultry meat prices above world levels. The main policy instruments applied in Russia include: (i) border protection; (ii) market intervention; (iii) payment based on output; (iv) concessional credit; (v) subsidies for inputs; (vi) leasing of machinery, (vii) equipment and livestock; (viii) per ha payments; (ix) tax preferences; and (x) concessions on repayment of arrears (OECD, 2016a). Russian agricultural producer support has imposed distortive measures on trade and production. MPS, as a result of the difference between domestic and international prices or the revenue transferred from food consumers to producers, is the predominant kind of support (two-thirds of the total); and budget support directly connected to production and inputs accounts for the other third. Less than 1% of support measures meet WTO 'green box' criteria which can be considered as 'decoupled income support'; and the share of agricultural tariff lines or imports with 10% or less import tariffs are significantly lower than in the USA, EU, Ukraine, or Canada (Sedik *et al.*, 2013).

Applied policy measures had a positive impact on agricultural production and trade. As can be seen from Fig. 2.4, in the period 2000–2014 cereal and pulses dominated in the structure of crops area, and after an increase of 4.9% they reached 58.8% in 2014. Areas under oil-bearing crops also more than doubled during that time, from 6.5% in 2000 to 14.3% in 2014. In contrast to this, the participation of fodder crops in the crops structure dropped by 36% in the observed period.

Production of wheat, as the most important cereal accounting for 32.2% in the structure of crops area, more than doubled, from 34.5 million tonnes in 2000 to 72 million tonnes in 2016. In the same period, sunflower seed production increased almost threefold, from 3.7 million tonnes to 10 million tonnes. The biggest growth – tenfold – can be observed in the production of soybeans and maize. Soybean output increased from 0.3 million tonnes to 3 million tonnes, while maize production increased from 1.5 million tonnes to 14.5 million tonnes (Fig. 2.5).

The increase of harvested area under maize and soybean and their output is of particular importance in relation to transgenic technology. Since Russia is not a producer of genetically modified crops, all its soybean and maize exports are considered to be non-genetically modified commodity export. This implies a big advantage for

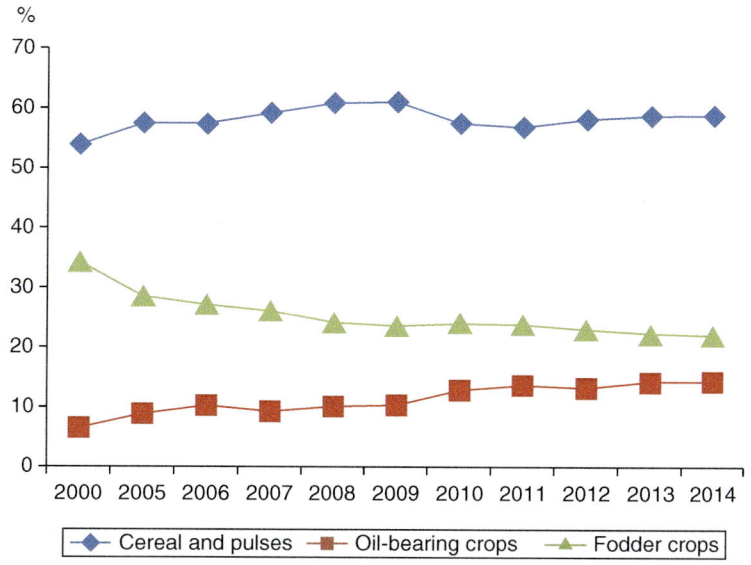

Fig. 2.4. Russia: Structure of crops area by main species, 2000–2014. (Adapted from Rosstat, 2015.)

Russia, and serious competitiveness problems for countries that export genetically modified soybean and maize, especially the USA and Brazil. After overtaking the USA as the world's largest wheat exporter for the first time in 2016, Russia has been taking an increasing share of the maize export market. An expansion of Russian non-genetically modified maize and soybean exports can be seen from Fig. 2.6 and Fig. 2.7. Maize export increased 96 times from 2007 and reached 4.7 million Mt in 2016. Soybean oilseed and soybean meal exports increased in the same period 90 and 9 times, respectively, both reaching 4.5×10^5 Mt (Fig. 2.7 and 2.8).

In earlier years, Russia exported its maize to the UK, Germany, Greece, Israel, Iran, Ireland, Spain, Italy, Cyprus, Lebanon, Libya, the Netherlands, Romania, Tunisia, Turkey, Azerbaijan, Armenia, and South Korea. The major destinations were Turkey and South Korea; but by November 2016 the top purchasing countries had become Iran and Japan. 2016 was also important because Russia expanded its export market by delivering maize for the first time to Vietnam and Bangladesh (APK Inform, 2016). Unlike Russian maize with its numerous buyers, almost all of the soybean export is directed towards China.

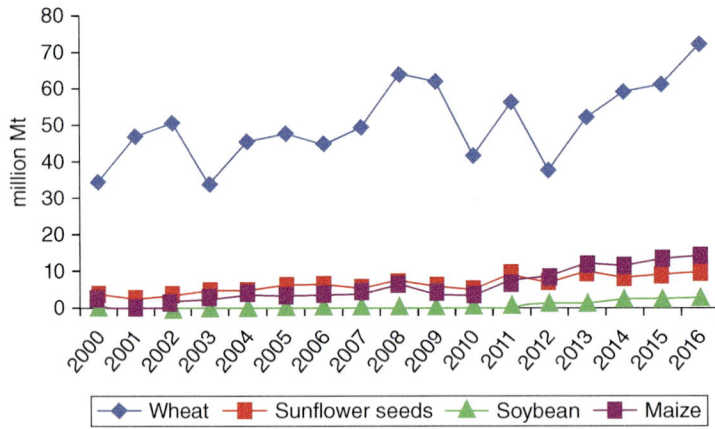

Fig. 2.5. Production of major crops in Russia, 2000–2016. (Adapted from IndexMundi (https://www.indexmundi.com/).)

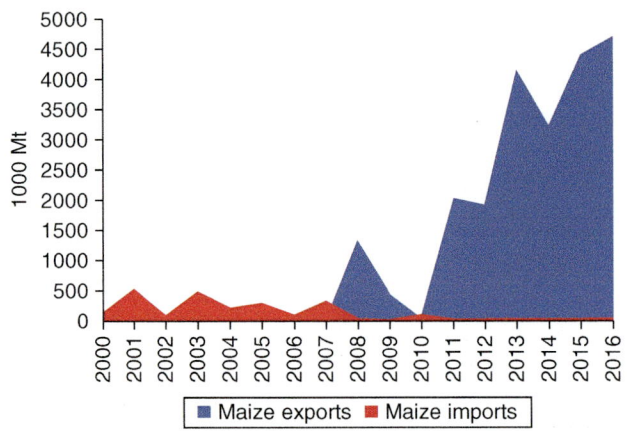

Fig. 2.6 Russia: Maize exports and imports, by year. (Adapted from IndexMundi (https://www.indexmundi.com/).)

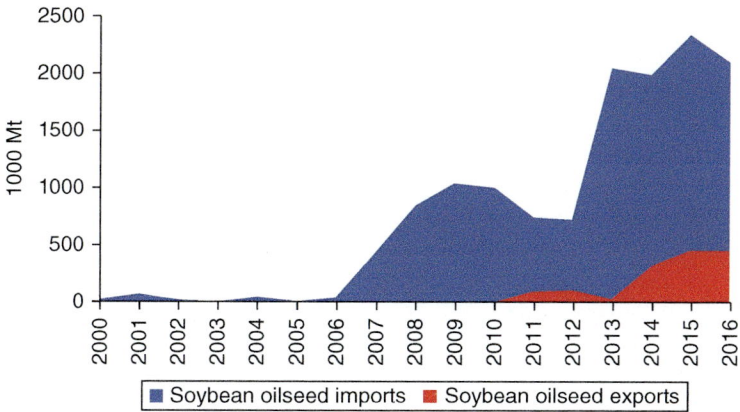

Fig. 2.7. Russia: Soybean oilseed exports and imports, by year. (Adapted from IndexMundi (https://www.indexmundi.com/).)

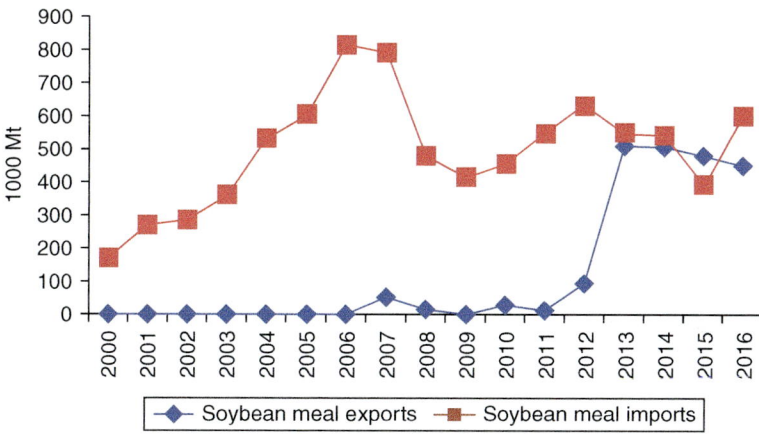

Fig. 2.8. Russia: Soybean meal exports and imports, by year. (Adapted from IndexMundi (https://www.indexmundi.com/).)

Although Russia has significantly increased its grain and oilseeds crops production, this is still not enough to obtain self-sufficiency in feedstuffs, because the increase in the area under feed crops, primarily soybeans, does not keep up with the enlargement of livestock and meat production (Fig. 2.9).

That is why a certain amount of transgenic crops and components, other than planting seeds, are legally entering the Russian territory through import. As can be seen from Figs 2.7 and 2.8, Russia imports huge quantities of soybean oilseed and a smaller quantity of soybean meal each year. There was a structural break in imports after 2007/08, when the country turned to more protectionist measures, as already explained. Some smaller quantities of transgenic maize are also imported into Russia. For example, in 2015 Russia imported 3.4×10^3 Mt of maize from the USA. However, even though the Sodruzhestvo Company from Kaliningrad maintains a separate facility for genetically modified soy, illegal contamination by this crop, as well as by maize and sugar beet, has been noted several times (Russia Beyond, 2014). In other words, the legal import of transgenic feedstuffs is the main source of illegal GMO contamination in Russia.

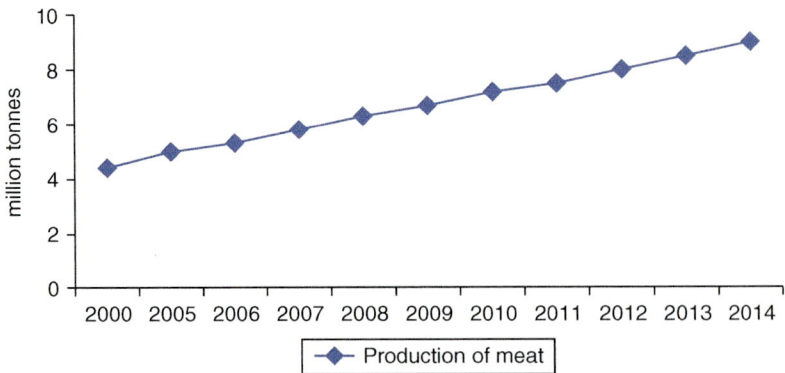

Fig. 2.9. Production of meat in Russia, by year. (Adapted from Rosstat, 2015.)

Russian transgenic neoregulation is a good example of how general agricultural policy affects specific national policy on GMOs. Self-sufficiency, underlined as the main agricultural goal, has caused agriculture transformation, reflected in a large increase of grain and meat production. The country has improved its position on the world food market and has greatly reduced dependence on food imports. In accordance with its self-sufficiency goal as well as the will of its citizens, Russia has adopted one of the world's strongest laws related to GMOs. However, the *de jure* prohibition of transgenic crops cultivation does not mean that the regulation has no weak points. The weakest point is toleration of GMO imports. Apart from the fact that Russia is obliged to follow WTO rules that consider GMOs as any other commodity, another reason for its liberal attitude to imports of feedstuffs is quite simple: Russia needs them. Feed crops production is not following livestock growth at the same pace, and the huge quantity of soybeans that Russia needs is difficult to find on the international market, since the largest exporting countries are producers of transgenic soybeans. If we take into consideration the projection that Russian dependence on meat imports will decreased significantly by 2025, especially pork and beef imports, which will decrease by 43% and 13.6%, respectively (USDA, 2016), it can be concluded that the attitude to transgenic feedstuffs imports is not likely to be changed. Of course, this does not mean that

there is no attempt to achieve self-sufficiency in soy production. For example, a major Russian meat producer region, Belgorod Oblast, has declared itself a GMO-free zone and has increased its own production of soybeans, and purchases only non-GMO feed. However, the public's attitude has helped the Russian government not only to adopt de jure prohibition of GMO cultivation, but also to tolerate transgenic feedstuffs imports. On the one hand, Russians have an extremely negative attitude towards GMO cultivation. On the other hand, they are totally indifferent to transgenic feed ingredients.

Brazil

Brazil is the second largest producer of transgenic crops in the world, just after the USA, planting transgenic soybean, maize, and cotton on 44.2 million ha in 2015 (James 2015). As of November 1, 2016, 58 GMOs – 34 maizes, 12 cottons, 10 soybeans, 1 dry edible bean, and 1 eucalyptus – were approved for cultivation (GAIN, 2016d). The country's position on GMO issues is quite similar that of the USA. The biggest differences lie in labelling, international agreements, and IPR protection. Brazil does not fully adhere to the concept of substantial equivalence, since products containing more than 1% of transgenic ingredients should be labelled. Unlike the USA, Brazil ratified the Cartagena Protocol in 2003, but is most often supportive of the positions advocated by the USA. It 'also opposed to the strict liability...

of an operator… and mandatory use of insurance or other financial instruments for the shipment of living modified organisms (LMOs)' (GAIN, 2016d). Brazil became a member of UPOV in 1999 by signing the 1978 version of the Act. This meant weaker protection than the USA and Russia adopted, since 'UPOV 1978 retains exemptions for farmers and plant breeders to save seeds for their own use' (Pechlaner and Otero, 2008, p. 360), while the 1991 version provides IPR 'on plants for 20 years, but it does not strictly require the use of patents, which would restrict seed saving. Rather, UPOV 1991 leaves it to national prerogative whether to adopt patents on plants or another system that would still allow for farmers and plant breeders to be exempted from restriction on seed saving. The USA chose to adopt patents and forgo the continuation of these exemptions' (Pechlaner and Otero, 2008, p. 357). Brazil lags behind the USA in patenting transgenic innovations, since it is still holding to the prohibition against patenting any naturally occurring product 're-gardless of isolation, purification or other chemical modification involved during the synthesis of an analogous substance in the laboratory' (De Souza, 2011, p. 62). However, this issue is still unresolved, because the Brazilian Intellectual Property Association (ABPI) puts continuous pressure on some scientists and congressmen to amend Article 10 of Law No. 9,279, which excludes from patentability any 'natural living beings, in whole or in part, and biological material, including the genome or germplasm of any natural living being, when found in nature or isolated therefrom, and natural biological processes' (De Souza, 2011, p. 62). This does not mean that the multinational companies Monsanto, Syngenta, and BASF, which have licensing agreements with the Brazilian Agriculture and Livestock Research Enterprise, have any problems in protecting their rights. On the contrary, they are collecting royalties with the support of the state.

Both Brazil and the USA have the same aspiration to the removal of trade barriers, guided by the same agricultural policy goal: to expand export. Thus, both countries have adopted weak regulatory oversight of transgenic technology.

As in the USA, in Brazil the public attitude has also helped the government to adopt a weak GMOs law. Citizens are not interested in GMOs – three-quarters have never heard of them; they are much more interested in food quality, prices, and expiry dates (GAIN, 2016d). The logical consequence is that, lacking awareness, they cannot be an engine of GMO resistance. The large food processors and retailers, especially the French hypermarkets, are the pillars of GMO rejection in Brazil. However, already in Brazil the diffusion of transgenic crops had not occurred without challenges forced mainly by Greenpeace and the Brazilian Institute for Consumer Defense (IDEC). In 1997, Greenpeace activists blocked the shipment of transgenic soybeans with the slogan, 'Frankensoy: Don't Swallow It'. 'In addition to media-designed protest events, Greenpeace engaged in legal mobilization, motivating the public prosecutor to undertake public civil actions', because 'it was the quickest way to stop genetically modified crops from arriving in the country, until they started a public campaign' (Motta, 2016, p. 86). These actions led to the establishment of a judicial moratorium, the Federal Court 'drew on the precautionary principle in the Constitution of 1988', asking for environmental impact assessment, even for open field trials, crop segregation, and labelling (Bauer, 2006). The moratorium lasted until the legalization of transgenic production by a presidential decree in 2003. The first legal import of transgenic soybeans in 2004 was greeted with the activists' slogan, *'Brazil melhor sem transgênicos!'* ['Brazil is better without transgenic food crops!'] (Greenpeace, 2004). The campaign under this slogan is still going on and is supported by 'certain environmental and consumer groups, including government officials within the Ministry of Environment, some political parties, the Catholic Church, and the Landless Movement' (GAIN, 2016d).

Although Brazil adopted GMOs relatively late, nowadays three of its ten most important crops (Fig. 2.10) are transgenic, with a high adoption rate. These are soybeans with a 93% adoption rate, maize with 83%, and cotton with 67% (GAIN, 2016d).

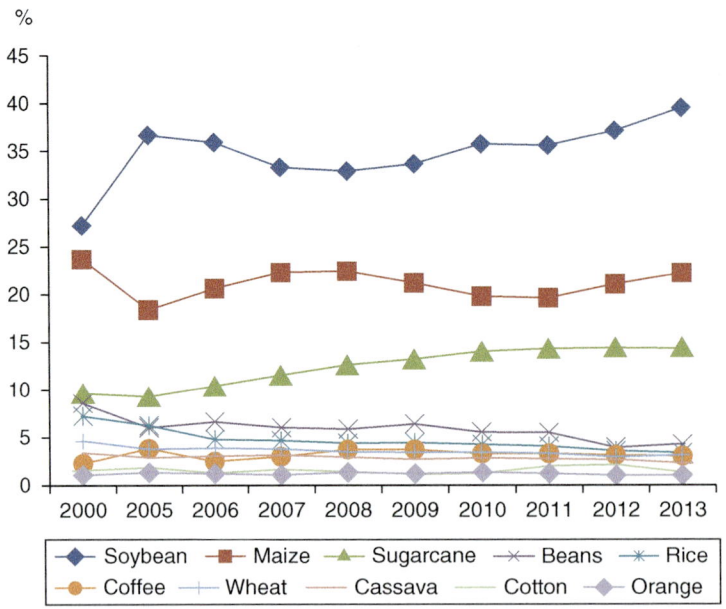

Fig. 2.10. Brazil: Structure of crops area by main species, 2000–2013. (Adapted from Rosstat, 2015.)

The participation of soybeans in the structure of crops area increased from 27.2% in 2000 to 39.5% in 2013, while maize and cotton, with some oscillations, dropped from 23.7% to 22.2% and from 1.6% to 1.3%, respectively. If we exclude coffee and sugarcane, production of Brazil's traditional crops such as beans, rice, wheat, and oranges have decreased. The increase of the sugarcane area from 9.6% to 14.4% in the period 2000–2014, as well as of the soybeans area, clearly indicates the country's direction toward biofuels production. Production of soybean oilseed reached 102 million tonnes in 2016 – 34 times more than Russian production, by way of comparison. And production of maize was 83.5 million tonnes or 6 times higher than in Russia. Brazil ranks as the second largest producer country of soybeans, just after the USA, while Russia ranks tenth. Besides, Brazil is much better positioned globally in the production of maize, occupying third place after the USA and China, while Russia is ranked ninth. However, the highest output in Brazil is achieved in sugarcane production, recorded at a level of 689.9 million tonnes in 2014 (Fig. 2.11).

Brazil produces almost three times more meat than Russia, achieving a greater increase in meat production than Russia in the period 2000–2014 (Fig. 2.12).

Although the production of oilseeds crops is continuously increasing, most of it is consumed in the domestic market (Fig. 2.13). In 2016, for the first time, oilseed exports exceeded domestic consumption, and half of the soybeans produced were exported. Brazil has also specialized in the crushing sector and produced huge quantities of soybean meal and oil both for the domestic market and for export (Fig. 2.14) (OECD/FAO, 2015).

An OECD/FAO projection stated that soybean will continue to be the most important agricultural Brazilian product and that there is a possibility of narrowing the gap with the USA, the world largest producer, with Brazil as the second largest by 2025, owing to the Brazilian potential to expand the area under this crop. However, the same projection emphasizes that Brazilian exports depend on China's economic performance, as China is the largest buyer (OECD/FAO, 2015). If this optimistic scenario becomes a reality and China's demand for oilseed imports increases by 2.9 Mt, half of the demand will be met by Brazil, which will be

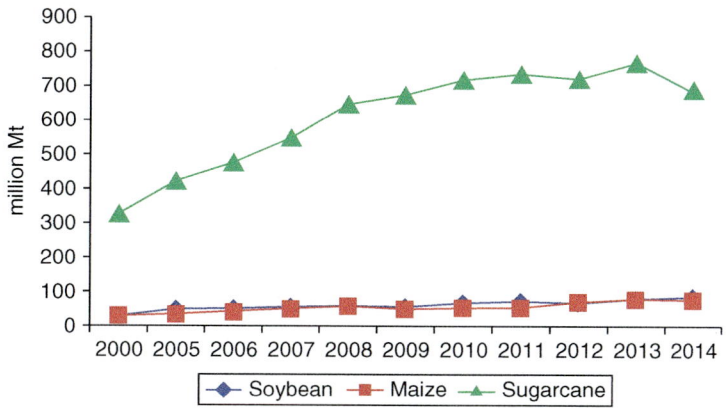

Fig. 2.11. Brazil: Production of top three farm products, 2000–2014. (Adapted from Rosstat, 2015.)

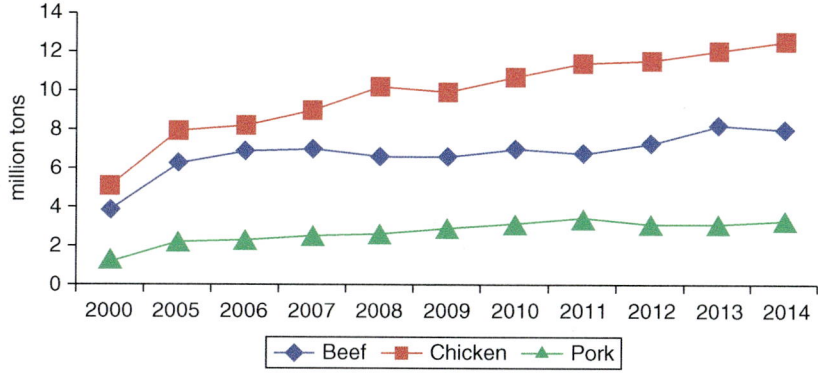

Fig. 2.12. Production of meat in Brazil, 2000–2014. (Adapted from Rosstat, 2015.)

stimulated to increase its areas under oil-seeds crops. If the opposite occurs, i.e. if China's economic growth gets worse than in the baseline, Brazil will decrease its production and exports to China as well as to other countries will drop by 3.2% and 2.1%, respectively.

As previously stated, the reasons for a country's policy toward GMOs should be sought in general agricultural performances and agrarian policy. Unlike China and India, Brazil has a significantly lower participation of the rural population in the total population (14.3%), even lower than the US and Spain. This has not occurred owing to farm size increase as in the USA and EU. In the period of the 1970s to 2000s, the USA increased their plots from 157.6 ha to 178.4 ha, while the same process in Spain resulted in a plot-size increase from 17.83 ha to 23.9 ha (Chand *et al.*, 2011). The specific issue with Brazil is that neither an increasing nor decreasing trend in the average size of farms can be defined, since the size of farms has fluctuated over decades: in the 1960s it was 74.9 ha, in the 1970s 60 ha, in the 1980s 70.7 ha, and in the 1990s 64.5 ha (Lowder *et al.*, 2016). However, the number of agricultural holdings in Brazil (5,175,000) is 38.7 times lower than in China, and 26.6 times lower than in India. Brazil has a much more favourable structure of farms than China and India. Farms of more than 10 ha account for one half of all properties (large farms more than 50 ha account for 19%) (FAO, 2015).

Apart from relatively favourable farm size (medium average) and available land per capita (0.39 ha), the reasons for the transgenic

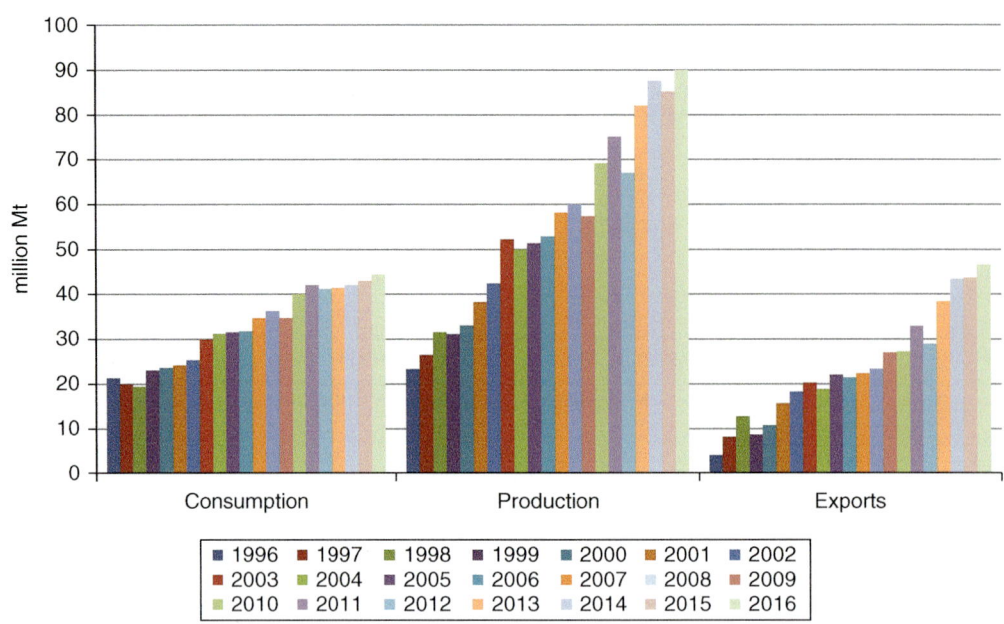

Fig. 2.13. Production, consumption, and exports of oilseeds in Brazil, 1996–2016. (Adapted from OECD/ FAO (2015).)

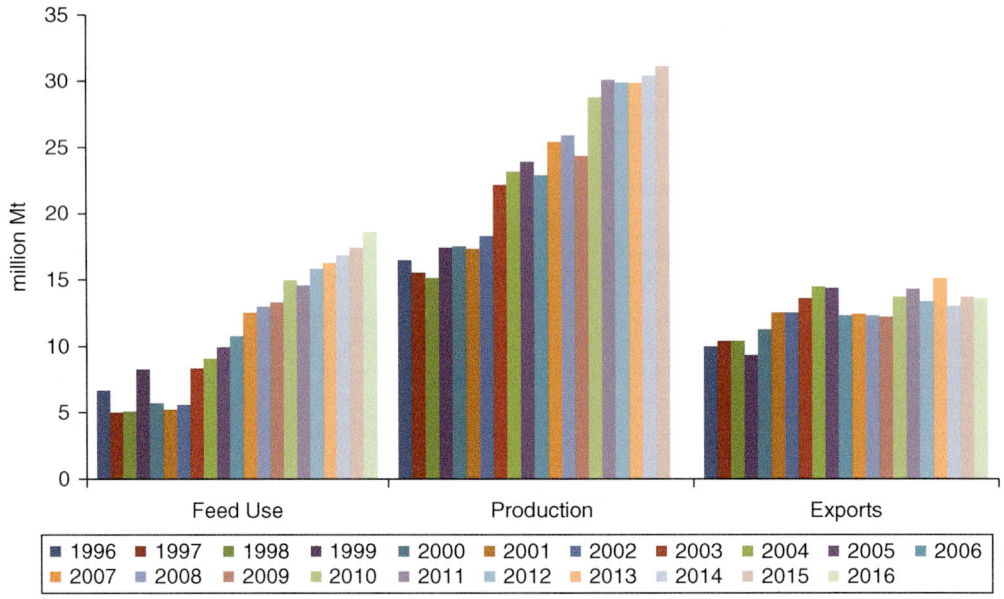

Fig. 2.14. Protein meal production, feed use, and exports in Brazil, 1996–2016. (Adapted from OECD/FAO (2015).)

technology diffusion is also found in the country's primary goal of agricultural policy: to further increase access to foreign markets. Brazil has huge bargaining power in agricultural trade liberalization negotiations. It has been a WTO member since January 1995 and a GATT member since July 1948. As of March 1990, the government adopted a comprehensive

programme of liberal reform, including abolition of import prohibition and tariff cutting
(Abreu, 1988). Unlike Russia, Brazil provides
a relatively low level of support to individual
farmers, but the budget for infrastructure
and research has grown considerably over
time. The level of PSE was 3.69% of gross
farm receipts (GFR), more than four times
lower than in Russia in 2014. The main element of the general services support estimate
(GSSE) in Brazil is payment of land restructuring for small family farms, while in Russia
this kind of support is generally directed to
the agricultural knowledge system. In both
countries PSE is the dominant part of TSE,
at about 80% in Brazil and 85% in Russia
(OECD, 2016b). The main policy instruments
in Brazil are: (i) minimum guaranteed prices
as an element of market price policy covering
a wide range of products including grain, livestock products, industrial and some traditional products; (ii) deficiency payments, and
maintaining a difference between market and
minimum prices; (iii) rural credits at prefer-
ential interest rates and debt rescheduling;
and (iv) crop insurance subsidies covering all
agricultural and livestock products. In addition, there are programmes which support organic farming and other sustainable practices

meeting environmental criteria, as well as
programmes supporting biofuel production,
infrastructure development, extension of irri-
gated areas, and improvement of milk competitiveness. All these instruments are applied
with the primary aims of further increasing access to foreign markets and improving
farmers' income.

The great success of the agrarian policy in
this respect can be seen from Fig. 2.15. Brazilian agricultural exports were twice as high
in 2013 as in 2007, and accounted for 9% of
the world's total exports. The most successful export products, soybeans, accounted for
23% of the total agricultural exports and
were worth US$23 billion in 2013 (OECD/
FAO, 2015).

Following a change in demand on the
world market, Brazilian export destinations
have also changed over time. In the 2000s
the main destinations were Europe and Central Asia; by the 2010s East Asia and the Pacific had assumed primacy, primarily because
of China's demand (Fig. 2.16).

Unlike the unexpected success of transgenic soybeans in Brazil, production of
transgenic cotton varies from year to year
(Fig. 2.17). In recent years world cotton
prices have been under pressure from

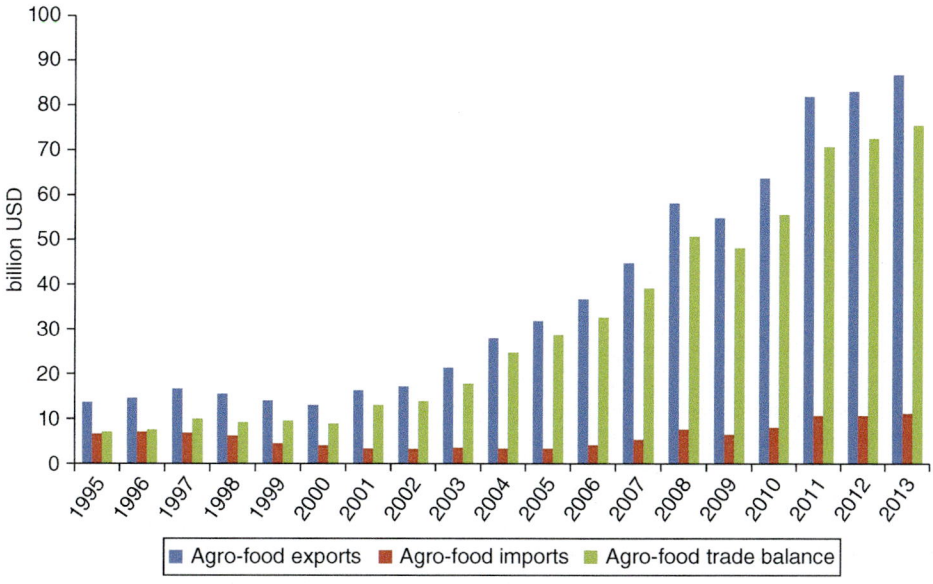

Fig. 2.15. Brazil: Agro-food trade 1995–2013. (Adapted from OECD/FAO (2015).)

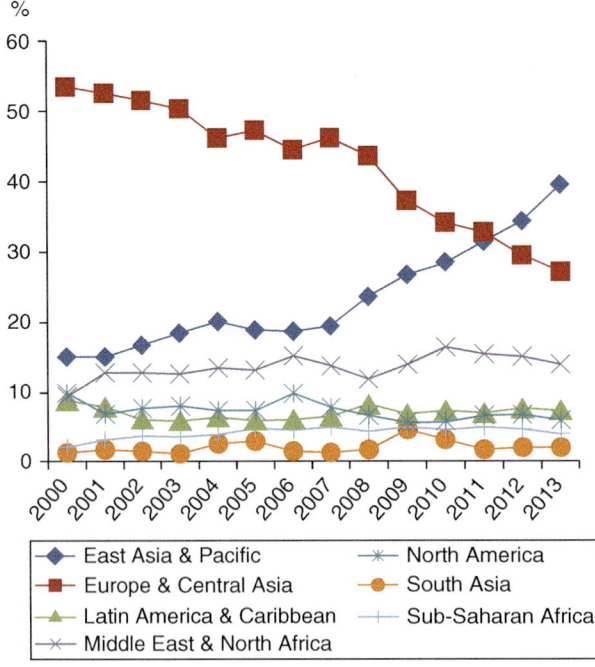

Fig. 2.16. Destination of Brazilian agricultural exports, 2000–2013. (Adapted from OECD/FAO (2015).)

Legend:
- East Asia & Pacific
- Europe & Central Asia
- Latin America & Caribbean
- Middle East & North Africa
- North America
- South Asia
- Sub-Saharan Africa

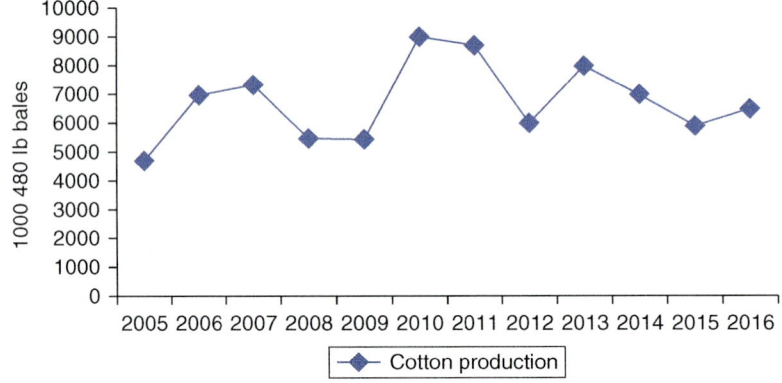

Fig. 2.17. Brazil: Cotton production, by year. (Adapted from IndexMundi (https://www.indexmundi.com/).)

decreased prices of synthetic fibre driven by lower oil prices. However, it is predicted that prices will reach a period of stability and Brazil will become the world's second largest cotton exporter by 2025 (OECD/FAO, 2015). The country has all the preconditions to achieve this. Having a long tradition in cotton production, Brazil has been a major world cotton supplier since the 19th century. It also has a favourable climate, large-scale farming, a cotton yield more than

double the world average, and targeted government support in the form of production, marketing and investment loans, and periodical debt rescheduling programmes (Kiawu *et al.*, 2011).

Brazil is a rather good example of how a country can respond to higher global demand by increasing agricultural production, and of how quickly a country can shift its production in a desired direction. The same rules apply both to Brazil and Russia: GMO policy is determined by the objectives of the overall agricultural policy. The main Brazilian goal is to expand export, thus Brazil quickly adapts to the demands of the world market. Regarding GMOs, Brazil proves that the lack of awareness and a weak social movement, except for sporadic events such as the stopping of shipments in 1997, leave the door open for all options. Brazil is an example of a paradox surrounding transgenic diffusion on the national and sub-national levels. Originally defined as GMO-free, the national policy was the very opposite of the real government intention to grow transgenic plants. On the other hand, the Rio Grande do Sul region declared itself a *'zona livre dos transgênicos'* ['GMO-free region'], as a result of the sub-national government initiative against transgenic crops. However, this initiative failed and Rio Grande do Sul became the most important transgenic producing region, while some regions without a GMO-free policy remain GMO-free (Bauer, 2006).

China

China produced transgenic crops on 3.7 million ha[ii] in 2015, which made this country the sixth largest producer in the world (James, 2015). Although China has approved for commercialization domestically developed transgenic tomato, cotton, petunia, sweet and chili pepper, papaya, rice, and maize, it only produces one edible transgenic crop, the papaya, on a smaller area (7000 ha), while the rest of the land is dedicated to transgenic cotton. Almost all cotton produced in the country is transgenic, since its adoption rate has reached 96%. Unlike Brazil, China produces only transgenic crops obtained from its own research, meaning that no foreign crops have been approved for commercialization. China is a unique case because it has banned foreign transgenic research and keeps the research and development (R&D) sector under state control. The Communist Party and President Xi support agricultural transgenic research, thus the government heavily invests in the R&D sector. But the official attitude to commercialization is much more careful. In addition to cotton, import permissions are issued for soybean, maize, and rape, but only as raw materials for processing.

This attitude is in line with their agricultural policy goal: to achieve food security through self-sufficiency in grain production. The objective is quite understandable if we bear in mind that: (i) out of the world's 570 million farms, 35% are located in China; (ii) 63% of its population is rural; (iii) national crop production is in the hands of the 200 million small family farms that are still important in dairy industry and swine production, despite the respectable development of large-scale livestock; (iv) the average farm size decreased in the period 1985–2010 from 0.7 to 0.6 ha; (v) agriculture's share of income decreased from 66.3% to 29.1%; and (vi) its share of capital dropped from 76.1% to 69.5% (FAO, 2015). Thus, the China experience mirrors an assertion of development economics theory which suggests that 'the gradual evolution of farm structure is paralleled by increased activity of factor markets as an efficiency-improving institution in resource reallocation' (Huang *et al.*, 2012, pp. 16–17).

Agricultural policy goal is underlined in 'the 13th five-year plan for the economic and social development of the People's Republic of China' issued in August, 2016 (State Council, 2016). The 13th Plan emphasizes, inter alia, that China will: (i) build a modern seed industry by establishing national seed production centres in Hainan, Gansu, Sichuan provinces and 100 regional superior food production centres; (ii) establish functional and protected zones for wheat and rice in order to guarantee stable acreage for basic crops; (iii) obtain high-yield green grain; (iv) promote sustainable agricultural development; (v) obtain zero growth rate in the use of pesticides and chemical fertilizers; (vi) progressively increase green-box subsidies and reduce trade-distorting domestic supports in the amber box; (vii) establish farming-based areas for livestock production; and (viii) support large family farms as a new type of agribusiness. Judging by this document, China will continue to work toward its goal to achieve self-sufficiency through domestic research innovation. Regardless of whether or not it becomes a producer of transgenic maize, which is a hot issue, the government will also promote eco-friendly and organic production.

China became a member of the WTO in December 2001. China's entry into the WTO was accompanied by much concern, since a lot of poor people are involved in the agricultural sector. Anyway, support to agricultural producers has continued to grow in recent years (OECD, 2016c). 'China is perhaps the most prominent example of a developing country that has shifted from taxing to subsidizing its agricultural sector' (Gale, 2013). The value of the subsidies in China 'doubled between 2008 and 2013' (Hejazi and Marchant, 2017) and in this respect Chinese support growth seems to be globally the highest. The government is implementing the following measures: (i) minimum guaranteed prices for wheat and rice and, from time to time, for other commodities such as maize, soybean, pork, rapeseed, cotton and sugar; (ii) intervention purchases by the China Grain Reserve Corporation; (iii) subsidies for improved seed varieties; (iv) subsidies for agricultural machinery purchasing; (v) compensation payment for inputs; and (vi) an agricultural insurance programme. Starting from 2014/15, China introduced a pilot target price programme for soybean and cotton and a 'tree subsidy' programme, as a combination of different types of payments. China proved that one of the goals of its policies is to reduce pollution by adopting the Environmental Protection Act and Zero-Growth Action Plan for Chemical Fertilizers and Pesticides. Regarding the structure of agriculture support, MPS remains the main contribution to the total support. TSE accounts for about 3.1% of GDP in recent years, three times more than in Russia and nine times more than in Brazil. The GSSE payment is mainly dedicated to the development of infrastructure costs of public stockholding, and the agricultural knowledge and innovation system. As in Russia and contrary to Brazil, domestic prices in China are above world prices – 23% on average, meaning that the level of price distortions is very high.

The 13th Plan foresees the establishment of a strict, smooth and transparent IPR system. Currently, IPR protection is very weak, mostly due to a fragmented seed industry that includes over 5,000 registered companies. An industry report stated that, 'over 50% of seeds sold in China are counterfeit, and for some varieties the percentage climbs to 80%' (GAIN, 2014). 'Under Article 25 of the Patent Law, animal and plant varieties as well as methods for the diagnosis and treatment of diseases cannot be patented. (Wong and Chan, 2016, p. 135), but seeds can be patented and the number of seed-related patents is growing over time (GAIN, 2017). Along with Brazil, China became a member of UPOV in 1999, under the weaker 1978 Act. The country ratified the Cartagena Protocol in 2005 with its extension to the Hong Kong Special Administrative Region in 2011.

China has strong labelling requirements supplemented with prevention of misleading advertising (GAIN, 2016e). All approved transgenic products are subject to mandatory labelling. These are, as cited in the Global Agricultural Information Network (GAIN) document: soybean seeds, soybeans, soybean powder, soybean oil, soybean meal, maize seeds, maize, maize oil, maize powder, rapeseed for planting, rapeseeds, rapeseed oil, rapeseed meal, cottonseed, tomato seed, fresh tomato, and tomato paste. Misleading advertising policy measures means prohibition of the non-GMO label on peanuts and sesame as well as on all other products not approved in China, because they are all already non-GMO. In other words, the non-GMO label is allowed and mandatory only on products for which a transgenic version is approved. It is believed that China uses the most sensitive PCR tests in order to detect GMO ingredients in shipments and refuses each container containing more than 0.1% (or sometimes less) of unapproved GMOs. Such an approach creates huge financial risks for foreign companies that consider this a trade-distorting measure.

Despite being a producer, China is considered to be a country whose policies slow down further diffusion of transgenic crops.

Strong public rejection has an essential role in this process of slowing down. The Ministry of Agriculture is trying to prepare the public for further expansion of GMOs with some campaigns, but at the same time opponents are increasing their activities (GAIN, 2016e). Activists have sued the Ministry several times for hiding information about GMOs. Very active in China is Greenpeace East Asia, that 'has made huge strides by convincing many supermarkets in China and Hong Kong such as Carrefour, Auchan and Walmart to stop selling genetically engineered rice, as well as foods farmed with the more hazardous pesticides' (Greenpeace, 2017). Greenpeace is also putting pressure on the government to investigate the presence of banned seeds, claiming that Liaoning, a seeds growing province, is heavily contaminated with illegally planted GMO seeds (Qin and Hao, 2016). Any state indication of further transgenic crops expansion has caused public reaction. Soon after Chemchina's US$44 bn bid for Syngenta, when Beijing, the largest grain growing province, proclaimed its intention to grow transgenic maize and soybean, China's north-eastern Heilongjiang province announced a five-year ban on transgenic maize, soybeans, and rice production, based on survey results (90% of its citizens rejected GMOs) (Financial Times, 2016). It is expected that the tension between the government and citizens will continue because the state does not want to abandon its plans. As Director-General of the Department of Science, Technology and Education, Liao Xiyuan has stated that China has a prudent attitude to GMOs and cannot afford to be left behind. Thus China will continue to push forward with its independent innovation efforts. As underlined in the 13th Plan, China will make non-edible cash crops its priority, with further expansion of transgenic cotton its top priority, but will also 'push forward with the commercial cultivation of pest resistant, genetically modified corn [maize] over the next five years'. In parallel, 'China will plant more non-GMO soybean and improve yields to fill the gap between supply and demand... By 2020, China will expand its soybean planting area to 140 million mu (9.3 million ha), and

yield per mu will be raised by 15 kg to 135 kg' (MOA, 2016).

China has adopted a regulatory oversight of transgenic technology that is both weak and strong at the same time. It is strong in relation to foreign seed entrance, weak in relation to importation of transgenic cotton and of maize and soybeans used as feed and for processing. As in the case of Russia and Brazil, the reason lies in the production and trade of agricultural commodities. As can be seen from Fig. 2.18, the largest part of the agricultural area in China is dedicated to rice, wheat, and maize as basic foods. From the same figure it is clear why the 13th Plan asks for the establishment of protected zones for wheat and rice: in the period 2000–2014, participation of rice and wheat in the total structure decreased by 0.9% and 2.6%, respectively. At the same time, areas under maize increased from 14.8% to 22.4%. The smallest part is dedicated to cotton, whose participation decreased from the maximum 3.9% in 2007 to 2.6% in 2014 returning to the level of 2000.

China produces more meat and grains than other BRICS countries. Meat production in China is 3.6 times higher than in Brazil and 9.6 times higher than in Russia, and reached the level of 87.1 million tonnes in 2014. China's cereal output was 557 million tonnes or about five times higher than in Russia and Brazil (Fig. 2.19). In the period 2000–2014, China achieved a slight increase in meat and cereal production in comparison to Brazil and Russia. Brazil increased its meat output by 130.5% and cereal production by 115.6%; Russia increased meat production by 103.5% and cereal production by 61.5%; while China increased meat production by 44.8% and cereal output by 37.5%.

China is the world's second largest producer of maize, after the USA, and the world's largest producer of non-transgenic maize. Its production in 2015 was 215.6 million tonnes, or 21% of global maize production (Statista, 2017). China is perhaps the main source of uncertainty in the world maize trade,

'swinging from the second-largest exporter in some years to an importer of large quantities in other years. China's corn [maize] exports are largely a function of

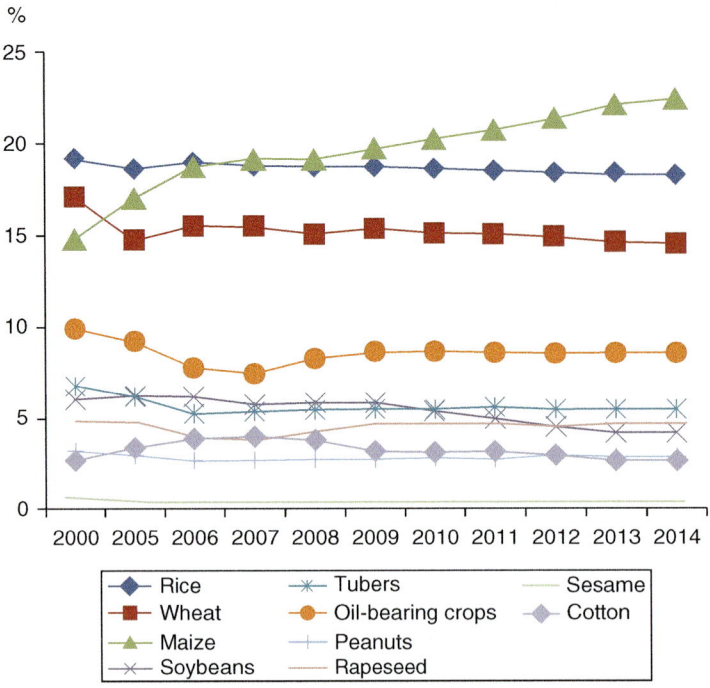

Fig. 2.18. China: Planting structure of farm crops by ten main species, 2000–2014. (Adapted from Rosstat, 2015.)

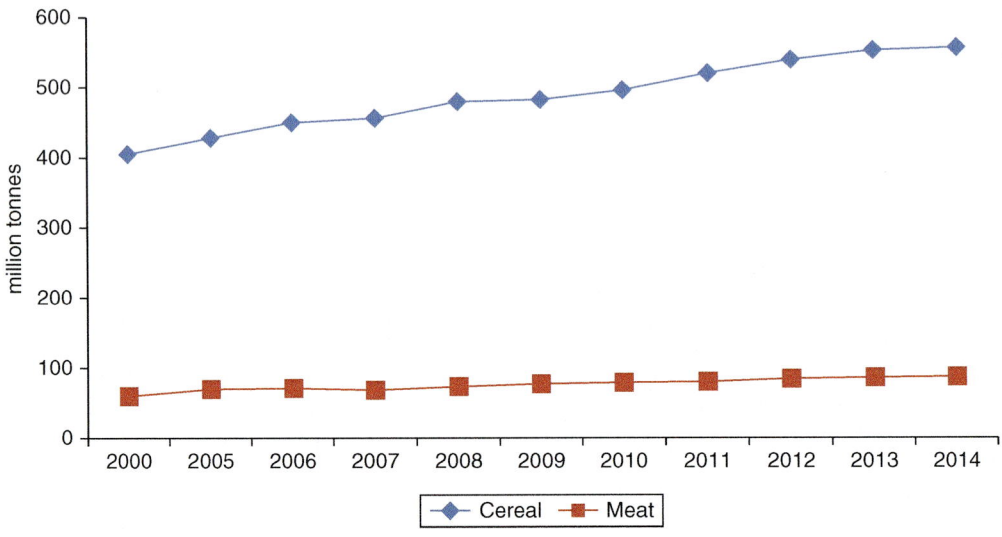

Fig. 2.19. China: Cereal and meat output, 2000–2014. (Adapted from Rosstat, 2015.)

government export subsidies and tax rebates, because corn [maize] prices in China are mostly higher than those in the world market. Large corn [maize] stocks are expensive for the government to maintain, and Chinese corn [maize] trade policy fluctuates with little relationship to the country's production, making China's corn [maize] trade difficult to predict'

(ERS USDA, 2017).

The foregoing is illustrated in Fig. 2.20. Since 2005, when the government started to purchase cheaper foreign maize under a stockpiling scheme, China has become a large importer. This is going to change, because the government has responded to excessive maize stocks by promoting the consumption of domestic maize and tightening restrictions on imports. Out of 11 countries with import permissions (Thailand, the USA, Peru, Laos, Argentina, Ukraine, Bulgaria, Brazil, Chile, Germany and Myanmar), China prefers Ukraine because of its non-transgenic production. US maize and multinational companies such as Syngenta are facing problems due to China's strategy to diversify suppliers and the possibility of shipment rejections in the case of the presence of unapproved transgenic varieties. In addition, the abandonment of the price support policy had by 2016 caused a fall in maize prices and made future import less attractive. Besides, in 2016 the government gave permission to two state-owned companies to sell maize abroad, which may enable China to become an important maize exporter again. If this comes true, a tectonic disturbance on the world maize market will occur and China's competitor countries, Brazil and the USA, will face difficulties.

However, China's growing livestock sector, the recovery in pork production, and steady growth in the poultry sector have increased demand for industry feed and protein meal. Sixty per cent of large domestic maize production is used as animal feed (Worldatlas, 2017) but there is still a feedstuffs shortage compensated by imports. When it comes to soybeans, China is less productive and more dependent on imports, as presented in Fig. 2.21. China imports about 90% of its soybean oilseeds to satisfy demand (MOA, 2016). For example, in 2016 China produced 12.5 million Mt of soybean oilseed, while domestic consumption was 100.8 million Mt.

As in the case of maize, China is a source of high uncertainty in the world cotton trade. From being a net cotton importer in the 1970s, in the following years China became a net exporter, and then again shifted to net imports (Fig. 2.22). Since 2011, China has been accumulating cotton in state stockpiles. 'Defending a high domestic cotton price during 2011–13, China has driven world cotton stocks to nearly double the average levels and 45 percent above the previous record for the years since 1950. With world stocks ending at near 90 percent of use in 2012 and 2013, global cotton markets face a difficult and costly transition if policy

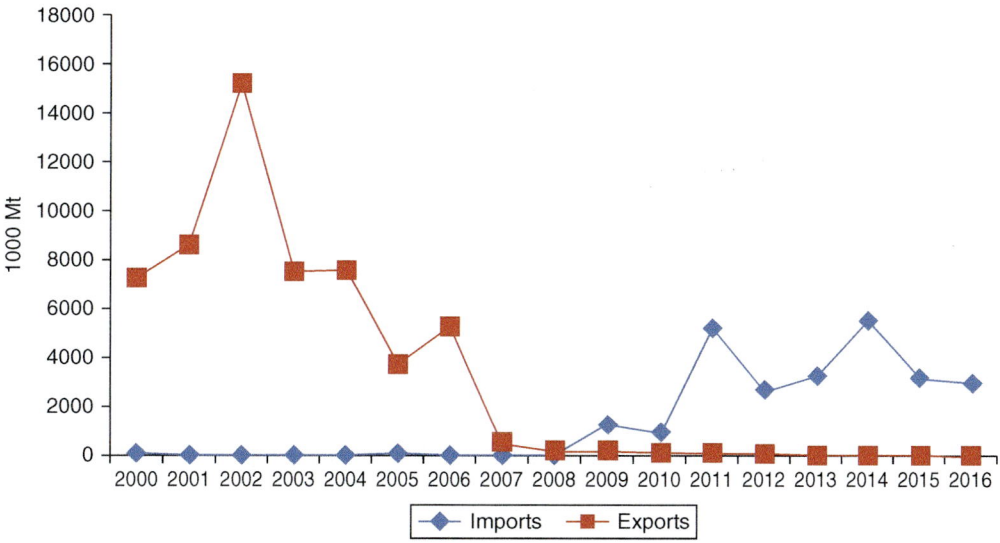

Fig. 2.20. China: Maize exports and imports, by year. (Adapted from IndexMundi (https://www.indexmundi.com/).)

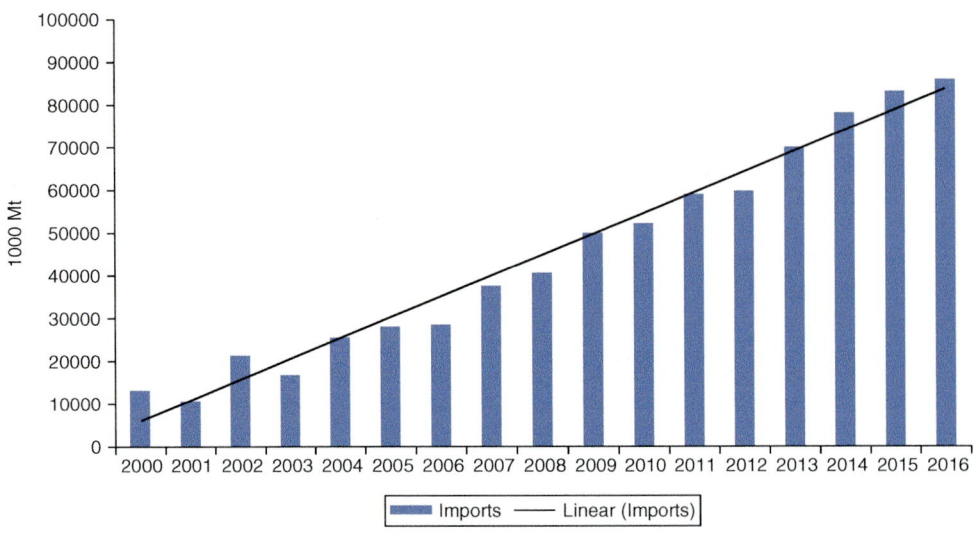

Fig. 2.21. China: Soybean oilseeds imports, by year. (Adapted from IndexMundi (https://www.indexmundi.com/).)

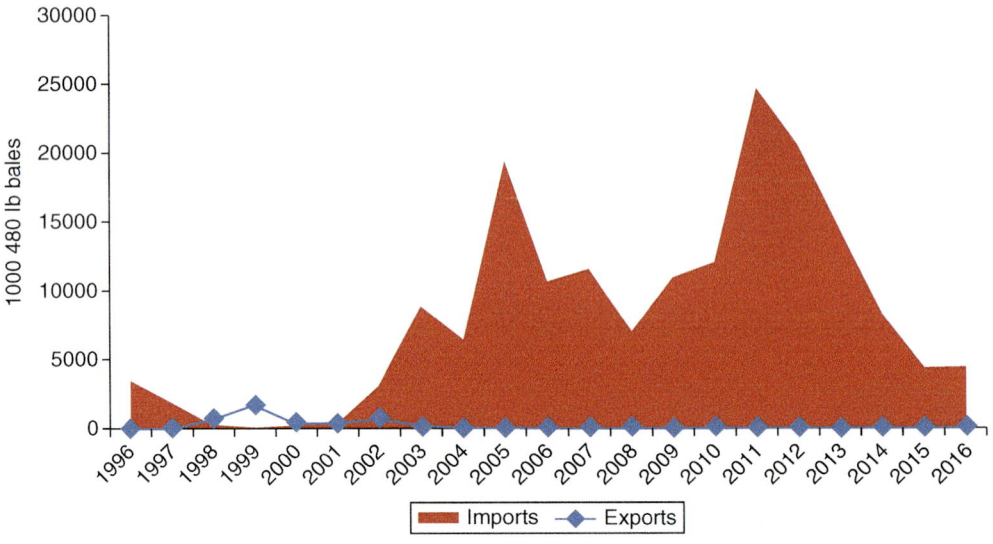

Fig. 2.22. China: Cotton imports and exports, by year. (Adapted from IndexMundi (https://www.indexmundi.com/).)

shifts in China return world stocks to normal levels with anything other than a long period of transition' (MacDonald *et al.*, 2015, p. 2). China's production of cotton of about 4.6 million tonnes in 2016 was lower than in 1997 (Fig. 2.23 and Fig. 2.24).

Regarding production of transgenic crops, China has adopted weaker neoregulation than Russia. However, if entrance of foreign companies is taken into account, China's regulation is stronger than Russia's. China and Russia have the same overall agricultural goal: to increase self-sufficiency. Nonetheless, there are huge differences between those two countries, a very important one being arable land (Table 2.2) and its

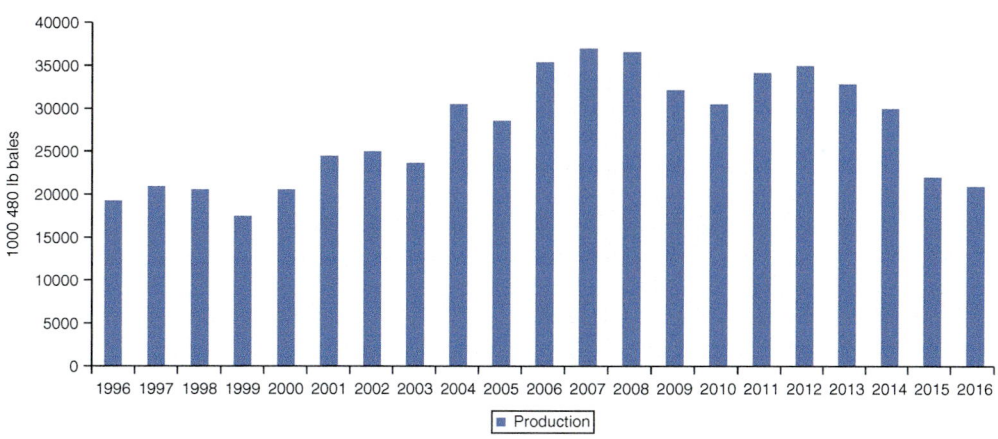

Fig. 2.23. China: Cotton production, by year. (Adapted from IndexMundi (https://www.indexmundi.com/).)

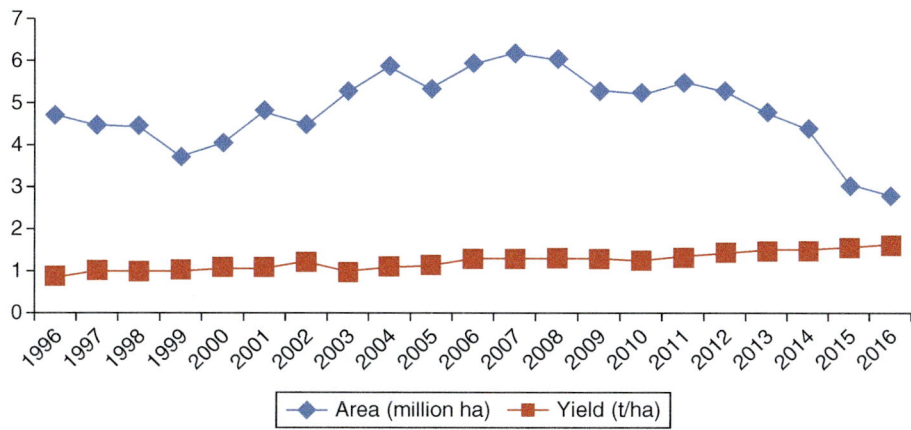

Fig. 2.24. China: Cotton area and cotton yield, by year. (Adapted from IndexMundi (https://www.indexmundi.com/).)

quality. Russia disposes of 0.86 ha per capita, while China disposes of 0.08 ha per capita. Russia has some of the world's most fertile agricultural lands, best reflected in the 'Chernozem Belt'. These lands were saved from destruction by chemicals because the chemical industry was directed towards Russian defence needs during the Cold War. Chinese farmers were for decades encouraged to use pesticides and fertilizers in large quantities, and as a result one-fifth of their farmland is polluted. This means that Russia can easily turn to organic and other ecological production, while it may be harder for China to do so because of poor land quality and high contamination with transgenic seeds. It seems that the Russian government has sincerely decided to protect its agricultural production from transgenic innovations, accepting transgenic animal feed as a 'necessary evil'. The Chinese government essentially supports transgenic technology, but has postponed its further expansion until 'all the prices are aligned'. China does not want foreign seeds and the position will be hard to change in the future, since the country is 'self-sufficient in rice, corn [maize], wheat, cotton, and soybean seeds and produces 80% of the vegetable and fruit seeds it uses', representing the world's second largest seed market (GAIN, 2014). China provides a good example of tensions between the government and citizens in relation to GMOs. In order to realize its true intention – to expand its own transgenic crops – China will have to overcome public resistance. Certainly, it will be interesting to consider these relations in the following period. Since China is a very unpredictable partner on the market, it will be also interesting to observe to what extent its partners and competitors are ready to adapt to sudden changes.

India

India is the fourth largest producer of transgenic crops in the world. Cotton, the only transgenic crop approved for commercial cultivation as seeds, fibre and feedstuff, is produced on 11.6 million ha with the adoption rate of 95% in 2015 (James, 2015). In addition, India has issued import permissions for soybean oil and canola oil. Like China, India invests heavily in transgenic R&D. Unlike China, however, both public and private sectors have developed several transgenic crops. Unlike China, India has allowed foreign companies' entrance into the seed market. Most of the transgenic cotton produced comes from Monsanto, thanks to a joint venture between Mahyco, an Indian company, and Monsanto. In this way, Mahyco is sublicensing Monsanto's transgenic cotton. Besides, India is flexible in relation to import shipments that may contain unapproved transgenic ingredients, mostly 'due to lack of testing facilities at the ports of entry/exit', so 'there has not been any known instance of interception of import consignments containing unapproved events' (GAIN, 2016f). As a supporter of mandatory labelling in international discussions, in 2013 India adopted rules stipulating that 'every package containing genetically modified food shall bear at the top of its principal display panel the word [genetically modified] GM' (GAIN, 2016f). The Indian government is trying to maintain price control by prescription of a maximum sale price for transgenic cotton. This causes a lot of pressure from industry, and 'price notifications

have been challenged in Indian courts by various industry stakeholders arguing that the order is unconstitutional and exceeds the authority granted'; so the Ministry of Agriculture and Farmers Welfare 'are likely to drop the draft licensing guidelines, and are exploring the possibility of introducing the licensing regulations through the provisions of the Protection of Plant Variety and Farmers Rights Act' (GAIN, 2016f).

Life forms are not patentable in India, meaning that no living thing occurring in nature can be protected by patent. India 'adopted a more restrictive European approach on patents by opting to utilize the morality clause in TRIPS' (Singh, 2015, p. 108). The Indian Patents Act provides that 'an invention, the primary or intended use or commercial exploitation of which would be contrary to public order and morality or which causes serious prejudice to human, animal or plant life or health or to the environment is not patentable' (Singh, 2015, p. 107). In international discussions together with an African group of countries, India insists that a review of Article 27.3 of TRIPS should be conducted in order to prohibit patents of life, including those of microbiological processes. In short, plants including seeds, varieties as well as essentially biological processes for the production of plants such as plant breeding and tissue culture techniques, are not patentable. However, there are interpretations that 'although genetically modified plants or seeds are not patentable in India, processes for the genetic modification of plants are patentable' (Yadavand and Kultshreshtha, 2017). Unlike the USA, EU, Russia, Brazil, and China, India is not a member of UPOV, but initiated a procedure for acceding to the UPOV Convention in 2002. The current Indian status in UPOV is unclear. Anyway, the initiative has caused a great resistance among leading civil society groups. India has decided to use the possibility of TRIPS to develop a sui generis system for protection of plant varieties. The Gene Campaign resulted in adoption of the Protection of Plant Varieties and Farmers' Rights Act in 2001, and for the first time in history India recognized the breeder's rights in order to stimulate the seed industry, but 'the proposed Plant Varieties Protection Authority, under the Act, is obliged to register [see Plant Authority, 2007] new strains of plant varieties developed by the farmers alongside the professional breeders The PVPA is also required to ensure equitable benefit sharing with the farmers' (Priyanka, 2005). The foregoing is in compliance with CBD and Cartagena, that India ratified in 1994 and 2003, respectively. But, as critics predicted, the 'extent to which farmers will be able to make use of the registration option may remain quite limited' (Antons, 2010, p. 120), even if it occurs, and will 'fall short of establishing real property rights of farmers to their knowledge and instead make them dependent on the national authority for most benefit sharing and compensation claims' (Antons, 2010, p. 138).

However, India is an example of resistance to international pressure in order to protect indigenous peoples. In favour of this cause, India organized the first internationalized protests against the WTO. Guided by Vandana Shiva, a famous environmentalist, 'in 1993, half a million farmers participated in a historic Bija Satyagraha rally at Bangalore's Cubbon Park...to keep seed in farmer's hands and to not cooperate with IPR Laws that make seed a corporate monopoly and make seed saving and seed sharing a crime' (Shiva, 2017a). The Navdanya movement also participates in monitoring GMO activities in India; for example, it has sued Monsanto and the Indian government for illegal introduction of GMOs into the country. More recently, on 8 September 2016 Vandana Shiva sent a letter to the Indian Prime Minister as stated on a website:

India's laws have been systematically violated by Monsanto which illegally introduced the first Bt cotton GMO crop in India. Monsanto is trying to subvert our Patent Law, our Plant Variety and Farmers Rights Act, our Essential Commodities Act, our Anti-Monopoly Act (Competition Act). It is behaving as if there are no Parliament, no Democracy, no Sovereign Laws in India to which it is subject. Or, it simply does not have any regard for them. The Bt cotton story in India is a story of illegality from the very beginning. Monsanto-Mahyco illegally imported Bt cotton in 1995 and illegally started open field trials in 1998... Monsanto started to illegally collect royalties from our farmers and rapidly established a monopoly. Today, 95% of cotton is Monsanto's Bt cotton. The government has had to take action to regulate the illegal and unlawful activities of Monsanto in royalty collection...Even while Monsanto's illegitimate claim to IPRs in the areas of seeds and plants is now out in the open, a new IPR scam is in the making...Although banned in India, Bayer finds ways to sell glufosinate, to the tea gardens of Assam and the apple orchards of Himachal Pradesh, illegally...Even though patents on seeds are not allowed, for more than one and a half decades Monsanto has extracted illegal royalties from Indian farmers, trapping them in debt, and triggering an epidemic of farmers' suicides. The systemic violation of Indian laws must be stopped immediately. We hope that as the Prime Minister of India you will take strict and urgent action on what has become a national emergency and a serious crisis of governance in the area of the seed, the first link in our food system.

(Shiva, 2017b)

The foregoing gives a picture of an India torn between the ambition to improve its position in the global transgenic arena, and the obligation to meet the needs of a numerous rural population (67.3%) (FAO, 2015) which became an important international ethical problem after the announcement by Prince Charles that commercial GMO propaganda had caused massive numbers of suicides. What applies to the other countries, goes for India too. Its transgenic national policy is more understandable if we look at the structure of production and trade relations. As in China, small family farms also predominate in India's farms structure, since the share of farms less than 0.5 ha is 47% (FAO, 2014). How unfavourable the farm structure is in India can be seen from the fact that India and the USA have about the same amount of arable land, but India has a 62.5 times higher number of agricultural holdings. Another similarity between India and China can be seen from the decreasing trend of farm size: while in 1971/1972 an average farm size was 2.3 ha, this had changed to

1.2 ha by 2000. In parallel, the number of agricultural holdings and the number of members of the rural population increased at the same rate (1.76%) since the 1970s (Chand *et al.*, 2011). Apart from the similarities, there are important differences between agricultural performances in China and India. China successfully increased agricultural growth and productivity and consequently significantly reduced the level of poverty. In India, food insecurity and malnutrition still exist on a large scale: 24.7% of the population are living on less than US$1.25 per day, and 60.6% of the population on less than US$2 per day; 46.6% of children in rural areas are underweight, while 33.9% of children in urban areas are underweight (FAO 2015). In 1987 the poverty headcount ratio in India was 44.76%, while in 2011 it was 21.23%. In China, the poverty ratio was 60.87% in 1987, but 7.9% in 2011; and it has continued to decrease according to the latest data (World Bank, 2013) to 1.85%.

India has been a WTO member since 1 January 1995 and a member of the GATT since 8 July 1948. The Indian position at WTO negotiations as well as the Indian approach to domestic support can be described as both offensive and defensive. Demanding to 'maintain significant security margins in trade policy to minimize the impact of external forces on Indian agriculture', India agreed with tariff reduction for developed countries and flexibility for developing countries. Also, 'while calling for limits on some green-box (decoupled) payments, India has expressed interest in relaxing the criteria governing relief to natural disasters' (Gopinath and Laborde, 2008, p. 3). India protects its heavily agricultural market, and its tariff rates are among the highest in the world – between 50% and 150%, on average, and 114% in 2009, several times higher than in Brazil (36%), China (16%), or the USA (34%, top ten agriculture markets). For example, as of April 2009 the bound products tariffs (%) were as follows: vegetable fat and oils 227; alcoholic beverages 150; oil seeds 130; grains 113; processed fruit and vegetables 111; and fresh and dried fruits, vegetables and nuts, excluding almonds, 100; the applied

tariffs were 24, 133, 30, 40, 30, 30, respectively (USITC, 2009). In regard to tariffs, India remains one of the world's most protectionist countries after WTO accession, but even with popular policies, the AMS is not likely to exceed the de minimis limits (10% of value of production) (Gopinath and Laborde, 2008). The main Indian agrarian policy objectives are to ensure food security by attaining food self-sufficiency, with a focus on wheat and rice, and to support farmers' incomes because of the previously described characteristics of Indian agro sector. To that aim, the government applies the following instruments: (i) minimum support price; (ii) input subsidies (fertilizer, electricity, irrigation, and seeds); (iii) regulated market; and (iv) food subsidies for consumers and trade policies.

Thus it is not surprising that India has dedicated larger areas to the production of cereals (Fig. 2.25).

Over time, the area under rice, peanuts, and vegetables has decreased, the area under fruit and tea has remained stable, while the areas under wheat, maize, soybeans, and sugarcane have increased. The area dedicated to transgenic cotton increased from 4.6% to 6.2% in the period 2000–2011. India has not succeeded in increasing agriculture productivity as China has, so the total cereal output in 2014 of 238.64 million tonnes was less than half of China (Fig. 2.26).

India is the world's largest milk producer, contributing 13% to global production. Most of the milk comes from buffaloes on small unorganized farms. Also, India is the world's largest exporter of buffalo meat. Nevertheless, when it comes to total meat production, India ranks fifth, after the USA, Brazil, the EU, and China. It produces about 6.3 million tonnes or 3% of world production. Of India's total meat production, buffaloes and cattle contribute 31% each, poultry 11%, goats and pigs 10% each, and sheep 5% (Apeda, 2017). Although meat production increased more than in other BRICS countries in the period 2000–2012, it is still nearly 14 times lower than in China, or 4 times more than in Brazil. The main obstacles to meat production in India are: inadequate measures of agrarian policy, the

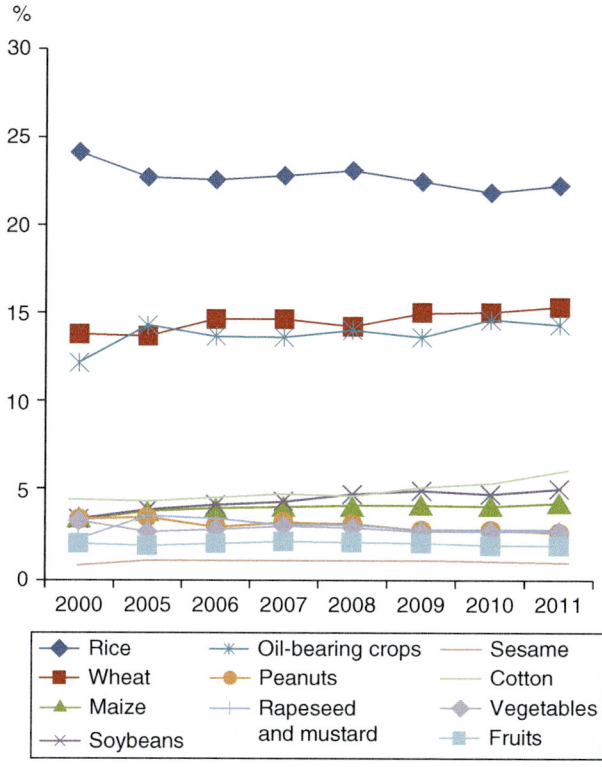

Fig. 2.25. India: Planting structure of farm crops, by year. (Adapted from Rosstat, 2015.)

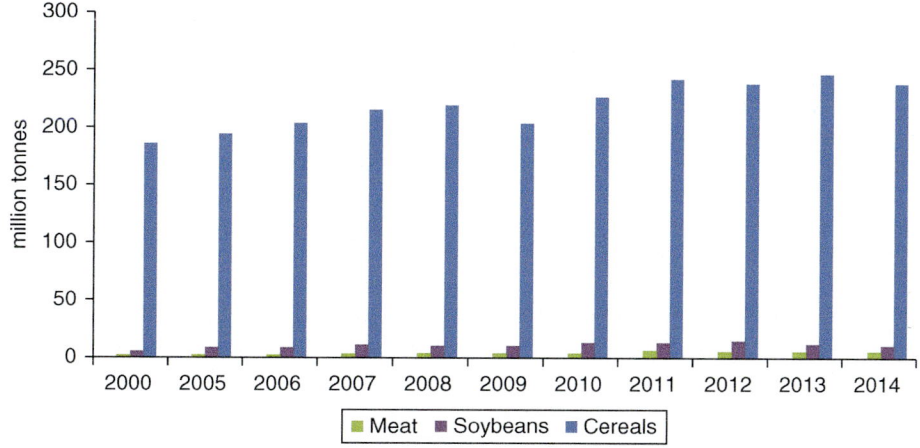

Fig. 2.26. India: Cereal, soybean, and meat outputs, by year. (Adapted from Rosstat (2015) and FAOSTAT (http://www.fao.org/faostat/).)

backwardness and poverty of rural society, technological lagging, and improper hygiene measures in slaughterhouses (Apeda, 2017).

Unlike China and Russia, India does not depend on soybean oilseed imports. For example, in 2016 the country produced 9.7 million Mt and consumed 9.3 million Mt domestically. On the other hand, India is heavily dependent on soybean oil import (Fig. 2.27). In 2016, its domestic consumption was 5.4 million Mt,

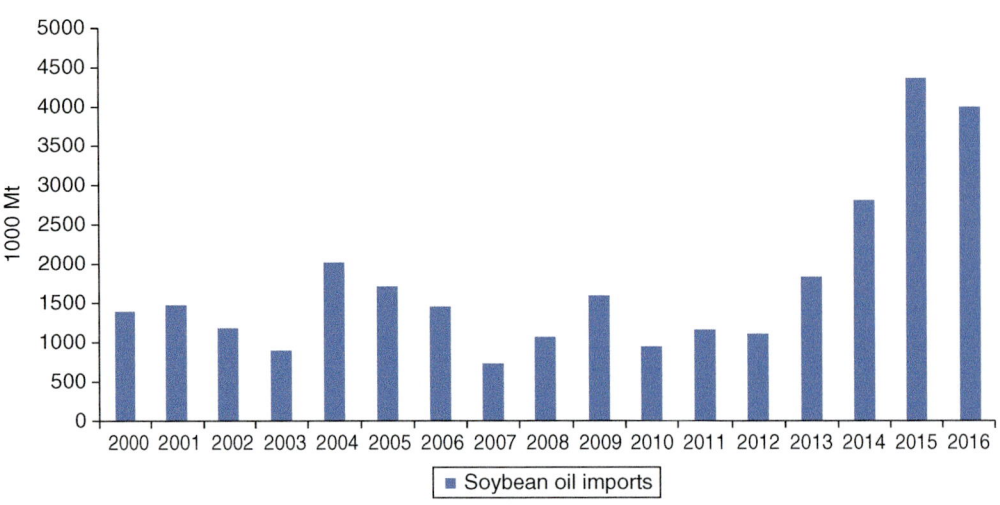

Fig. 2.27. India: Soybean oil imports, by year. (Adapted from IndexMundi (https://www.indexmundi.com/).)

while the country produced just 1.3 million Mt. Soybean oil is mainly used as food for the growing population, and only in small part as feedstuffs for poultry. India's demand for edible oils 'has been rising consistently with the compound annual growth rate (CAGR) of 2.7% in the last 3 years, and around 5.5% in the last 5 years', from 11.6 million Mt in 2003/04 to 17.5 million Mt in 2012/13 (NSA, 2015). Because of that, and stimulated by the price drop in 2012, India has become the leading net importer of soybean oil, exceeding China. According to the latest data, it is estimated that India imported 40 million tonnes in 2016. Indian consumers 'favored it for cooking samosas, dosas, and curries, but the relatively high price of soy oil was a deterrent for many consumers in the country' (The Wall Street Journal, 2016). India's import of soybean oil of about 3.5 million tonnes from Argentina, Brazil, and Paraguay as well as a small quantity of canola oil from Canada, reflects poor processing facilities and decreasing production of oil seeds. The problems can be overcome only by reforming oilseed cultivation.

The shortfall between oilseeds supply and demand is not a new issue for India. New varieties of soybeans, hailed as a miracle that would increase farmer's income by 88%,

were introduced in India in the 1960s, but soy production failed to repeat the US success (Chand, 2007), owing to instability in yield, instability in prices and the associated risk to farmers' incomes, as well as underdeveloped crop insurance. 'Expansion of the soybean planted area took place without large and clear increases in productivity over time' and 'there was no element in soybean cultivation that favored large-size holdings' – as much as 60.12% of soybean cultivation has been on small farms (Chand, 2007). In addition, consumers' demand for soy products was quite low, except for one product, nutria-nuggets. 'Soybean oil was found to be more lucrative in India than using soy for other kinds of food', and '90% of soybeans produced in this country are used for oil extraction' (Chand, 2007, p. 7). Failing to achieve a significant soybean production improvement that would keep up with the domestic demand growth, India has become the largest net importer of vegetable oils since the 1980s. Market liberalization and decline in international prices have provided the stimulus for the rapid rise of soybean oils imports.

In contrast to the obvious failure in soybean production, 'India has recently become the world's largest exporter of cotton yarn, and by 2024 will be closing in on China to

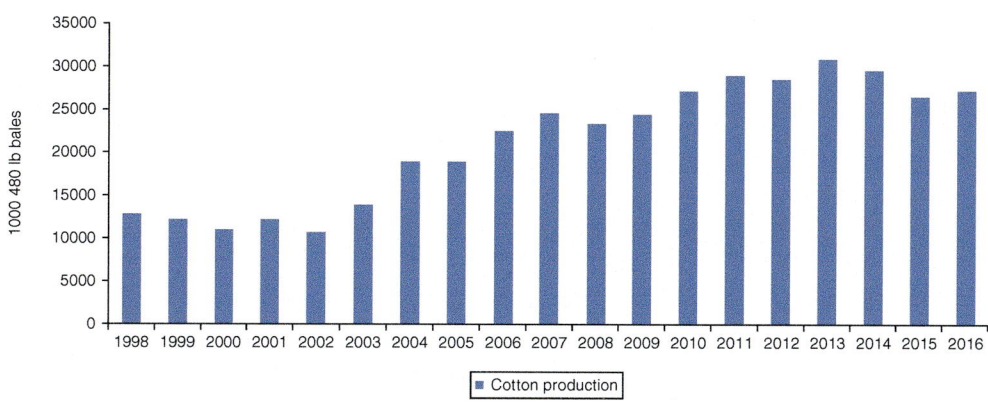

Fig. 2.28. India: Cotton production, by year. (Adapted from IndexMundi (https://www.indexmundi.com/).)

have the world's largest domestic market in terms of population. India's textile industry has been the largest beneficiary of China's shift away from processing cotton fibre into textiles during 2012–2014' (OECD/FAO, 2015). Since 2015, India, with its cotton production of 5.7 million tonnes, has become the world's leading cotton producer country, leaving China behind – by 2006, India had displaced the USA to the third position. The importance of the Indian textile industry based on cotton can be seen from the following facts: (i) it contributes 4% to the GDP; (ii) it directly employs about 45 million people; and (iii) it is a major contributor to foreign exchange earnings (Ministry of Textiles, 2015). Apart from some oscillations, India doubled cotton production in the period 1998–2016 (Fig. 2.28). As the OECD/FAO projection states, differences in cotton production between China and India over the years will increasingly grow in favour of India. It is expected that in 2024 India will account for 30% of the world's cotton output, producing 9 million tonnes of cotton, significantly more than China (Fig. 2.29) (OECD/FAO, 2015).

Certainly, yield growth and increase of harvested area are prerequisites for fulfilling this prediction. Until now, India has obtained a significantly lower yield than the global average of 800 kg/ha, although the yield increased by 81.8% in the period 1998–2016 (Fig. 2.30). In general, cotton yield has been very unstable, reaching the maximum

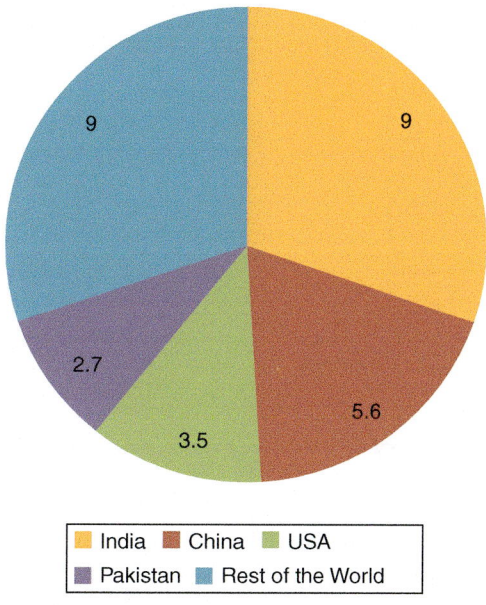

Fig. 2.29. Projection: World's cotton production in 2024, by major producers, million tonnes. (Adapted from OECD/FAO, 2015.)

of 577 kg/ha in 2013, then falling to 500 kg/ha in the following year, and again increasing to 549 kg/ha in 2016. Also, the harvested area has fluctuated over time, increasing by 15.2% in the last year compared with 1998, but still not reaching the level of 2014 (Fig. 2.31).

Regarding exports, it is predicted that India will remain the second largest exporter after the USA. 'India frequently imposed export

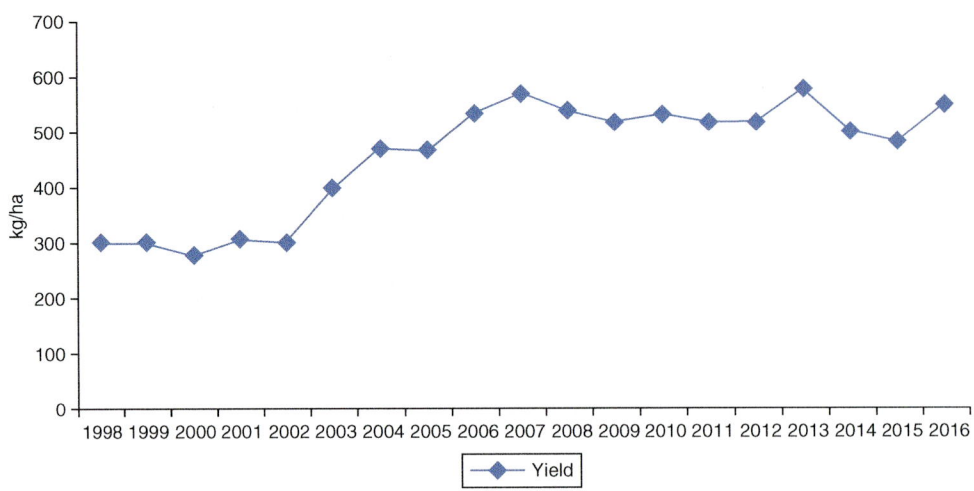

Fig. 2.30. India: Cotton yield, by year. (Adapted from IndexMundi (https://www.indexmundi.com/).)

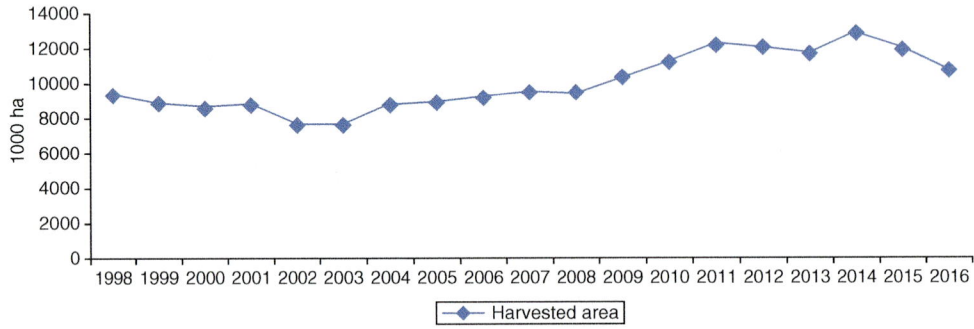

Fig. 2.31. India: Cotton area harvested, by year. (Adapted from IndexMundi (https://www.indexmundi.com/).)

quotas to maintain low cotton prices for its textile industry, and it was a net importer for 7 consecutive years between 1998 and 2004' (OECD/FAO, 2015), as presented in Fig. 2.32. After that time, India became a net exporter but with a huge amount of fluctuation. The highest export quantity was obtained in 2011, 11.08 million 480 lb bales, but in 2016 exports were lower than the level of 2006.

Since its illegal adoption in 1998, transgenic cotton in India has remained a controversial issue. Both opponents and proponents have different views about yield instability and general production performance. For example, one opponent's view is: '[*Bacillus thuringiensis*] *Bt* cotton had promised higher yield, low fertilizer use, and tolerance to pests, but 15 years on, it has failed on all counts. As pests develop resistance, farmers are forced to increase pesticide use (Yadav, 2016). Proponents blame the loss in production on a faulty decision to grow the wrong type of cotton: 'India is cultivating a longer maturing variety (more than 180 days' duration), and this gives insects that turn up in November a chance to attack the crop' (Yadav, 2016). However, the possibility that India will shift away from transgenic cotton is very small. Most probably, it will introduce new cotton varieties to overcome problems. This is clearly stated in the Government of India's Twelfth Five Year Plan for 2012–2017 (Planning Commission, 2013). While

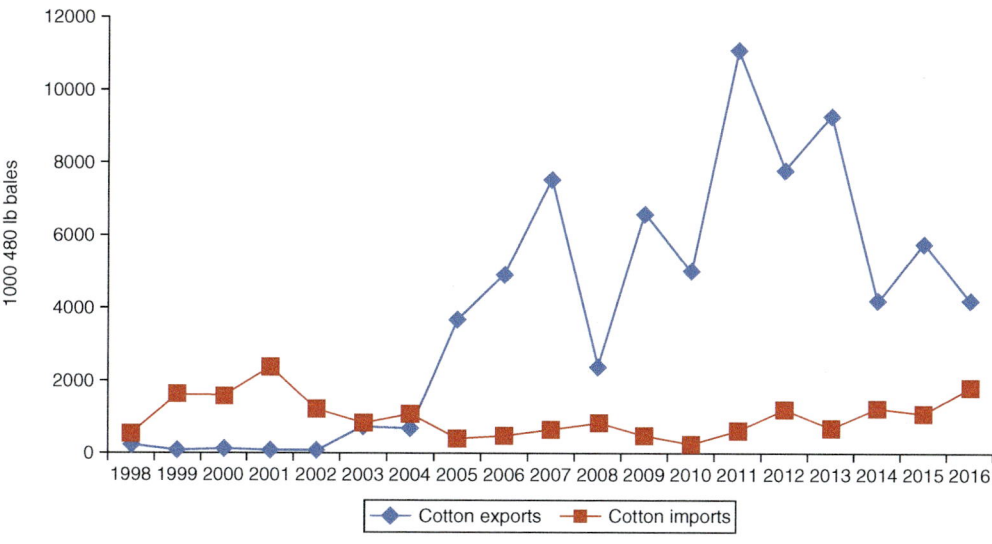

Fig. 2.32. India: Cotton exports and imports, by year. (Adapted from IndexMundi (https://www.indexmundi.com/).)

acknowledging the GMO controversy and the legitimacy of complaints of non-availability of non-*Bt* seeds in some regions, as well as the fact that advances in biotechnology do not necessarily involve GMOs, the Plan emphasized it is 'necessary to remain abreast with the latest advances in biotechnology' and proclaimed the need to 'put in place scientifically impeccable operational protocols and a regulatory mechanism to permit GMOs only when they meet rigorous tests that can outweigh misgivings'. Furthermore, it can be expected that India will start to produce transgenic edible crops such as brinjal (aubergine). A relaxed attitude to field trials speaks in favour of this. Huge farmer and environmental group protests in 2010 led to a brinjal ban, 'but under the government of Prime Minister Narendra Modi, voted into power a year ago, India has quietly changed course on [genetically modified] GM field testing' (Kumar, 2015, p. 138).

India's approach to GMOs can be assessed as ambivalent. On the one hand, the country is trying to keep up with the global transgenic R&D, on the other hand, it seeks to meet the needs of a numerous rural population. India has allowed both public and private sectors to get involved in transgenic R&D, and allowed foreign companies to enter the seed market. In order to ensure equitable benefit sharing with the farmers, India has developed a *sui generis* system for protection of plant varieties, and is trying to protect farmers' incomes by prescription of maximum sale prices. However, it often withdraws decisions under pressure from industry. Striving to achieve the food self-sufficiency goal, India has not allowed production of edible transgenic crops. On the other hand, since it is facing significant poverty, it is constantly increasing imports of transgenic foods in the form of vegetable oils. India's decision to import large quantities of soybean and canola oil is governed solely by its affordable price.

However, social movements in India are better organized than in China; they take the shape of large street protests and attract international attention. But since India is weaker than China or Brazil in many agricultural indicators, transgenic technology has made a stronger indirect impact on its agricultural social structure.

South Africa

South Africa is the first and the largest producer of transgenic crops in Africa and the ninth producer of GMOs in the world. It produced transgenic maize, cotton, and soybean on 2.3 million ha in 2015 (a drop from 2.9 million ha in 2013) (James, 2015), with an adoption rate of 89%, 100%, and 95%, respectively (GAIN, 2016g). Unlike the above discussed countries which use GMOs mainly as feedstuffs or processed ingredients, South Africa is a unique example because 'it has allowed the country's staple foods to be genetically modified' (Jaffer, 2017).

All transgenic crop varieties commercially produced in South Africa are developed in the USA. Monsanto, Pioneer, and Syngenta are three companies that have obtained permission for general release of their maize, cotton, and soybeans for commercial plantings, food and feed use, import and export. In addition, three animal vaccines from Intervet and Ceva Animal Health have been approved. Six crops have commodity clearance or import permission as foods or feeds: maize, soybeans, canola cotton, rice, and rapeseed as developed by Du Pont Pioneer, Monsanto, Syngenta, DowAgrowScience, BASF, Pioneer Hi-Bred, and AgrEvo. However, until now South Africa has imported transgenic maize from Argentina, Brazil, and Paraguay, or conventional maize from EU or Zambia, but not from the USA owing to the non-synchronicity of the regulatory approval process. In 2015/16 a serious drought decreased the yield of South Africa maize by 40%, and in 2017 the country issued a permit for 1.3 million tonnes of US maize for the first time (Dzonzi and Crowlay, 2017). South Africa has a liberal policy in relation to the performance of trials. Apart from 178 open field agricultural trials, the Triclinium and Wits companies are currently conducting trials for the *human immunodeficiency virus* (HIV) vaccine (GAIN, 2016g). Its own research is conducted on vegetables, ornamental plants and indigenous crops, but without great success.

South Africa has a GMO policy similar to the USA. In relation to labelling and testing, the neoregulation paradigm of substantial equivalence has found fertile ground in South Africa. Labelling is required only if a transgenic food product significantly differs from a non-transgenic counterpart or if it contains allergens or human/animal proteins.

There is no routine testing of imports in order to detect unapproved transgenic ingredients unless they are considered to involve health considerations. South Africa became a UPOV member in 1977 under the 1978 Act and 'is one of the few African countries that had a plant variety protection regime in place prior to the adoption of the TRIPS Agreement'...The Plant Breeders' Rights Act 'provides for protection of new varieties of plants, both conventionally bred and genetically modified'... 'the Patents Act excludes patents for both plant and animal varieties', but 'this exclusion does not extend to a variety developed through a microbiological process, such as plants modified through genetic engineering' (Bedasie, 2012, pp. 126–127). In other words, IPR is strong, and transgenic products can be protected through both patent and plant breeders' rights. Thus, company fees are collected in the same way as in the case of US farmers who are obliged to sign a one-year licensing agreement, which at once pays for the seed and technology fees. South Africa signed the Cartagena Protocol in 2003. However, its implementation will be carried out in phases. Analyses of public attitudes to GMOs have shown a 'major increase in public awareness of biotechnology, and a major increase in the attitude that favours the purchasing of [genetically modified] GM food' in the period 2004–2015. The proportion of the public that would purchase GM foods on the basis of health considerations has increased from 59% to 77%, on the basis of cost considerations it has increased from 51% to 73%, and on the basis of environmental considerations it has increased from 50% to 68% (Gastrow *et al.*, 2016). In general, 49% of the population in South Africa believes that it is safe to eat transgenic food. Resistance to GMOs exists in this country, but has a weak influence. For example, mandatory labelling is still on hold although 75% of the population consider it necessary, and it is stipulated by the 2011 Consumer Protection Act.

Regardless of the weak regulatory oversight and strong IPR, diffusion of GMOs in this country is not going smoothly nor without resistance. South Africans have joined the international community in a global march against Monsanto. Greenpeace Africa is launching a campaign, 'Say NO to Release of GMO Vaccine in South Africa'. BioWatch South Africa fights for biodiversity, food sovereignty, and social justice and is supporting the training of small-holder farmers to strengthen agro-ecology practices and GMO awareness. As of June 2016, 25,000 people have signed a petition opposing Monsanto's field trial on drought-tolerant crops under the campaign slogan, 'Monsanto profiting from climate change: NO to bogus drought-tolerant GMOs' (Seed Freedom, 2016).

Maize is the most important crop in South Africa and is produced in some years on over half of the sown area (Fig. 2.33). As already mentioned, most of the maize produced in the country is transgenic. Since 1997, the first year of transgenic maize approval, the adoption rate has progressively increased from below 5% to about 90% in 2016 (GAIN, 2016g). Maize production is characterized by fluctuations in both the harvested area and the output (Fig. 2.34). From more than 4 million ha in 1996, the harvested area of maize dropped to 2.2 million ha in 2015, while production dropped from more than 10 million Mt to less than 8 million Mt. Maize production recovered in

2016, and South Africa ranked as the world's tenth largest producing country, and the ninth largest exporting country. Despite the country's importance on the world trade market, export performances are still unstable (Fig. 2.35). Most often, South Africa is an exporter country and the main destinations for its maize are African countries. Yet, in 2006, 2014, and 2015 South African maize production was below the annual national demand, and from being an exporter it became a huge importer of corn. Despite transgenic production, the severe drought that affected the region in some years caused irreversible damage to the crops and lowered yield. In 2006 South Africa imported 1.1 million Mt of maize, while in 2015 it imported 3 million Mt.

Some oscillations in transgenic soybeans production are also present, but a clear linear growth trend can be observed in both harvested area and output (Fig. 2.36). The planted area under soybean reached a record of 700,000 ha in 2016, between 90–95% of which was estimated to be transgenic seed. This is an impressive growth compared to 1996, when 87,000 ha were planted with soybeans. Transgenic soybeans were introduced in South Africa for the first time in 2001. Even though South Africa was facing

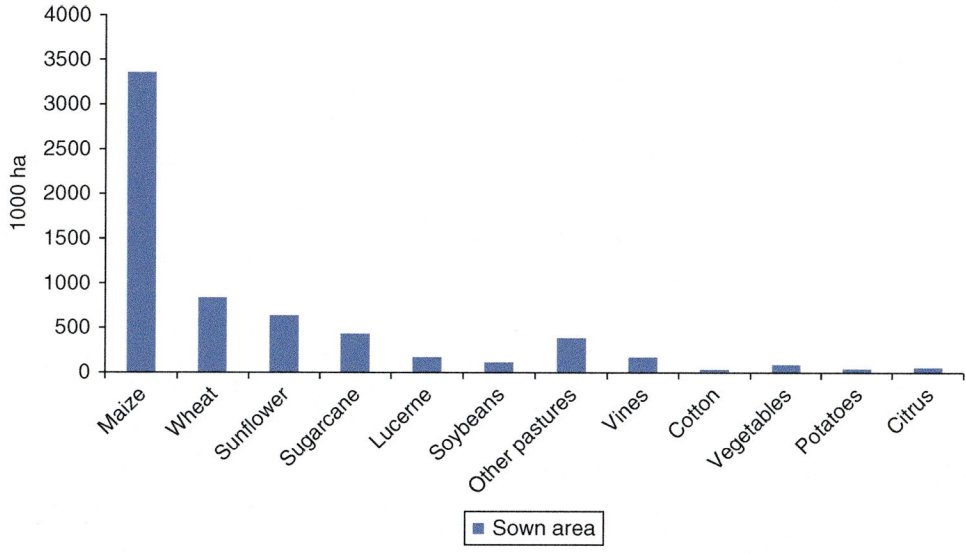

Fig. 2.33. South Africa: Structure of crops area by main species. (Adapted from FAO, 2005.)

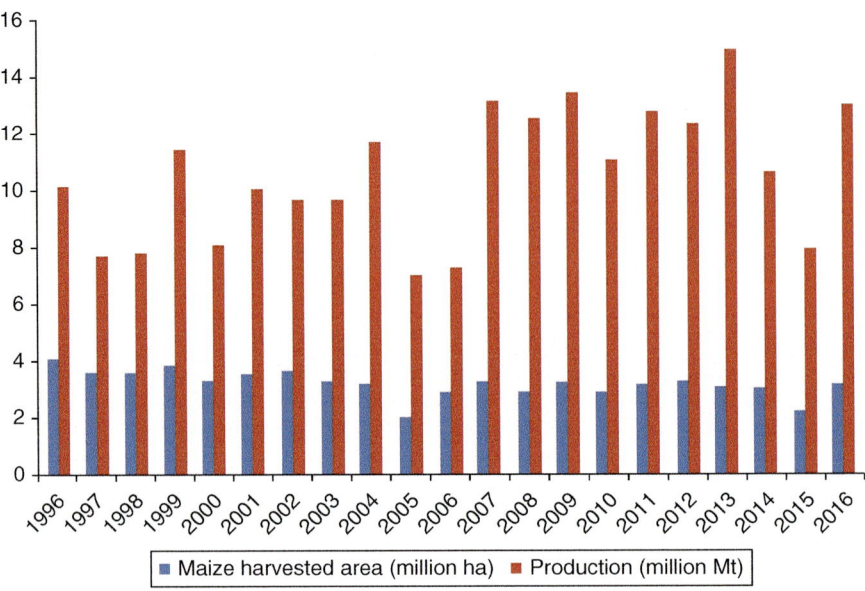

Fig. 2.34. South Africa: Maize production and area harvested, by year. (Adapted from IndexMundi (https://www.indexmundi.com/).)

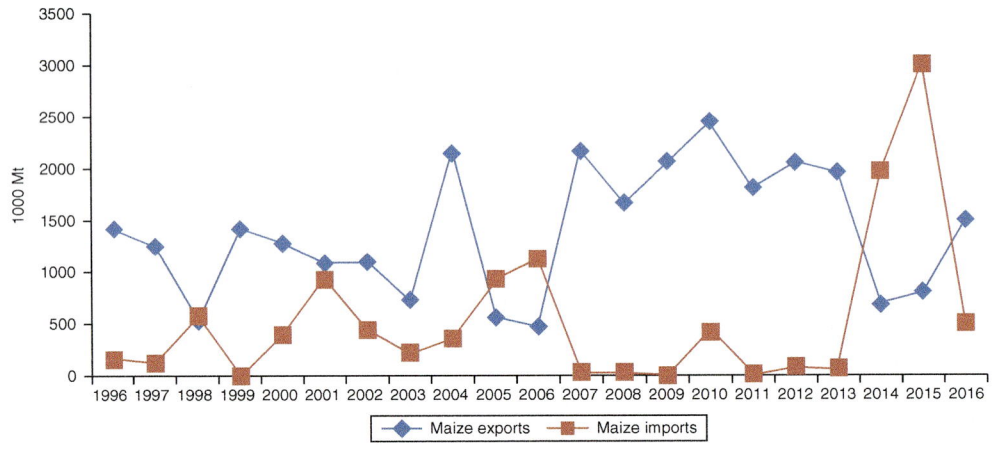

Fig. 2.35. South Africa: Maize exports and imports, by year. (Adapted from IndexMundi (https://www. indexmundi.com/).)

food insecurity in some years, areas with soy were increasing because of large investments in oilseed processing capacity that is now estimated at 2.2 million Mt per annum (GAIN, 2016g), or twice as much as domestic production.

Significant growth in demand for soybeans is mainly driven by the increasing demand for animal feed. As can be seen from Fig. 2.37, the biggest increase in livestock industry is recorded in poultry production, from 870,000 tonnes in 2000 to about 1.6 million tonnes in 2013. In the same period mutton production increased by 41%, while beef and veal production increased by 44%. Although South Africa, the largest importer of soybeans in sub-Saharan Africa, accounts for an average of 72% of import demand

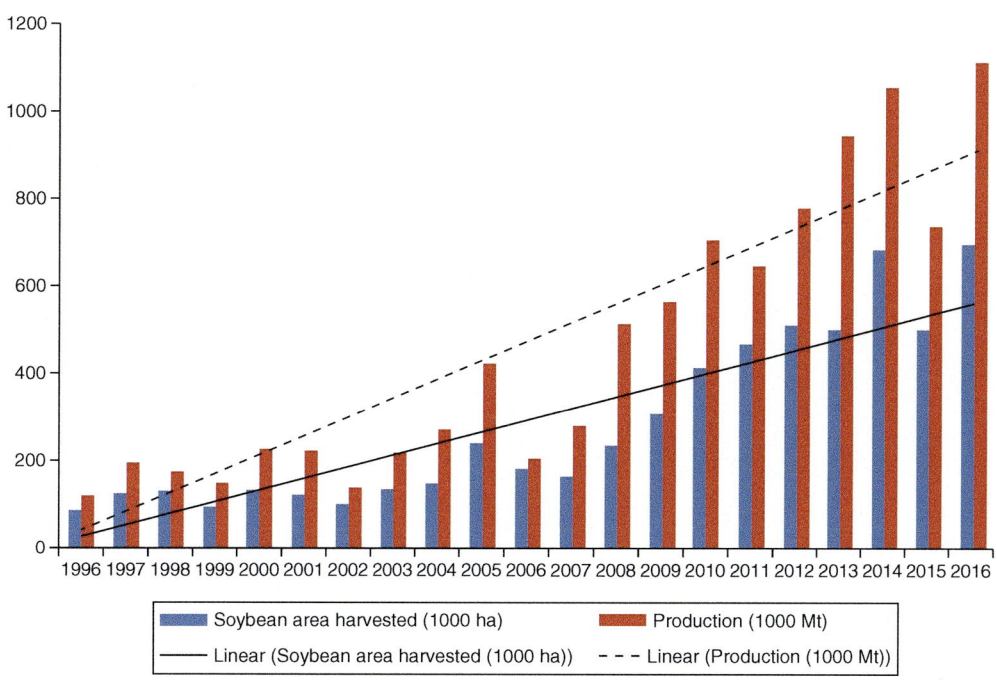

Fig. 2.36. South Africa: Soybean area harvested, by year. (Adapted from IndexMundi (https://www.indexmundi.com/).)

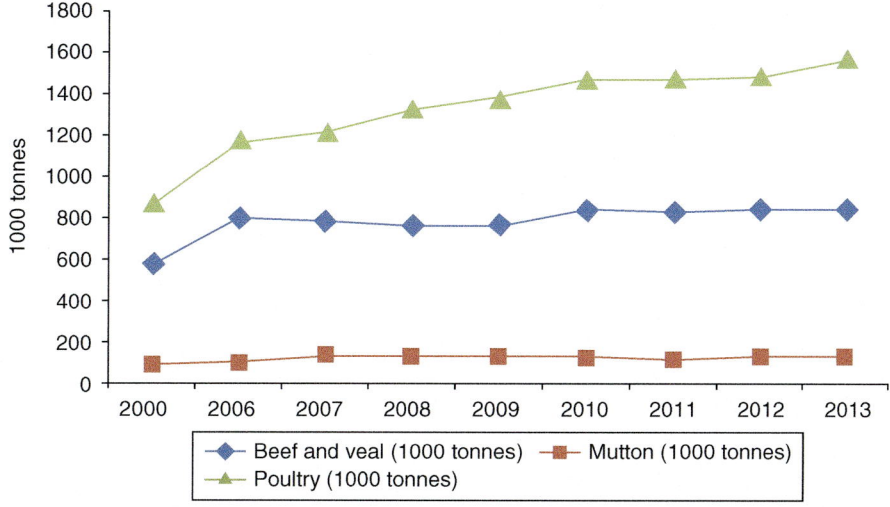

Fig. 2.37. South Africa: Meat production, by year. (Adapted from Rosstat, 2015.)

(Sihlobo and Kapuya, 2016), in seeking to reduce dependence on soybean imports, domestic production cannot meet the demand (Fig. 2.38). Despite net exports in the periods 2003–2004 and 2007–2011, since 2012 South Africa has established its position as a net importer of soybean oilseeds.

Production and trade performances of cotton, South Africa's third transgenic crop, are worse than in the case of maize and soybeans. From the first introduction of transgenic cotton in 1998 until 2016, cotton areas dropped 13 times, while production decreased 4.5 times (Fig. 2.39). South Africa's farmers are increasingly giving up this product, mainly owing to a negative movement in cotton prices (GAIN, 2016g). Regardless of cotton prices, in recent years production has never been at the required level for the country to become a net

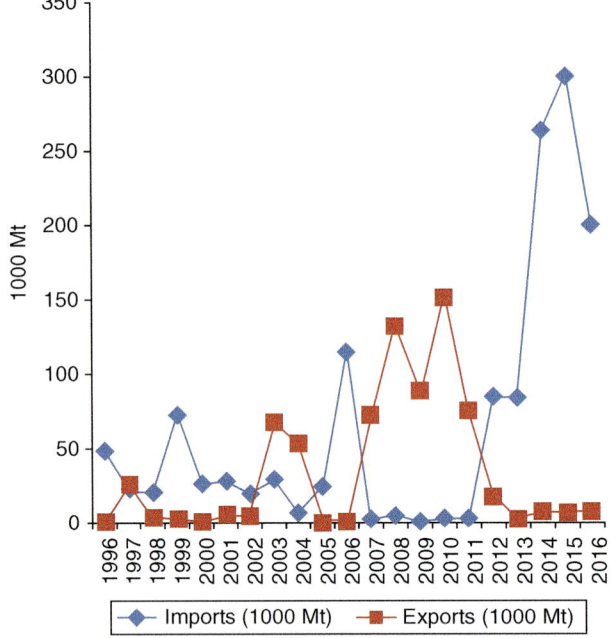

Fig. 2.38. South Africa: Soybean exports and imports, by year. (Adapted from IndexMundi (https://www.indexmundi.com/).)

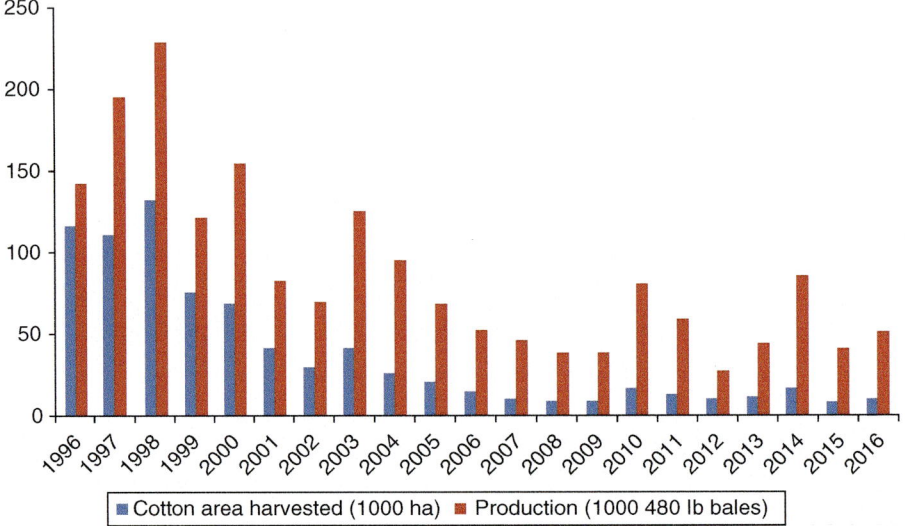

Fig. 2.39. South Africa: Cotton production and harvested area, by year. (Adapted from IndexMundi (https://www.indexmundi.com/).)

exporter of cotton. In the entire observed period South Africa was a net importer of cotton, although the ratio between export and import had decreased tenfold by 2016 compared to 1996 (Fig. 2.40).

In general, from the case of South Africa it can be seen once more that an overall agricultural goal determines national GMO policy. In contrast to the small farms' domination in China and India, and that of the medium average farm in Brazil, in South Africa it is the domination of large farms, even several times larger than in the USA, that characterizes the farm structure. South Africa has farms of a size above all international standards. These trends towards increasing size were evident by the 1990s. 'The average sized farm increased from 738 ha per farm in 1953, to 867 ha in 1960, to 988 ha in 1971, and to 1339 ha in 1981, but declined to 1280 ha per farm in 1988' (Van *et al.*, 1995, p. 9). However, South African agriculture 'is dualistic in nature', consisting of a less developed subsistence sector and well-developed commercial farms which 'produce more than 95 percent of the total

marketed agricultural output' (FAO, 2005, p. 2). One of South Africa's greatest agricultural disadvantages is a shortage of arable land (9.8% of its territory, of which just 3% is considered truly fertile), because of which the country lags behind other BRICS countries. This shortcoming defines the direction of production, since the majority of land is used for grazing and livestock farming. However, South Africa has a significantly lower number of holdings than other BRICS countries (1,093,000) but a higher percentage of people living in rural areas (35.2%) than Brazil and Russia (FAO, 2015).

Like Brazil, South Africa became a member of the GATT in 1948 and a member of the WTO in 1995. Both countries follow policies that lead to low price distortions. Thus domestic prices with some exceptions are aligned with world prices. Furthermore, both countries dedicate a similar amount of the GDP to the TSE (about 0.3%), and in both countries support to farms is below 5% of the GFR. After reforms in the mid-1990s agricultural intervention in the market has been reduced, and export subsidies and

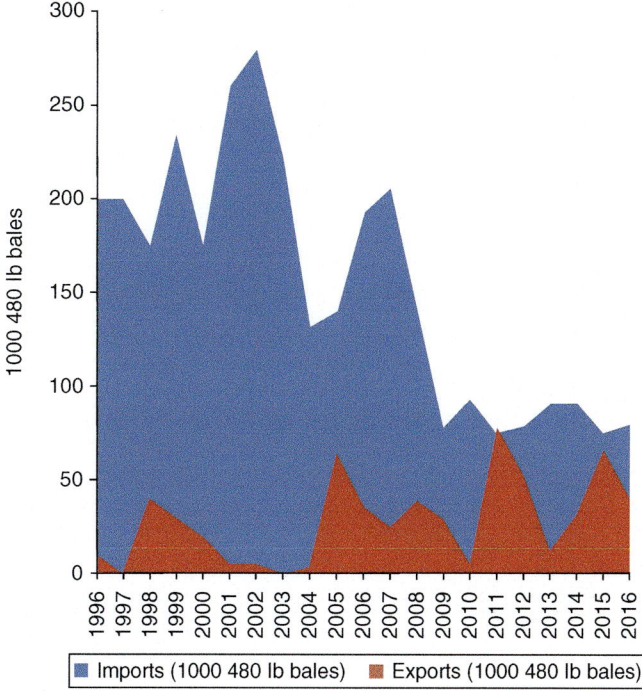

Fig. 2.40. South Africa: Cotton exports and imports, by year. (Adapted from IndexMundi (https://www.indexmundi.com/).)

market support interventions have been completely eliminated (OECD, 2016d). Land reform, the key issue of agricultural policy, predestined the applied measures targeted at smallholders: (i) support based on output and inputs; (ii) production loans for new farmers; (iii) training and skills development; (iv) capacity building; (v) business development and support; and (vi) upgrading of the irrigation scheme. The applied measures are aligned with the country's three major agricultural policy goals: (i) building an efficient and internationally competitive agricultural sector; (ii) supporting the emergence of a more diverse structure of production with a large increase in the smallholders' enterprises; and (iii) conservation of natural resources (Ministry for Agriculture and Land Affairs, 1988).

Since the country has given agriculture the main role in building a strong economy and has been trying to build an efficient and internationally competitive agricultural sector, it has adopted a very weak regulatory approach to GMOs, quite similar to the USA. However, unlike the USA, South Africa has failed to significantly improve its transgenically produced food export. On the contrary, with some variations, South Africa has remained dependent on imports of all the three transgenic crops commercially produced in the country. South Africa's agricultural productivity has remained stagnant in the 21st century. 'Productivity grew rapidly at 3.98% between 1981 to 1989, owing to mechanization and use of fertilizer, herbicides, pesticides…From 1989–1994 growth of productivity declined to as little as 0.28%, due to inflation. But after 1994 the growth was positive due to a positive net farm income… then it became stagnant due to declining output growth and increasing use of inputs around 2008' (Ramaila et al., 2011, pp. 6–7). Despite their active presence in South Africa, it seems that NGOs have much less influence on the country's decisions than in other BRICS countries. This should be understood within the post-apartheid transition context and the infancy of social movements. 'South African politics is characterized by a dominant political party which gains considerable legitimacy and

electoral support by virtue of its past position and success, rather than as a result of its current policies.' In such circumstances, social movements' 'efforts appear to be directed towards the central state and there is little investment in more decentralized and devolved authorities and decision-making structures… to draw a simple analogy, most movement organizations spend more time trying to get a share of the cake than in challenging the size of the cake and who has the right to set the rules that govern access to the cake' (Mitlin and Mogaladi, 2009).

However, South Africa is good proof that a strong neoliberal stance in respect to GMOs is not a guarantee for success. The most important parameters are still (i) climate; (ii) land quality; (iii) proper selection of cultivars; (iv) proper use of agrochemicals and mechanization; and (v) proper selection and application of agricultural policy measures – and these have all been important for centuries.

2.3.4   Serbia

Serbia, an upper middle income country located at the crossroads of Central and Southeast Europe, and an official candidate for membership of the EU, has unique GMO policies compared with those of the previously discussed countries. In Serbia, the production and commercialization as well as importation of transgenic crops and products is strictly forbidden by a 2009 law. Currently, there are no foreign companies involved in transgenic experimental research. It has one of the most turbulent histories of recent times among the countries formerly under socialistic regimes. The civil war, the North Atlantic Treaty Organization (NATO) airstrikes, international economic sanctions, the refugee influx, and one of the world biggest hyperinflations contributed to the Serbian economy of 2000 being half the size it was in 1990 (Lovre and Brankov, 2016). In addition, Serbia is still struggling to preserve its territorial integrity since the province of Kosovo unilaterally declared independence from Serbia in February 2008.

All these hardships would be more difficult to handle if the country were not to a large extent achieving self-sufficiency in food production. Unlike all the countries discussed so far, Serbia is not a member of the WTO. The accession process that started officially in 2005 is well under way, but with at least one major issue pending, i.e. the harmonization of Serbian GMO regulations with WTO principles of trade.

Serbia is also unique because 80% of cities and municipalities (135 out of 169) have declared themselves GMO-free. A huge campaign, 'Serbia without GMOs', has been underway for several years. In 2016, the 'March against Monsanto' was organized by the Beekeepers Association and with the support of the leading anti-GMO portal PPNS (http://www.prviprvinaskali.com/) in the city of Kragujevac. The public attitude towards GMOs is extremely negative. A study has shown that rejection of transgenic food is 'mostly associated with possible adverse effects on human health, together with moral and ethical issues and distrust in companies that produce transgenic food' (Brankov et al., 2013a, p. 8295). Despite a difficult economic situation, only 19.7% of respondents would buy transgenic food if it tasted the same as traditional food, but was cheaper. Apart from the perceived risks and benefits, lack of confidence in institutions has a direct impact on readiness to accept transgenic food in Serbia (Brankov et al., 2013b, p. 117). The analysis showed that in Serbia there is great consumer interest in organic products (Brankov et al., 2013c, p. 158). A desire to maintain personal health and the environment is the main motivation for the great consumer interest in organic food products (Brankov et al., 2014, p. 9605). The root of the huge distrust of all those involved in GMOs, as well as the huge concern with health, can be found in the NATO bombing of Serbia with depleted uranium in 1999. The explosion of malignant tumours because of uranium dust after the 10-year latency period is evident in Serbia (Jovanovic et al., 2012, p. 188). Furthermore, international sanctions against Serbia in the periods 1991–1995 and 1998–1999 gave a lesson about the importance of self-sufficiency in food production. After the 'democratic changes' in the 2000s, the poorly executed *privatization* and *restructuring* of *agrokombinates* in a suspicious manner led to the establishment of several large landowners, the biggest Serbian agro-businessmen (Brankov, 2018). This has additionally deepened the public's distrust of government decision-making bodies. On the wings of a desire to preserve 'what was left', Serbian citizens are united on the issue of resistance to GMOs. Such an attitude creates problems for the political elite who, exposed to frequent elections, have postponed the decision on amending the rigorous GMO law for several years, despite pressure from the USA, the WTO and the EU.

Serbia is a very good example of the importance of social movements in relation to the GMO issue. Montenegro, the smallest country in the Balkan region, is a completely opposite example. In absolute silence, without any social movement reactions, under Djukanovic's regime the country adopted a GMO law aligned with EU regulations in 2008, just 2 years after separating from the State Union of Serbia and Montenegro.

The Serbian government's ambivalent attitude is a reflection of the desire of the political elite to join the WTO and thus secure a substantial improvement of IPR protection and a revision of the law in line with TRIPS standards (GAIN, 2016h). This has been evidenced in the insufficiently effective destruction of illegal GMO plantations, as the media have often reported. On the other hand, Serbia ratified the CBD in 2002 and Cartagena in 2006, and promoted organic soybean production by signing the Danube Soya Declaration in 2013. This ambivalence is a reflection of competing import and export interests, since grain farmers and trader organizations are not united on the issue.

According to the latest 2012 Agricultural Census, available land comprises 68.9% of the total territory of Serbia (77,592 sq. km without Kosovo), of which about 3.8 million ha is agricultural land, although almost one eighth is not utilized (Sevarlic, 2015). Since the 2000s, the large-scale horizontally and vertically integrated *agrokombinats* from the socialist era have been in the process of being privatized and reconstructed. *Agrokombinats* are in large part now transformed into enterprises which hold 17.8% of the utilized agricultural area, and possess 207.4 ha on average. The remaining 82.2% of utilized agriculture area is owned by family farms of 4.5 ha on average. Available arable land per capita in Serbia is 0.36 ha and higher than in China, India and South Africa, unfortunately because of a bleak demographic situation reflected in a negative natural growth and a significant aging of the population. According to the 2011 Census of the Population, Serbia's population had decreased by 5% since 2002 and stood at 7.18 million people, 40.6% of whom live in rural settlements (RZS, 2011). As presented in Table 2.8, 77.4% of farms are smaller than 5 ha, while farms larger than 50 ha account for about 1% of the total number of holdings. Unlike other socialist countries, Serbia never fully collectivized agriculture. For example, in 1989 'the private sector accounted for 83% of total maize output, 59% of total wheat output, 48% of total beef output, and nearly 80% of all pork output' (EC, 2006, p. 8). The primary goals of agricultural policy after market liberalization are production growth and stability of producers' incomes, as well as an increase in commercial family farms' competitiveness. Market support measures such as export refunds, intervention purchase and support to storage are being withdrawn and replaced by direct payments 'based on output (price supplements), payments per ha, and animal and input subsidies (refunds, subsidized interest rates and insurance premiums, etc.). On average, 70% of the total agricultural budgetary supports (42% in 2006 and 91% in 2013) were allocated to direct payments and variable input subsidies...' (Bogdanov and Rodic, 2014:163). Anyway, as Lovre put it: 'The

declared attempt to increase the supply and production efficiency of agricultural products has not materialized in the measures of agricultural policy, as is well illustrated by the inconsistency of the structure of agricultural policy measures. It is indisputable that liberalization of market agricultural and food products has in actual fact been twisted, with extremely negative effects on the size of supply' (Lovre, 2013, p. 21).

In Serbia, agriculture contributes more to GDP than in Brazil, Russia, South Africa, USA and Spain, but less than in India and China (Table 2.7 and 2.9). The same relationship applies for the share of agricultural EAP in the total EAP. Serbian agriculture generates less income per worker than Brazilian, Russian, South African, US and Spanish agriculture. For example, it generates 19.6 times less income than US agriculture or 2 times less income than Brazilian and 2.3 times less than Russian agriculture, but 4.2 times more than India's and 2.8 times more than China's. Also, agricultural capital stock per worker is higher in Serbia than in India and China – 7.7 and 9.8 times, respectively. In general, the strongest side of the Serbian food system is food safety, diet diversification, and the low poverty headcount ratio (0.1%), while the biggest weakness is the GDP per capita PPP (Power Purchasing Parity) (Brankov and Milovanovic, 2015).

Contrary to the previously mentioned countries, in Serbia beef, pork, and poultry meat production has not yet reached the pre-transition level (Fig. 2.41). Production of beef in 2015 was 45% lower than in 1990, pork meat production decreased by 1.4%, while poultry production decreased by 17.3%. The data indicate inadequate state protection of domestic production against unfair competition from foreign producers after the market liberalization.

Serbia has still not fully recovered its meat production after the collapse in the

Table 2.8. Farms in Serbia, by size. (From Sevarlic, 2015.)

Number of holdings (thousands)	Share of holdings by farm size class (%)					
631.5	<2 ha	2–5	5–10	10–20	20–50	>50 ha
	48.1	29.3	14.3	5.2	2.1	1

1990s. On the other hand, it has significantly increased soybean production (Fig. 2.42), but not enough to be self-sufficient in feedstuffs. As a result, it imports a certain quantity of soybean meal almost every year (Fig. 2.43). Since 2013 and the signing of the Danube Soya Declaration, Serbia has significantly increased soybean and soybean oil export (Fig. 2.44 and 2.45), and for the first time since 2010–2011, produced in 2016 more soybean meal than was domestically consumed. In 2016, Serbia ranked 16th in world soybean oilseed production, and 29th in soybean meal production.

The foregoing means that Serbia is on the path to achieving stable self-sufficiency for

Table 2.9. Serbia agricultural GDP, income, and capital stock per worker.

Agricultural GDP (billion USD)[a]	Agriculture in GDP (%, 2014)[a]	Agricultural EAP (thousands, 2010)[b]	Agricultural EAP (share of total)[b]	Ratio of agricultural GDP to agricultural EAP (USD/person)[c]	Agricultural capital stock (million, constant 2005 USD)[b]	Agricultural capital stock, per worker (constant 2005 USD)[b]
3.497	8.2	617	13	5,667.7	7,409	10,554

[a]CIA (2015); [b]FAO (2012); [c]Authors' calculation

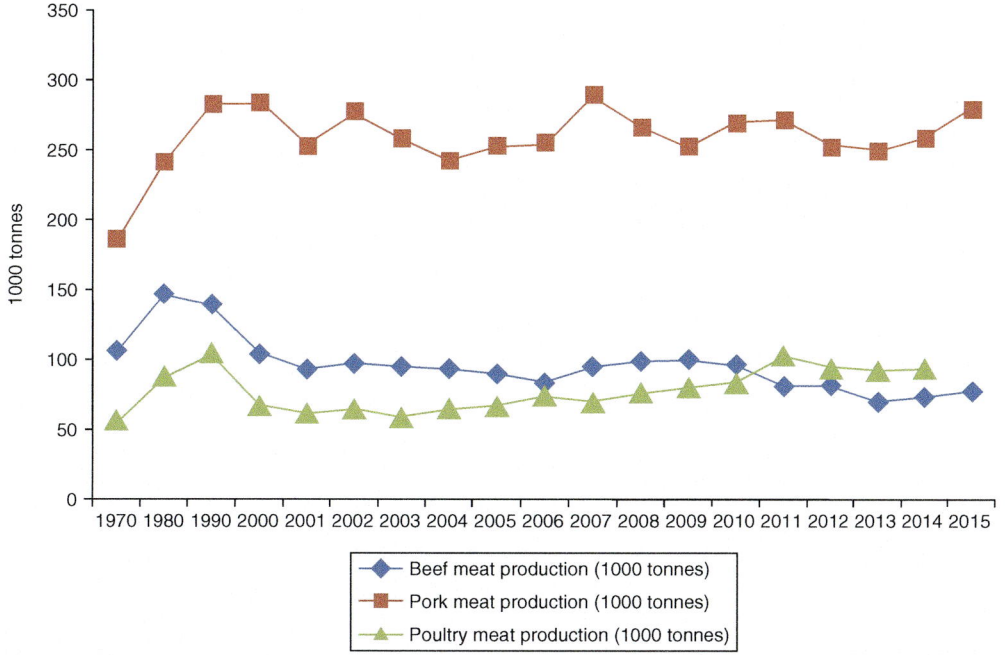

Fig. 2.41. Serbia: Meat production, by year. (Adapted from the Republic of Serbia Bureau of Statistics (http://www.stat.gov.rs/en-us/).)

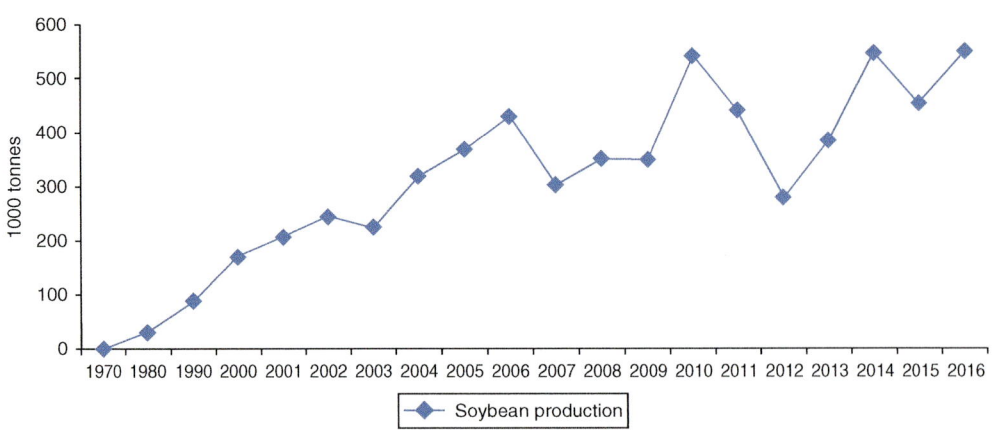

Fig. 2.42. Serbia: Soybean production, by year. (Adapted from the Republic of Serbia Bureau of Statistics (http://www.stat.gov.rs/en-us/).)

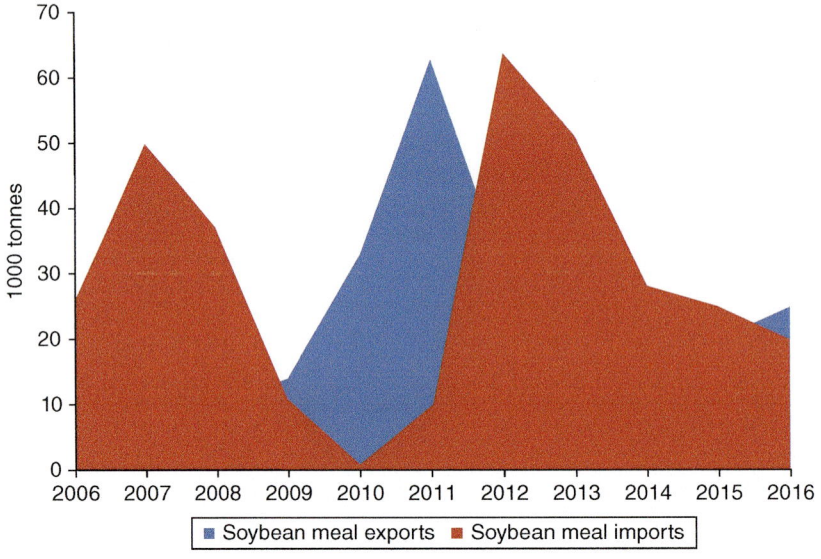

Fig. 2.43. Serbia: Soybean meal exports and imports, by year. (Adapted from FAOSTAT (http://www.fao.org/home/en).)

both the current level of livestock production and the projected increase. This is in line with research that has shown that soybean acreages should increase by 32% to satisfy the current needs, and soybean should be grown on 270,000 ha to satisfy future needs (Bosnjak *et al.*, 2012). Thanks to official government confirmation that the soybean produced is non-GMO, Serbian exporters have become more competitive

internationally (GAIN, 2016h). In addition, starting from December 2016, the first 'national quality label': *Srpski Kvalitet* [Serbian Quality], was launched. To qualify for the trademark, products must use Serbian raw materials. It guarantees, among other things, that food products such as milk, eggs, and meat are GMO-free through the entire production chain. The label has opened the door for adding value to Serbian food products

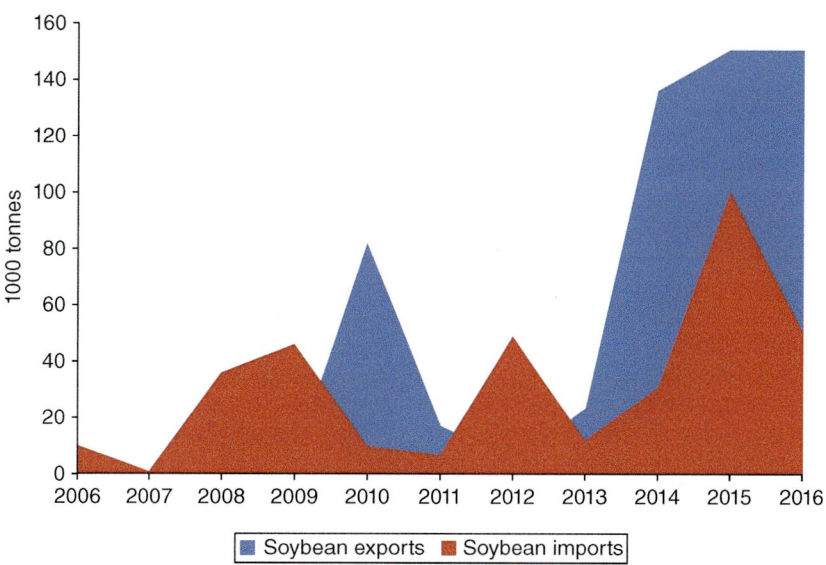

Fig. 2.44. Serbia: Soybean exports and imports, by year. (Adapted from FAOSTAT (https://www.fao.org/home/en).)

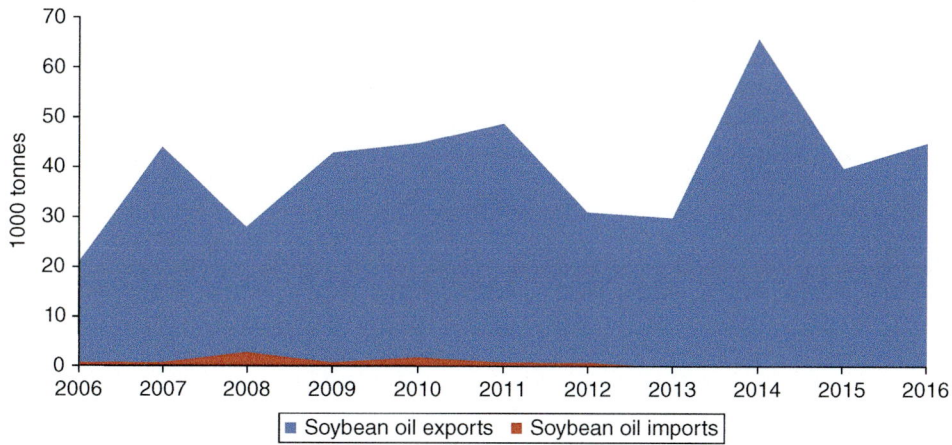

Fig. 2.45. Serbia: Soybean oil exports and imports, by year. (Adapted from FAOSTAT (http://www.fao.org/home/en).)

and increasing their competitiveness on the international market.

Unlike China and Russia, Serbia is facing no urgent need for GMO soybeans to feed livestock. A slow recovery of the sector after the collapse in the 1990s enabled two parallel processes: an increase of livestock productivity and an increase of non-GMO soybeans production. In other words, wisely guided, Serbia can improve its position on the international market as a non-GMO food producer. This is corroborated by the production of non-GMO maize. Serbia is one of the leading global maize exporters, ranking eighth in 2015, after the USA, Brazil, Ukraine, Argentina, EU, Russia, and Paraguay. Taking into consideration that the USA, Brazil, Argentina, and Paraguay are producers of transgenic maize, it can be noticed that Serbia is the fourth largest country exporter of non-GMO maize. Maize is the leading Serbian agricultural export commodity, mainly exported to neighbouring Mediterranean and North-African countries. However, thanks to its non-transgenic features it has found new markets in South Korea and Japan in recent years. The Serbian GMO policy is in line with agricultural goals to achieve stability of producers' incomes and increase competitiveness. Limited by area, Serbia cannot compete with the world's largest producers on the basis of economies of scale, but can compete in small niches such as non-GMO food. Although changes to the GMO law have not been completed yet, and will be continued after the next elections in 2018 or 2020, a well-organized social movement in a small country like Serbia 'gave a lecture' to the rest of the world.

2.3.5 Final remarks: Tying the case studies together

Our study of the different countries shows that national GMO policies should not be analyzed separately, but as integral parts of agricultural policy. Each of the discussed countries accepts or refuses transgenic products depending on its overall agricultural goals. The countries where the main goal is to achieve self-sufficiency have adopted strong regulatory oversight more or less successfully. The countries whose main goal is to expand exports have approved a weak regulatory approach to GMOs. Besides this, the countries in which small farms achieve a significant amount of agricultural output seek more to protect their own markets than the countries in which agricultural production is mainly in the hands of large producers. The very different success ratios of transgenic technology in Brazil and South Africa confirm that a neoliberal stance in regard to GMOs is no guarantee of success. Moreover, technology in itself is not a guarantee of satisfactory agricultural output; other parameters, historically important in agricultural production, have a much greater

impact. This study confirms a previous finding that 'less developed and less economically powerful countries...suffer the brunt of negative social impacts with the introduction of this new technology' (Pechlaner and Otero, 2008, p. 366).

Although the third food regime has yet to assume its final shape, it can be considered to have already solidified around a central axis of transgenic technology, with soy as its point of stability. As Friedmann and McMichael in their famous 1989 paper pointed out: 'the shift to soy, outside of Asia almost completely an American crop between World War II and 1973, reflected American power' (Friedmann and McMichael, 1989, p. 109). Since the creation in the 1930s of the conditions for soy to compete with other oil seeds, soy has played an important role in agriculture transformation. It is generally considered that 'soy was at the center of the postwar transformation of agriculture, and with it, major shifts in the international division of labor. Most of the story applies to meat, but its origins also lie' in soy's combined properties as processed soybeans vegetable oil initially important for the manufacture of margarine, and as an excellent source of animal feed (Friedmann and McMichael, 1989, p. 110). The same observation applies today, with soy in an even stronger position through IPR protection. Transgenic soybeans have broken through all the barriers of national regulations and found their way into all parts of the world. Regardless of whether a country adopted a *de jure* prohibition of cultivation, as Russia did, or accepted substantial equivalence approaches, like South Africa, all have accepted transgenic soybeans. Even Serbia, a country that has adopted the strongest legislation and has shown the strongest resistance, has imported certain quantities of transgenic soybean meal in some years of feeds shortage.

The final shape of the third food regime in relation to GMOs depends on the efforts of the resistance. Most probably, the benefit of the resistance will be the crystallization of a smaller or larger number of market niches for non-GMO products. It can hardly be expected that any kind of effort can fully eliminate products of transgenic technology.

Even Russia, with its real and potential power, will face remnants of GMO contamination if it decides to withdraw from the importation of any transgenic products. It is well known that farmers' dependence on corporations as suppliers of inputs, established during the second food regime, further deepened under the third regime for those who accepted the new technology. IPR protection has contributed to an increase of inequality between nations and has even created conditions for the restoration of colonialism, i.e. neocolonialism. In response to this, social movements have become better organized, compared to their beginnings in the 1960s, and farmers who have stayed out of GMO production have become symbols of freedom in the modern food system.

Notes

[i] 'National production activities are divided into activities that contribute directly or indirectly (intermediate consumption of goods and services) to the production of means of production (department 1), means of consumption (department 2), and exports (department 3), and imports are divided into imports of capital goods or intermediate goods for the production of capital goods and imports of consumer good' (Dunford, 1990, p. 311).

[ii] Data about the area in China significantly vary. Further in the text we apply USDA estimations.

References

Abreu, M. (1988) Brazil, the GATT, and the WTO: history and prospects. Available at: http://www.econ.puc-rio.br/uploads/adm/trabalhos/files/td392.pdf (accessed 30 May 2018).

Antonow, K. (2010) Patentability of life forms in Europe. Available at: http://www.ictsd.org/downloads/2010/03/antonow-patentability-of-life-forms-in-europe.pdf (accessed 30 May 2018).

Antons, C. (2010) *Sui generis* protection for plant varieties and traditional knowledge in biodiversity and agriculture: The international framework and national approaches in the Philippines and India. *Indian Journal of Law and Technology* 6, 89–139.

Apeda (2017) Animal product. Available at: http://apeda.gov.in/apedawebsite/six_head_product/animal.htm (accessed 30 May 2018).

APK Inform (2016) In November, Russia export record volumes of corn – SovEcon. Available at: http://www.apk-inform.com/en/news/1076266#.WKYeeDiZl_k (accessed 30 May 2018).

Bauer, M.W. (2006) Paradoxes of resistance in Brazil. In: Gaskell, G. and Bauer M.W. (eds) *Genomics and Society: Legal, Ethical and Social Dimensions*, Earthscan, London, United Kingdom.

Bedasie, S. (2012) The possible overlap between plant variety protection and patent: Approaches in Africa with particular reference to South Africa and Ethiopia. *Haramaya Law Review* 1, 125–136.

Beier, E. and Fiero, A.W. (2015) The US and Russian patent systems: Recent amendments and global implications for the protection of intellectual property rights. *The John Marshall Review of Intellectual Property Law* 14, 504–519.

Benjamin, R.M. (2011) The national prevention strategy: Shifting the nation's health-care system. *Public Health Reports* 126, 774–776.

Bernauer, T. and Aerni, P. (2008) Trade conflict over genetically modified organisms. Available at: http://citeseerx.ist.psu.edu/viewdoc/download?doi=10.1.1.524.3725&rep=rep1&type=pdf (accessed 30 May 2018).

Bernauer, T. and Meins, E. (2003) Technological revolution meets policy and the market: Explaining cross-national differences in agricultural biotechnology regulation. *European Journal of Political Research* 42, 643–683.

Bernstein, H. (2001) The peasantry in global capitalism: Who, where and why? *Socialist Register* 37, 25–51.

Bernstein, H. (2015) Food regimes and food regime analysis: A selective survey. Available at: http://www.plaas.org.za/sites/default/files/publications-landpdf/BICAS_WP_2-Bernstein.pdf (accessed 30 May 2018).

Bogdanov, N. and Rodic, V. (2014) Agriculture and agricultural policy in Serbia. In: Volk, T., Erjavec, E. and Mortensen, K. (eds) *Agricultural Policy and European Integration in Southeastern Europe.* FAO, Budapest, Hungary, pp. 153–171.

Borras, J.R., Saturnino M., Edelman, M. and Kay, C. (2008) Transnational agrarian movements: Origins and politics, campaigns and impact. In: Saturnino, M., Borras, J.R., Edelman, M. and Kay, C. (eds) *Transnational Agrarian Movements Confronting Globalization.* Wiley-Blackwell, Chichester, United Kingdom, pp. 1–37.

Bosnjak, D., Rodic, V. and Muncan, P. (2012) Soybean acreages needed to satisfy consumption of basic livestock products in Serbia. *Bulgarian Journal of Agricultural Science* 18, 539–544.

Brankov, T. (2018) Food policy in the republic of Serbia. In: Smithers, G. (ed.) *Reference Module in Food Science*. Elsevier, Amsterdam, the Netherlands, pp. 1–4.

Brankov, T. and Lovre, K. (2012) The role of international organizations in the spread of genetically modified food. *Zbornik Matice srpske za drustvene nauke/Social Science Collection* 138, 29–38.

Brankov, T. and Milovanovic, M. (2015) Measuring food security in the Republic of Serbia. *Ekonomika Poljoprivrede/Economics of Agriculture* 62, 801–812.

Brankov, T., Sibalija, T., Lovre, K., Cvijanovic, D. and Subic, J. (2013a) The impact of biotechnology knowledge on the acceptance of genetically modified food in Serbia. *Romanian Biotechnological Letters* 18, 8295–8306.

Brankov, T., Sibalija, T., Lovre, K., Cvijanovic, D. and Subic, J. (2013b) Structural equation modeling of consumer attitudes toward genetically modified food in Serbia. *Metalurgia International* 18, 114–119.

Brankov, T., Sibalija, T. and Cvijanovic, D. (2013c) Serbian public attitude towards 'green'. In: Radovic-Markovic, M., Vojteski-Kljenak, D. and Jovancevic, D. (eds) *Rural Entrepreneurship: Opportunities and Challenges*. Faculty of Business Economics and Entrepreneurship, Belgrade, Serbia, pp. 159–182.

Brankov, T., Sibalija, T. and Subic, J. (2014) Serbian consumers' willingness to buy food products produced without the use of pesticides. *Romanian Biotechnological Letters* 19, 9605–9614.

Brankov, T., Lovre, K., Popovic, B. and Bozovic, V. (2016) Gene revolution in agriculture: 20 years of controversy. In: Jamal, F. (ed.) *Genetic Engineering: An Insight into the Strategies and Applications*. InTech, Rijeka, Croatia, pp.1–22.

Buttel, F.H. (2001) Some reflections on late twentieth century agrarian political economy. *Cadernos de Ciência & Tecnologia/Notebooks of Science & Technology* 18, 11–36.

CBD (2017) Biodiversity-related Conventions. Available at: https://www.cbd.int/brc/ (accessed 30 May 2018).

Chand, R. (2007) Agro-industries Characterization and Appraisal: Soybeans in India. FAO, Rome, Italy.

Chand, R.P.A., Prasanna, L. and Singh, A. (2011) Farm size and productivity: Understanding the strengths of smallholders and improving their livelihoods. *Economic and Political Weekly* 46 (26–27), 5–11.

Chorev, N. and Babb, S. (2009) The crisis of neoliberalism and the future of international institutions: A comparison of the IMF and the WTO. *Theory and Society* 38, 459–484.

CIA (2007) The 2007 World Fact Book, by United States. Available at: http://www.freeinfosociety.com/media/pdf/4767.pdf (accessed 1 June 2018).

CIA (2015) The World Fact Book. Available at: http://statisticstimes.com/economy/countries-by-gdp-sector-composition.php (accessed 1 June 2018).

Clar, E., Martín-Retortillo, M. and Pinilla, V. (2016) The Spanish path of agrarian change, 1950–2005: From authoritarian to export-oriented productivism. Available at: http://repositori.uji.es/xmlui/bitstream/handle/10234/158884/DT%20SEHA%201602.pdf?sequence=1 (accessed 1 June 2018).

Connor, J.M. (1994) North America as a precursor of changes in Western European food-purchasing patterns. *European Review of Agricultural Economics* 21, 155–173.

Constance, D.H., Choi, J.Y. and Lara, D. (2015) Engaging the organic conventionalization debate. In: Freyer, B. and Bingen, J. (eds) *Re-Thinking Organic Food and Farming in a Changing World*, Springer, Dordrecht, Netherlands, pp.161–185.

Davison, J. (2010) GM plants: Science, politics and EC regulations. *Plant Science* 178, 94–98.

De Souza, E.P. (2011) Biopatents in Brazil. *ELNI Review* 2, 58–65.

De Visser, C.L.M., Schreuder, R. and Stoddard, F. (2014) The EU's dependency on soya bean import for the animal feed industry and potential for EU produced alternatives. *OCL – Oilseeds and Fats, Crops and Lipids* 21, D407.

Doctrine (2010) Doctrine on food security of the Russian Federation, Presidential Decree no. 120. Available at: http://kremlin.ru/acts/6752 (accessed 1 June 2018).

Doh, J.P. and Guay, T.R. (2006) Corporate social responsibility, public policy, and NGO activism in Europe and the United States: an institutional-stakeholder perspective. *Journal of Management Studies* 43, 47–73.

Dunford, M. (1990) Theories of regulation. *Environment and Planning D: Society and Space* 8, 297–321.

Dzonzi, T. and Crowlay, K. (2017) South Africa approves 1.3 million tonnes of US GMO Corn Import. Available at: https://www.bloomberg.com/news/articles/2017-01-25/south-africa-approves-1-3-million-tonnes-of-u-s-gmo-corn-imports (accessed 1 June 2018).

EC (2006) Serbia Country Report. Available at: http://ec.europa.eu/agriculture/sites/agriculture/files/external-studies/2006/applicant/serbia_en.pdf (accessed 1 June 2018).

EC (2015) Fact Sheet: Questions and Answers on EU's Policies on GMOs. Available at: http://europa.eu/rapid/press-release_MEMO-15-4778_en.htm (accessed 1 June 2018).

EC (2016) Genetically Modified Organisms – Traceability and Labelling. Available at: http://eur-lex.europa.eu/legal-content/EN/TXT/?uri=LEGISSUM:l21170 (accessed 1 June 2018).

EC (2018) GMO legislation. Available at: https://ec.europa.eu/food/plant/gmo/legislation_en (accessed 1 June 2018).

Eco Watch (2015) It's official: 19 European countries say 'No' to GMOs. Available at: https://www.ecowatch.com/its-official-19-european-countries-say-no-to-gmos-1882106434.html (accessed 1 June 2018).

Ecomercioagrario (2017) Spain, the European leader in industries that export pork to China. Available at: http://ecomercioagrario.com/espana-pais-europeo-con-mayor-numero-de-industrias-autorizadas-para-exportar-carne-de-porcino-a-china/ (accessed 1 June 2018).

ERS USDA (2017) Trade. Available at: https://www.ers.usda.gov/topics/crops/corn/trade/ (accessed 1 June 2018).

EU (2006) The EU Approach to the Millennium Round. Available at: http://trade.ec.europa.eu/doclib/docs/2006/december/tradoc_111111.pdf (accessed 1 June 2018).

Falkner, R. (2007) The political economy of 'normative power' Europe: EU environmental leadership in international biotechnology regulation. Journal of European Public Policy 14, 507–526. Available at: https://static1.squarespace.com/static/538a0f32e4b0e9ab915750a1/t/538db328e4b0aff4bbdaffa1/1401795368887/Falkner_2007_EU_Normative_Power_final_ms.pdf (accessed 1 June 2018).

FAO (2005) Fertilizer Use by Crop in South Africa. FAO, Rome, Italy.

FAO (2012) The State of Food and Agriculture 2012 – Investing in Agriculture for a Better Future. FAO, Rome, Italy.

FAO (2014) The State of Food and Agriculture – Innovation in Family Farming. FAO, Rome, Italy.

FAO (2015) The State of Food and Agriculture – Social Protection and Agriculture: Breaking the Cycle of Rural Poverty. FAO, Rome, Italy.

FAO (2016) The State of World Fishery and Aquaculture. FAO, Rome, Italy.

FAS (2018) U.S. Regulation of Genetically Modified Crops. Available at: https://fas.org/biosecurity/education/dualuse-agriculture/2.-agricultural-biotechnology/us-regulation-of-genetically-engineered-crops.html (accessed 1 June 2018).

FDA (2018) Labelling of Food Derived from Genetically Engineered Plants. Available at: https://www.fda.gov/food/ingredientspackaginglabeling/geplants/ucm346858.htm (accessed 1 June 2018).

Fernandez-Cornejo, J., Wechsler S.J. and Milkove, D. (2016) The Adoption of Genetically Engineered Alfalfa, Canola and Sugarbeets in the United States. Economic Information Bulletin 163, November 2016. Available at: https://www.ers.usda.gov/webdocs/publications/81176/eib-163.pdf?v=42697 (accessed 1 June 2018).

Fernandez-Wulff, P. (2013) Why and How Spain Became the EU's Top Grower of GMOs. Available at: https://ourworld.unu.edu/en/why-and-how-spain-became-the-eus-top-grower-of-gmos (accessed 1 June 2018).

Financial Times (2016) China province bans GMO crops for five years. Available at: https://www.ft.com/content/a221fb5e-c750-11e6-8f29-9445cac8966f (accessed 1 June 2018).

Fowler, C. and Mooney, P. (1990) The Threatened Gene: Food, Politics and the Loss of Genetic Diversity. The Lutterworth Press, Cambridge, UK.

Friedmann, H. (1993) The political economy of food: A global crisis. New Left Review 197, 29–57.

Friedmann, H. (2005) From colonialism to green capitalism: Social movements and emergence of food regimes. Research in Rural Sociology and Development 11, 227–264.

Friedmann, H. (2009) Feeding the empire: The pathologies of globalized agriculture. Socialist Register 41, 124–142.

Friedmann, H. and McMichael, P. (1989) Agriculture and the state system: The rise and fall of national agricultures, 1870 to present. Sociologia Ruralis 29, 93–117.

Friedmann, H. and McNair, A. (2008) Whose rules rule? Contested projects to certify Local production for distant consumers. Journal of Agrarian Change 8(2–3), 408–434.

GAIN (2014) Planting Seeds Annual 2014. Available at: https://gain.fas.usda.gov/Recent%20GAIN%20Publications/Planting%20Seeds%20Annual%202014_Beijing_China%20-%20Peoples%20Republic%20of_7-18-2014.pdf (accessed 2 June 2018).

GAIN (2015) China Grain and Feed Annual – 2015. Available at: https://gain.fas.usda.gov/Recent%20GAIN%20Publications/Grain%20and%20Feed%20Annual_Beijing_China%20-%20Peoples%20Republic%20of_5-11-2015.pdf (accessed 2 June 2018).

GAIN (2016a) Spain Agricultural Biotechnology Annual. Available at: https://gain.fas.usda.gov/Recent%20GAIN%20Publications/Agricultural%20Biotechnology%20Annual_Madrid_Spain_11-4-2016.pdf (accessed 2 June 2018).

GAIN (2016b) EU 28 Agricultural Biotechnology Annual 2016. Available at: https://gain.fas.usda.gov/Recent%20GAIN%20Publications/Agricultural%20Biotechnology%20Annual_Paris_EU-28_12-6-2016.pdf (accessed 2 June 2018).

GAIN (2016c) Russia Federation Biotechnology Annual Agricultural Biotechnology Annual 2016. Available at: https://gain.fas.usda.gov/Recent% 20GAIN%20Publications/Agricultural%20 Biotechnology%20Annual_Moscow_Russian%20 Federation_4-3-2017.pdf (accessed 9 June 2018).

GAIN (2016d) Brazil Agricultural Biotechnology Annual. Available at: https://gain.fas.usda.gov/Recent% 20GAIN%20Publications/Agricultural%20Biotech nology%20Annual_Brasilia_Brazil_11-22-2016.pdf (accessed 2 June 2018).

GAIN (2016e) *Agriculture Biotechnology Annual: China Moving Towards Commercialization of Its Own Biotechnology Crops*. Available at: https://gain.fas.usda.gov/Recent%20GAIN%20 Publications/Agricultural%20Biotechnology%20 Annual_Beijing_China%20-%20Peoples%20 Republic%20of_12-16-2016.pdf (accessed 2 June 2018).

GAIN (2016f) *India Agriculture Biotechnology Annual*. Available at:: https://gain.fas.usda.gov/ Recent%20GAIN%20Publications/Agricultural%20 Biotechnology%20Annual_New%20Delhi_ India_12-12-2016.pdf (accessed 9 June 2018).

GAIN (2016g) Agricultural Biotechnology Annual: Biotechnology in South Africa. Available at: https://gain.fas.usda.gov/Recent%20GAIN%20 Publications/Agricultural%20Biotechnology%20 Annual_Pretoria_South%20Africa%20-%20 Republic%20of_11-21-2016.pdf (accessed 2 June 2018).

GAIN (2016h) Serbia Agricultural Biotechnology Annual. Available at: https://gain.fas.usda.gov/ Recent%20GAIN%20Publications/Agricultural %20Biotechnology%20Annual_Belgrade_Serbia_ 12-14-2016.pdf (accessed 2 June 2018).

GAIN (2017) China's Planting Seeds Market Continues to Grow. Available at: https://gain.fas. usda.gov/Recent%20GAIN%20Publications/ China%27s%20Planting%20Seeds%20 Market%20Continues%20to%20Grow_Beijing_ China%20-%20Peoples%20Republic%20of_ 1-17-2017.pdf (accessed 2 June 2018).

Gale, F. (2013) *Growth and Evolution in China's Agricultural Support Policies*. U.S. Department of Agriculture, Economic Research Service, Washington, DC, United States.

Garzon, I. (2005) Multifunctionality of Agriculture in the European Union: Is There Substance Behind the Discourse's Smoke? Available at: https://arefiles. ucdavis.edu/uploads/filer_public/2014/03/27/ multifunctionality.pdf (accessed 2 June 2018).

Gastrow, M., Roberts, B., Reddy, V. and Ismail, S. (2016) Public Perception of Biotechnology in South Africa. Available at: http://www.pub.ac.za/ wp-content/uploads/2016/10/Public-Perceptionsto-Biotechnology.pdf (accessed 2 June 2018).

Gereffi, G. (1994) The organization of buyer-driven global commodity chains: How US retailers shape overseas production networks. In: Gereffi, G. and Korzeniewicz, M. (eds) *Commodity Chains and Global Capitalism*. Praeger, London, UK, pp. 95–122.

Goodwin, B.K., Marra, M. and Piggott, N. (2016) The cost of a GMO-free market basket of food in the US. In: Kalaitzandonakes, N., Phillips, P.W.B., Wesseler, J. and Smyth, S.J. (eds) *The Coexistence of Genetically Modified, Organic and Conventional Foods*. Springer, New York, pp. 363–378.

Gopinath, M. and Laborde, D. (2008) *Implications for India of the May 2008 Draft Agricultural Modalities*. International Centre for Trade and Sustainable Development, Geneva, Switzerland.

Greenpeace (2004) Activists Painting Transgenico and GMO on Cargo Ship in Brazil. Available at: http:// media.greenpeace.org/archive/Activists-Painting-Transgenico--and--GMO--on-Cargo-Ship-in-Brazil-27MZIF8J5LQ.html/ (accessed 2 June 2018).

Greenpeace (2017) The History of Greenpeace. Available at: http://www.greenpeace.org/eastasia/ about/history/ (accessed 2 June 2018).

Grzybowski, K. (1972) United States–Soviet Union Trade Agreement of 1972. *Law and Contemporary Problems* 37, 395–428.

Hartwick, E. and Peet, R. (2003) Neoliberalism and nature: The case of the WTO. *The Annals of the American Academy of Political and Social Science* 590, 188–211.

Hejazi, M. and Marchant, M. (2017) China's evolving agricultural support policies. *Choices* 32, 1–7.

Henneberry R.S. (2013) US export market development programs. In: Armbruster, W.J. and Knutson, R.D. (eds) *US Programs Affecting Food and Agricultural Marketing*. Springer, New York, pp. 195–227.

Huang, K.S. (1985) *US Demand for Food: A Complete System of Price and Income Effects*. USDA Economic Research Service, Washington, DC.

Huang, J., Wang, X. and Qui, H. (2012) *Small-scale Farmers in China in the Face of Modernisation and Globalization*. International Institute for Environment and Development and HIVOS, London and Hague, UK and Netherlands.

Huffman, W. (2017) The economics of organic and GMO farming systems (in the US): Interactions and how they might co-exist. *Economics Working Papers* 17035. Available at: https://lib.dr.iastate. edu/cgi/viewcontent.cgi?referer=https://scholar. google.com/scholar?hl=en&as_sdt=0%2C5&as_ ylo=2017&q=the+future+of+organic+and+ GMO+in+USA&btnG=&httpsredir=1&article =1035&context=econ_workingpapers (accessed 2 June 2018).

Hyatt, J.T. (1997) An overview of north/south trade in food products. *Journal of Food Distribution Research* 28, 1–6.

Jaffee, S. and Gordon, P. (1992) *Exporting High-Value Food Commodities: Success Stories from Developing Countries*. World Bank, Washington, DC.

Jaffer, Z. (2017) SA only country allowing GM staples: We have no choice. Available at: http://www.thejournalist.org.za/kau-kauru/gm-staples (accessed 2 June 2018).

James, C. (2015) ISAAA Brief 51-2015: Executive Summary. Available at: http://www.isaaa.org/resources/publications/briefs/51/executivesummary/default.asp (accessed 3 February 2017).

James, C. (2016) ISAAA Brief 52 – 2016: Executive Summary. Available at: http://www.isaaa.org/resources/publications/briefs/52/executive summary (accessed 16 December 2017).

Jovanovic, V., Petkovic, S. and Cikaric, S. (2012) Zlocin u ratu-Genocid u miru: Posledice NATO bombardovanja Srbije 1999. godine [The crime in war-Genocide in Peace: Consequences of NATO bombing of Serbia in 1999]. *Official Gazette*. Belgrade, Serbia.

Kiawu, J., Valdes, C. and MacDonald, S. (2011) Brazil's Cotton Industry: Economic Reform and Development. Available at: https://www.ers.usda.gov/webdocs/publications/35849/6125_cws11d01_1_.pdf?v=41055 (accessed 2 June 2018).

Kodras, J. E. (1993) Shifting global strategies of US foreign food aid, 1955–90. *Political Geography* 12, 232–246.

Korobko, I.V., Georgiev, P.V., Skryabin, K.G. and Kirpichnikov, M.P. (2016) GMOs in Russia: Research, society and legislation. *Acta Naturae* 8, 6–13.

Kumar, S. (2015) India eases stance on GM crop trials. *Nature* 521, 138–139.

La Via Campesina, (2017) Who Are We? Available at: https://viacampesina.org/en/who-are-we/ (accessed 8 June 2018).

Lerman, Z. and Shagaida, N. (2007) Land policies and agricultural land markets in Russia. *Land Use Policy* 24, 14–23.

Liefert, W.M. and Liefert, O. (2012) Russian agriculture during transition: performance, global impact, and outlook. *Applied Economic Perspectives and Policy* 34, 37–75.

Lovre, K. (2013) Policy of support to agriculture and rural development. In: Skoric, D., Tomic, D. and Popovic, V. (eds) *Agri-food Sector in Serbia*. Serbian Association of Agricultural Economists and Serbian Academy of Science and Art, Belgrade, Serbia.

Lovre, K. and Brankov, T. (2016) Multiple sustainability dimensions of retail sector in Serbia until 2013. *Industrija* [Industry] 44, 133–149.

Lowder, S.K., Skoet, J. and Raney, T. (2016) The number, size, and distribution of farms, smallholder farms, and family farms worldwide. *World Development* 87, 16–29.

Lucht, J. (2015) Public acceptance of plant biotechnology and GM crops. *Viruses* 7, 4254–4281.

Ludlow, P. (2005) The making of the CAP: Towards a historical analysis of the EU's first major policy. *Contemporary European History* 14, 347–371.

Luttrell, C.B. (1973) The Russian Wheat Deal—Hindsight vs. Foresight. Available at: https://www.staff.ncl.ac.uk/david.harvey/MKT3008/RussianOct1973.pdf (accessed 2 June 2018).

MacDonald, S., Gale, F. and Hansen, J. (2015) Cotton policy in China. Available at: https://www.ers.usda.gov/publications/pub-details/?pubid=36245 (accessed 2 June 2018).

Magnan, A. (2012) Food regimes. In: Plicher, J.M. (ed.) *The Oxford Handbook of Food History*. Oxford University Press, New York.

Majkovic, D., Borec, A., Rozman, C., Turk, J, and Pazek, K. (2005) Multifunctional concept of agriculture: just an idea or the real case scenario? *Društvena istraživanja: časopis za opća društvena pitanja* 14, 579–596.

Malko, A.M. (2006) Reproduction right of seed potatoes in Russia. In: Haase, N.U. and Haverkort, A.J. (eds) *Potato Developments in a Changing Europe*. Wageningen Academic Publishers, Wageningen, Netherlands, pp. 175–179.

Mascarenhas, M. and Busch, L. (2006) Seeds of change: intellectual property rights, genetically modified soybeans and seed saving in the United States. *Sociologia Ruralis* 46, 122–138.

McMichael, P. (2009) A food regime genealogy. *The Journal of Peasant Studies* 36, 139–169.

McMichael, P. (2011) The food regime in the land grab: Articulating 'global ecology' and political economy. *International Conference on Global Land Grabbing, 6–8 April 2011*, Institute of Development Studies, University of Sussex, Falmer, England, pp. 1–26.

Meyer, N. (2018) Donald Trump officially comes out in favor of GMOs, confirms organic movement's worst fears. *AltHealthWorks* 9 January 2018. Available at: https://althealthworks.com/14853/donald-trump-officially-comes-out-in-favor-of-gmos-confirms-organic-movements-worst-fears/ (accessed 2 June 2018).

Ministry for Agriculture and Land Affairs (1988) Agricultural Policy in South Africa. Available at: http://www.nda.agric.za/docs/Policy/policy98.htm (accessed 2 June 2018).

Ministry of Textiles (2015) From Farm to Fabric: The Many Faces of Cotton. Available at: https://www.icac.org/getattachment/mtgs/Plenary/74th-Meeting/Details/Documents/Country-Statements/India.pdf (accessed 2 June 2018).

Mintz, K. (2017) Arguments and actors in recent debates over US genetically modified organisms (GMOs). *Journal of Environmental Studies and Sciences* 7, 1–9.

Mitlin, D. and Mogaladi, J. (2009) Social Movements and Poverty Reduction in South Africa. Available at: http://hummedia.manchester.ac.uk/schools/seed/socialmovements/es/publications/reports/Mitlin_Mogaladi_SouthAfricamappinganalysis.pdf (accessed 2 June 2018).

MOA (2016) MOA holds press conference on GMO. Available at: http://english.agri.gov.cn/news/dqnf/201604/t20160414_168130.htm (accessed 2 June 2018).

Monostyrsky, O.A. (2004) Food safety of Russia: yesterday, today and tomorrow. In: Сороколетова, Н.Е., Аркадьевна, Л.Н. Кондратенко Е, Нетипанова Н.В. (2014) Современные аспекты использования генномодифицированных компонентов в продуктах питания и методы их обнаружения. Технологии пищевой и перерабатывающей промышленности АПК–продукты здорового питания [Modern aspects of genetically modified components in food and their methods of detection products]. *Technology and Food Processing Industry, Agribusiness Healthy Food* 4, 75–81.

Motta, R. (2016) Social Mobilization, Global Capitalism and Struggles over Food: A Comparative Study of Social Movements. Routledge, Abington, Pennsylvania.

Myers, J.S. and Sbicca, J. (2015) Bridging good food and good jobs: From secession to confrontation within alternative food movement politics. *Geoforum* 61, 17–26.

Nestle, M. (2016) Corporate funding of food and nutrition research: science or marketing? *JAMA Internal Medicine* 176, 13–14.

NSA (2015) One-and-a-half million tonnes of soybean oil imported by India yearly. Available at: http://nsai.co.in/news/92-oneandahalf-million-tonnes-soyabean-oil-imported-by-india-yearly (accessed 2 June 2018).

OECD (2011) Evaluation of Agricultural Policy Reforms in the United States. OECD Publishing, Paris, France.

OECD (2016a) Agricultural Policy Monitoring and Evaluation 2016: Russian Federation. Available at: http://www.keepeek.com/Digital-Asset-Management/oecd/agriculture-and-food/agricultural-policy-monitoring-and-evaluation-2016/russian-federation_agr_pol-2016-22-en#.WIRI6rmZl_k#page1 (accessed 2 June 2018).

OECD (2016b) Agricultural Policy Monitoring and Evaluation 2016: Brazil Available at: http://www.keepeek.com/Digital-Asset-Management/oecd/agriculture-and-food/agricultural-policy-monitoring-and-evaluation-2016/brazil_agr_pol-2016-7-en#.WIWtX7mZl_k (accessed 2 June 2018).

OECD (2016c) Agricultural Policy Monitoring and Evaluation 2016: China. Available at: http://www.keepeek.com/Digital-Asset-Management/oecd/agriculture-and-food/agricultural-policy-monitoring-and-evaluation-2016/china_agr_pol-2016-10-en#.WIWScbmZl_k (accessed 2 June 2018).

OECD (2016d) Agricultural Policy Monitoring and Evaluation 2016: South Africa. Available at: http://www.keepeek.com/Digital-Asset-Management/oecd/agriculture-and-food/agricultural-policy-monitoring-and-evaluation-2016/south-africa_agr_pol-2016-23-en#.WIWkPrmZl_k#page7 (accessed 2 June 2018).

OECD/FAO (2015) *OECD-FAO Agricultural Outlook 2015–2024.* OECD Publishing, Paris, France.

Palley, T. (2010) America's exhausted paradigm: macroeconomic causes of the financial crisis and great recession. *New School Economic Review* 4, 15–43.

Paul, H. and Steinbrecher, R. (2003) *Hungry corporations: Transnational Biotech Companies Colonise the Food Chain.* Zed Books, London, UK and New York

Pechlaner, G. and Otero, G. (2008) The Third Food Regime: Neoliberal Globalism and Agricultural Biotechnology in North America. *Sociologia Ruralis* 48, 351–371.

Pig Progress (2016) Spanish pig industry reaching record sizes. Available at: http://www.pigprogress.net/Finishers/Articles/2016/4/Spanish-pig-industry-reaching-record-sizes-2784529W/ (accessed 2 June 2018).

Planning Commission (Government of India) (2013) *Twelfth Five Year Plan 2012–2017: Volume II.* Sage Publications, New Delhi, India.

Plant Authority (2017) List of 147 crop species open for registration under new/extant/farmers variety. Available at: http://www.plantauthority.gov.in/List%20of%20147%20Crop%20species%20registration%20under%20Extantaandfarmers.htm (accessed 8 June 2018).

Pollack, M.A. and Shaffer, G.C. (2000) Biotechnology: The next transatlantic trade war? *Washington Quarterly* 23, 41–54.

Priyanka, R. (2005) UPOV and Rights of Farmers–An Indian Perspective. Available at: http://www.intelproplaw.com/Articles/cgi/download.cgi?v=1131337938 (accessed 2 June 2018).

Pritchard, B. (2009) The long hangover from the second food regime: A world-historical interpretation of the collapse of the WTO Doha Round. *Agriculture and Human Values* 26, 297–307.

Qin, L. and Hao, F. (2016) Illegal GM Crops found in China's breadbasket. Available at: https://www.chinadialogue.net/article/show/single/en/8535-Illegal-GM-crops-found-in-China-s-breadbasket (accessed 2 June 2018).

Raffaelli, M. (2009) *An Investigation Into the Breakdown of International Commodity Agreements*. Woodhead Publishing Limited, Cambridge, UK.

Ramaila, M., Mahlangu, S. and du Toit, D. (2011) *Agricultural Productivity in South Africa: Literature Review*. Directorate: Economic Services, Production Economics Unit, Department of Agriculture, Forestry and Fisheries, South Africa.

Rai, R. (2007) Introduction to Plant Biotechnology. Available at: http://www.daff.gov.za/docs/GenReports/AgricProductivity.pdf (accessed 16 February 2017).

Rausser, G. and Zilberman, D. (2014) Government agricultural policy, United States. In: Alfen, N.K. (ed.) *Encyclopedia of Agriculture and Food Systems, Volume 3*. Elsevier Academic Press, Waltham, Massachusetts, pp. 518–528.

Raynolds, L. and Wilkinson, J. (2007) Fair trade in the agrarian and food sector: Analytical dimensions. In: Raynolds, L.T., Murray, D.L. and Wilkinson, J. (eds) *Fair Trade: The Challenges of Transforming Globalization*. Routledge, Abington, Pennsylvania.

Reardon, T. and Berdegué, J.A. (2002) The rapid rise of supermarkets in Latin America: Challenges and opportunities for development. *Development Policy Review* 20, 371–388.

Robinson, D. (2007) Exploring Components and Elements of Sui Generis System for Plant Variety Protection and Traditional Knowledge in Asia. ICTSD, Geneva, Switzerland.

Rosset, P., Rice, R. and Watts, M. (1999) Thailand and the world tomato: globalization, new agricultural countries (NACs) and the agrarian question. *International Journal of Sociology of Agriculture and Food* 8, 71–94.

Rosstat (2015) *BRICS: Joint Statistical Publication 2015*. Available at: http://www.gks.ru/free_doc/doc_2015/BRICS_ENG.pdf (accessed 2 June 2018).

Rosstat (2016) Russia in Figures. 2016: Statistical Handbook. Rosstat, Moscow, Russia. Available at: http://www.gks.ru/free_doc/doc_2016/rusfig/rus16e.pdf (accessed 2 June 2018).

Russia Beyond (2014) Genetically modified crops enter Russia. Available at: https://in.rbth.com/economics/2014/01/16/genetically_modified_crops_enter_russia_32331 (accessed 2 June 2018).

RZS (2011) *2011 Census of Populations, Households and Dwellings in the Republic Serbia*.

Statistical Office of the Republic of Serbia, Belgrade, Serbia.

Sedik, D., Lerman, Z. and Uzun, V. (2013) Agricultural policy in Russia and WTO accession. *Post-Soviet Affairs* 29, 217–247.

Seed Freedom (2016) Unprecedented opposition to Monsanto's GMO trials in South Africa: Enough is enough. Available at: http://acbio.org.za/wp-content/uploads/2016/06/Joint-Press-Release_DTstack-.pdf (accessed 2 June 2018).

Sevarlic, M. (2015) *Poljoprivredno zemljiste* [Agricultural land]. Statistical Office of the Republic of Serbia, Belgrade, Serbia.

Shaw, J.D. (2007) *World Food Security: A History since 1945*. Palgrave Macmillan, Hampshire, England.

Shiva, V. (2017a) Bija Satyagriha. Available at: http://navdanya.org/campaigns/bija-satyagriha (accessed 2 June 2018).

Shiva, V. (2017b) Violation of India's IPR laws and Competition Act by MNCs. Available at: http://navdanya.org/news/604-violation-of-indias-ipr-laws-and-competition-act-by-mncs (accessed 2 June 2018).

Sihlobo, W. and Kapuya, T. (2016) South Africa's Soybean Industry: A Brief Overview. Available at: http://www.grainsa.co.za/south-africa-s-soybean-industry:-a-brief-overview (accessed 2 June 2018).

Singh, K.K. (2015) *Biotechnology and Intellectual Property Rights*. Springer, New Delhi, India.

Skocpol, T. and Finegold, K. (1982) State capacity and economic intervention in the early New Deal. *Political Science Quarterly* 97, 255–278.

State Council (2016) The 13th Five-year Plan for Economic and Social Development of the People's Republic of China. Available at: http://en.ndrc.gov.cn/newsrelease/201612/P020161207645765233498.pdf (accessed 2 June 2018).

Statista (2017) World Corn Production by Country. Available at: https://www.statista.com/statistics/254292/global-corn-production-by-country/ (accessed 2 June 2018).

The Wall Street Journal (2016) How India's taste for soy oil has fueled a surge in imports. Available at: https://blogs.wsj.com/indiarealtime/2016/10/07/how-indias-taste-for-soy-oil-has-fueled-a-surge-in-imports/ (accessed 2 June 2018).

TNS Opinion & Social (2010) Biotechnology, Eurobarometer 73.1. Available at: http://ec.europa.eu/commfrontoffice/publicopinion/archives/ebs/ebs_341_en.pdf (accessed 2 June, 2018).

UN (2015) World Urbanization Prospects. The 2014 Revision. Available at: https://esa.un.org/unpd/wup/publications/files/wup2014-report.pdf (accessed 2 June 2018).

USDA ERS (2013) *Farm Size and the Organization of US Crop Farming*. Economic Research Report, 152. US Department of Agriculture, Economic Research Service. Available at: https://www.ers.usda.gov/webdocs/publications/45108/39359_err152.pdf (accessed 2 June 2018).

USDA (2016) USDA Agricultural projections to 2025. Available at: https://www.usda.gov/oce/commodity/projections/USDA_Agricultural_Projections_to_2025.pdf (accessed 2 June, 2018).

USDA (2017) How the Federal Government Regulates Biotech Plants. Available at: https://www.aphis.usda.gov/aphis/ourfocus/biotechnology/SA_Regulations/CT_Agency_Framework_Roles (accessed 2 June 2018).

USDA (2017a) World Agricultural Production. Available at: https://apps.fas.usda.gov/psdonline/circulars/production.pdf accessed 2 June 2018).

USITC (2009) *India: Effects of Tariffs and Nontariff Measures on US Agricultural Exports*. United States International Trade Commission, Washington, DC.

US Spending (2018) US Health Care Spending History from 1900. Available at: https://www.usgovernmentspending.com/healthcare_spending (accessed 2 June 2018).

Van, Z.J., Binswanger, H.P. and Thirtle, C.G. (1995) The Relationship Between Farm Size and Efficiency in South African Agriculture. Available at: http://documents.worldbank.org/curated/en/127831468760155107/pdf/multi-page.pdf (accessed 2 June 2018).

Wehrheim, P., Serova, E.V., Frohberg, K. and von Braun, J. (2000) Introduction and overview. In: Wehrheim, P., Serova, E.V. Frohberg, K. and von Braun, J. (eds) *Russia's Agro-food Sector: Towards Truly Functioning Markets*. Kluwer Academic Publishers, Dordrecht, Netherlands, pp. 1–8.

Wilson, D. and Purushothaman, R. (2006) Dreaming with BRICs: The Path to 2050. Available at: http://media.library.ku.edu.tr/reserve/resspring10/intl532_mgec632_ZOnis/3_The_challenge_of_BRICs.pdf (accessed 5 April 2017).

WIPR (2016) Russia: still a good place for IP protection. Available at: http://www.worldipreview.com/contributed-article/russia-still-a-good-place-for-ip-protection (accessed 2 June 2018).

Wong, A.Y.T. and Chan, A.W.K. (2016) Genetically modified foods in China and the United States: A primer of regulation and intellectual property protection. *Food Science and Human Wellness* 5, 124–140.

Worldatlas (2017) World Leaders in Corn (Maize) Production, by Country. Available at: http://www.worldatlas.com/articles/world-leaders-in-corn-maize-production-by-country.html (accessed 2 June 2018).

World Bank (2013) Poverty Headcount Ratio at $1.90 a Day (2011 PPP) (% of Population). Available at: https://data.worldbank.org/indicator/SI.POV.DDAY (accessed 9 June 2018).

World Bank (2017) Poverty. Available at: http://www.worldbank.org/en/topic/poverty (accessed 2 June 2018).

Yadavand, C. and Kultshreshtha, G. (2017) Patenting in biotechnology – the Indian scenario. Available at: http://www.iam-media.com/intelligence/IAM-Life-Sciences/2016/Articles/Patenting-in-biotechnology-the-Indian-scenario (accessed 2 June 2018).

Yadav, S. (2016) Fly in the face of Bt cotton. Available at: http://www.thehindubusinessline.com/blink/know/fly-in-the-face-of-bt-cotton/article8561303.ece (accessed 2 June 2018).

Zerbe, N. (2004) Feeding the famine? American food aid and the GMO debate in Southern Africa. *Food Policy* 29, 593–608.

Zobbe, H. (2002) The Economic and Historical Foundation of the Common Agricultural Policy in Europe. *Xth EAAE Congress Exploring Diversity in the European Agri-Food System*, Zaragoza, Spain, pp. 1–17. Available at: http://ageconsearch.umn.edu/bitstream/24867/1/cp02mo01.pdf (accessed 2 June 2018).

3 Does Transgenic Food Production Affect World Food Prices?

Food prices are of special concern to poor countries and poor people. As previously stated, the second food regime mechanism failed to solve world hunger, leading the FAO to call a World Food Summit in 1974 to examine global food production and consumption. To be sure, this was not the first attempt to solve hunger, since the Great Depression in the 1930s with its disastrous effects on 'consumer purchasing power and on the incomes of primary producers, underlined the need for some form of intergovernmental arrangement for staple food-stuffs... In the early 1930s, Yugoslavia proposed that in view of the importance of food for health, the Health Division of the League of Nations should disseminate information about the food position in representative countries of the world. Its report was the first introduction of the world food problem into the international political arena' (Shaw, 2007, p. 6).

It is well known that agricultural commodity prices are determined by the interaction of supply and demand as market fundamentals and by exogenous shock such as adverse weather conditions. 'Long-run changes in food demand are primarily the result of population and income growth, but are also influenced by relative price changes and the evolution of dietary patterns. Demand for agricultural raw materials such as rubber is related to economic growth more generally. Long-run expansion in supply is primarily driven by technological progress, which reduces costs. In the past, technological progress reduced costs and induced supply expansion at a faster rate than population and income growth expanded demand leading to a long-run relative decline in agricultural commodity prices' (FAO, 2009, p. 11). 'The widely shared view was that low food prices were a curse to developing countries and the poor' (Swinnen, 2010, p. 2). But, after the 2007/08 price rise – caused mainly by 'the demand for grains and oilseeds as biofuel feedstocks' and the 'demand shocks by decreasing supply responsiveness' (Gilbert, 2010, p. 3) to nutritional transition in China and throughout Asia – the vast majority of reports state that 'high food prices have a devastating effect on developing countries and the world's poor' (Swinnen, 2010:3). However, historical lessons have shown that not all people are affected in the same way, 'the urban and rural poor and people in food import-dependent countries are most vulnerable to international commodity price increases, when these are transmitted to local markets, because they spend the largest proportions of their incomes on food... smallholder farmers, many of whom are also poor and food insecure, can be enabled to benefit from higher food prices and become part of the solution by reducing price spikes and improving overall food security' (Da Silva *et al.*, 2012). Thus, the problems of high food prices need to be doubly resolved: in the short run to help the world's poor and in the long run by proper direction of production, consumption and trade in an age of increasing population, demand and climate change. Excluding sub-Saharan Africa, the Green Revolution has caused manifest effects with more or less success in many parts of the world. For this reason we hereafter assess the manner in which the new Gene Revolution has affected agricultural commodity prices.

In our work we use the premise that international price increases are predominantly driven by price changes in the US market (Headey and Fan, 2010). The USA was for a long time the largest exporter of maize and wheat, and the second largest exporter of soybean oilseeds. In 2016, US exports accounted for 39% of the world maize trade, 15% of the wheat trade and 40% of the world soybean oilseeds trade. US wheat exports' dominance was overtaken by Russia and fell from 50% in the 1970s via 30–40% in the 1980s to 15% in 2016. Similarly, in the 1970s the USA dominated the world in unprocessed oilseeds, with a global market share of more than 70%, but in recent years with Brazil and Argentina's involvement, its share has fallen below 50%. However, there is no doubt that the USA plays a very important role in soybean, wheat and maize markets, and consequently it has an important role in their price formation. 'Hence US grain prices are typically quoted as international prices', the only exception is rice, 'where Thai prices are typically quoted' (Headey and Fan, 2010, p. 7). The reason for the exclusion of rice lies in the fact that the USA lags significantly behind the Asian countries India, Thailand, Vietnam and Pakistan, and accounted for just 9.2% of rice world exports in 2016/17. In addition, rice prices are more volatile than prices of other grains due to the specificity of the rice market, reflected in the small amount of export of total production (8.1% in 2016) and highly inelastic demand as it is the major staple for Asian people. Another specificity of rice is the dominance of small farmers and traders in production and the market. Since there is no commercially important production of transgenic rice and wheat, we exclude these crops from specific analysis further on in the text. We focus on the two most important transgenic edible crops, maize and soybean.

Analysis has convincingly demonstrated that the US biofuel policies represent a new episode of the distorting effects of agrarian policies, which not only cause the instability of global prices but also have a negative impact on the welfare of consumers in food-importing countries. It has also been mentioned that the agrarian lobby, which advocates various forms of support for the ethanol production, is very aggressive. The rise of maize prices and biofuel production, the decline in the stock/use ratio, and the increase of areas under HT and *Bt* maize in the USA, all changed at about the same time. Regardless of whether the cause is transgenic technology or not, there can be no doubt that the USA had significantly increased maize production in the previous period. But, unfortunately this escalation was invested in biofuels production, without any contribution to saving the world from hunger. In any case, transgenic technology, primarily through seed prices, has contributed to rising food prices. Seed costs have increased on a per kilogram produced basis, on a percentage of operating cost basis and on a percentage of revenue basis. Increases in seed costs per kilogram of soybeans and maize produced, indicate that yield increase did not keep pace with seed cost increase. Thus, it seems reasonable to conclude that the significant increases in maize and soybean seed prices mostly occurred owing to the introduction of transgenic varieties into production. The rise in food prices, being in close correlation with their instability, disproportionately affects underdeveloped countries and poorer population categories. There is no doubt that the most important effect is manifested by the reduction in real income and the growth of income volatility. However, the very unfavourable situations in least-developed countries (LDCs) cannot be attributed directly to transgenic technology. LDCs are neither significant producers nor significant importers of transgenic crops.

3.1 Long-run Trends in International Prices

In the last 45 years the world food system has collapsed twice, first in 1972/74 and then again in 2007/08. Both crises signalled the end of cheap food periods that lasted for almost 30 years each. Although the first crisis reached its peak in the 1970s with the US–Soviet Union wheat transfer, real food prices began to rise in the 1960s. The second food crisis occurred after a steady decline in food prices during the 1980s and 1990s, and after reaching the lowest point ever in the 2000s. The period after 2007/08 is characterized by price volatility; after a moderate decline in the 2010s prices of key staple foods have remained higher than before the crisis (Fig. 3.1).

The price level of rice, soybeans, wheat, and maize in the second food crisis were in real terms as high as in the late 1970s or early 1980s. In both crises the biggest jump was recorded in the price of rice. The rice price shot up by more than 200% in 1974 compared with 1971 and by 86.4% in 2009 compared with 2006. Wheat prices increased more in the first crisis than in the second

one, by 88.8% and 48.4%, respectively. Interestingly, the prices of soybeans and maize whose transgenic varieties have been grown in significant quantities since 1996 rose more in the second food crisis than in the first one. The price of soybeans increased by 43% in 1974 compared with 1971, while in 2009 the soybean price was higher by 51.6% than in 2006. The price of maize was 60% higher in 2009 than in 2006, and 45.7% higher in 1974 than in 1971. With the exception of wheat, whose price in both post-crisis 8-year periods dropped by about 45%, price recovery after the second crisis has been slower than after the first one. Eight years after the first crisis the prices of rice, soybeans, and maize decreased by 68.7%, 47.5%, and 50.5%, respectively; 8 years after the second crisis the real prices of rice, soybeans, and maize decreased by 33.3%, 15.1%, and 21.9%, respectively.

Compared with 1997, the real prices of soybean, maize, and rice in 2016 increased by 25.6%, 24.3%, and 19.3%, respectively while wheat prices decreased by 4.5% in the same period (Fig. 3.2). Considering that more than 80% of soybeans and about 30% of maize in the world is transgenic, the rice

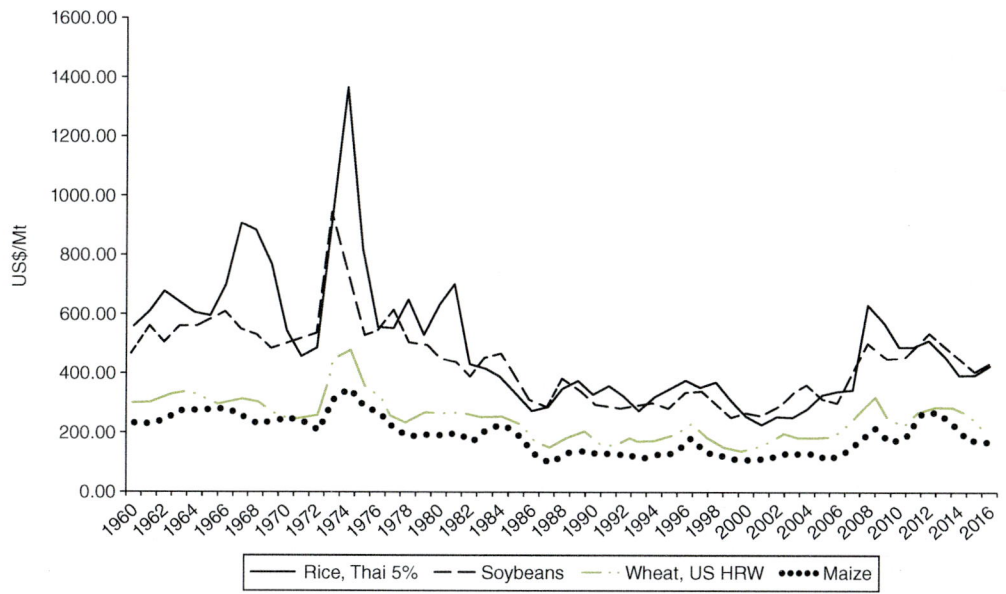

Fig. 3.1. Trends in real international prices of key staple foods, 1960–2016. (Adapted from World Bank, 2017.)

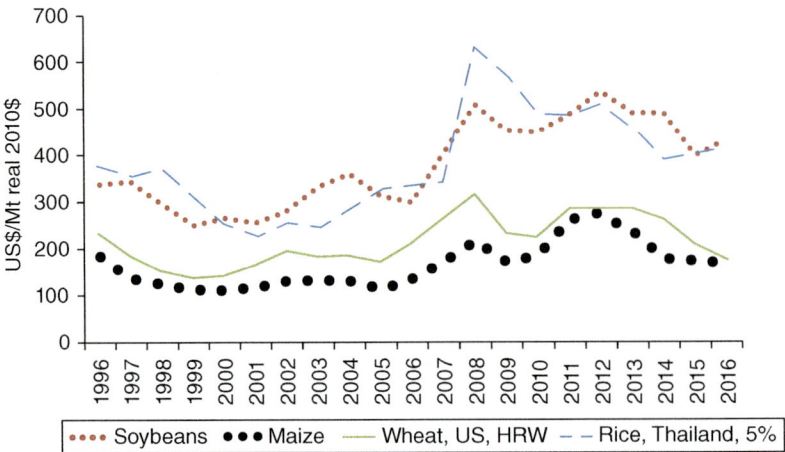

Fig. 3.2. Trends in real international prices of key staple foods since the commercialization of transgenic crops in 1996. (Adapted from World Bank, 2017.)

export market is small and segmented and 'preliminary estimates suggest that export restrictions, and the resulting panic buying, explain about 50 percent of the recent increase in prices' (Lipsky, 2008), and that there is no commercially grown transgenic wheat variety, we can roughly estimate that transgenic technology has contributed to the rise in food prices.

Regarding the timing of price changes, short-term price data presented in Fig. 3.3 show that in the period before the commercialization of transgenic crops, as well as in the period just before the second crisis, maize prices started to rise first. In the same periods nominal wheat prices remained the most stable, while the price of soybeans increased the most. It is well known that oil prices increased sharply before the 1972/74 crisis and the same is true for the second 2007/08 crisis (Fig. 3.3 and Table 3.1). All energy prices grew significantly, not just petroleum, but there is a difference in that 4 years after the second crisis petroleum prices rose, while natural gas and coal prices decreased. With the exception of urea, all other fertilizer prices, the price of diammonium phosphate (DAP), triple superphosphate (TSP) and potassium chloride were higher in real terms in the second crisis than in the first. Furthermore, chicken meat prices were higher during the second crisis. On the eve

of the creation of GMOs in 1995, prices of soybeans, maize, rice, and wheat were between 60% and 70% lower than in 1974. In addition, food non-staples and other agricultural commodities were cheaper: beef was 50% cheaper, chicken 13%, and cotton 41%. It was a period of inexpensive fertilizers; prices of phosphate and nitrogen fertilizers were about 75% lower and potassium fertilizers about 20% lower than in 1974. After the second food crisis and after the commercialization of transgenic crops, the commodity prices changed significantly. In the last observed year 2016 – 20 years after GMOs cultivation and 8 years after the second crisis – the prices of the most important transgenic crops, soybean and maize, were 53% and 26% higher, respectively. Contrary to this, the price of the third most important transgenic crop, cotton, was 26% lower than in 1995. However, since cotton is a non-edible crop, the decrease of cotton prices is of less importance than the increase of soybean and maize prices. As important feedstuffs they certainly contributed to the sharp rise in beef and chicken prices, which went up 100% between 1995 and 2016. This leads us to another less harsh estimation that transgenic technology has contributed to the rise in food prices.

Certainly, it should not be ignored that prices of fertilizers and energy have had an

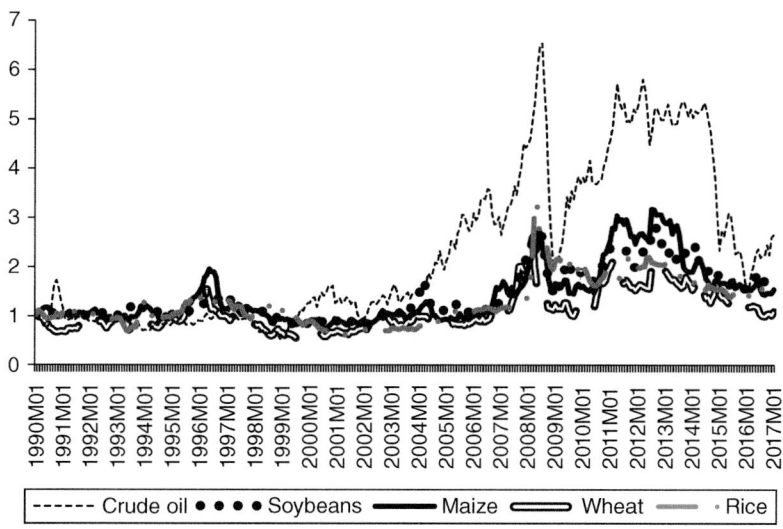

Fig. 3.3. Trends in nominal prices of staple crops and oil, January 1990–January 2017, price index January 1990. (Adapted from World Bank, 2017.)

Table 3.1. Changes in international prices across commodity groups, the 1972/1974, 1991/1995, 2007/2008 crises and today (percentage change of prices measured in real 2010 US$). (Adapted from World Bank, 2017.)

Commodity	1974	1978 (%)	1995	1999 (%)	2008	2012 (%)	2016
Food staples (US$/Mt)							
Soybeans	753.7	−33	282.0	−11.3	508.4	+5.6	431.8
Maize	348.6	−45.6	134.3	−16.7	217.0	+24.8	169.4
Rice	1374.0	−52.7	349.2	−11.7	632.3	−19.2	421.7
Wheat	477.4	−49.7	192.5	−27.8	317.0	−10.3	177.4
Food non-staples (US$/kg)							
Beef	4.2	−4.8	2.1	+9.5	3.1	+22.6	4.2
Chicken meat	1.5	−13.3	1.3	+23.1	1.8	+5.5	2.6
Other agricultural commodities (US$/kg)							
Cotton A index	3.9	−23.1	2.3	−34.8	1.5	+20.0	1.7
Fertilizers (US$/Mt)							
DAP	883.7	−70.4	235.6	−6.4	940.5	−47.9	367.6
TSP	806.5	−77.1	162.8	+17.7	855.1	−51	309.2
Urea	759.2	−71.5	203.7	−59.4	479.1	−23.2	212.1
Potassium chloride	160.7	−34	128.1	+17.8	554.4	−24.9	261.4
Energy							
Crude oil (US$/bbl)	29.2	−16.8	18.7	+19.8	94.3	+1.1	45.6
Coal Australia ($/Mt)	44.9	+21.8	42.8	−25	123.6	−29.2	70.1
Natural gas (2010=100)	31.2	+44.9	39.4	+25.4	174.8	−48.5	60.2

impact on the cost of agricultural production. In the observed periods (1995–2016) phosphate fertilizer prices grew between 50% and 90%, potassium fertilizer prices by 104%, urea by 4%, while crude oil, coal, and natural gas prices increased by 144%, 63.8%, and 52.8%, respectively. However, it is very difficult to estimate effects of rising energy prices on food prices, as Headey and Fan (2010, pp. 25–28) put it: 'cross-country evidence suggests that substantial variations exist across countries. US agricultural production in particular is almost solely dependent on oil for energy use. To rising fuel costs we also need to add the enormous surge in price of fertilizers, most of which are made from energy products... energy costs can constitute up to 90 percent of the cost of fertilizer production... the bulky nature of grains means that agricultural prices are strongly influenced by transport costs... Hence, at least for US production costs, we find that rising energy costs are a strong factor... which at least suggests the possibility that rising energy prices caused food prices to increase, rather than reverse. However, we also attribute a large role to demand-side factors that would have interacted with supply-side factors affecting production costs. If the supply curve alone had shifted upward because of rising fuel prices, the profits of farmers and food wholesalers would not normally be expected to rise much unless demand was very inelastic. Because US farmers (and major firms such as Cargill) experienced sharply rising profits in 2006, 2007, and 2008 it can safely be inferred that demand factors are also important contributors to rising food prices.' Summing up, 'a major effect of rising energy prices was a consequent surge in demand for biofuels' (Headey and Fan, 2010, p. xiii).

After the 2007–2008 food crisis, and especially after the escalation of food prices in 2010–2011, extensive economic studies took into consideration the causes and impacts of high and unstable food prices in international trade. Explaining the causes of the extreme increase in food prices, most analyses (e.g. Gilbert and Morgan, 2010; Abbott et al., 2011; Gilbert and Morgan, 2011; Tadesse et al., 2016) found the main causes in the following simultaneous factors: rapid economic growth in China, India, and other Asian countries; long-term disinvestment in agriculture; decline in inventories of most agricultural and food products; poor harvest in some regions of the world caused by climate disorders (drought in Australia, floods in Northern Europe, and a

heatwave in Southern Europe); depreciation of the US dollar; expansion of biofuel production; and financialization (speculative activities) of the food system. Von Braun and Tadesse (2012) categorized all these global factors into three groups: (i) root causes – extreme weather events, increasing biofuel feedstock demand, increasing volume of futures trading in commodity markets; (ii) intermediate causes – concentration of food production in few areas, lack of information on world food, economic growth in emerging markets; and (iii) immediate causes – export restrictions, aggressive food imports, decline in world food reserves.

The present study has no ambition to analyse the effects of each of the individual factors, but confirms the empirical research with a strong correlation between long-term (trend), medium-term and short-term impacts of food price changes (volatility and spikes). Last but not least, it is worth noting that the rise in food prices, being in close correlation with their instability, most strongly affects underdeveloped countries and poorer population categories. There is no doubt that the most important effect is manifested by the reduction in real income and the growth of income volatility. The extent of reduction and instability of income is conditioned by a number of factors, among which the most important are: (i) the share of the food sector in income generation; and (ii) the share of food in total budget expenditures. Based on the share of the income of the food sector and the share of food in total expenditure, it is possible to identify three groups of stakeholders according to the effect of growth and instability of food prices: (i) urban consumers – these consumers do not produce food, so the increased food prices do not affect their nominal income, but reduce their real income. The size of the decrease depends on the share of food costs in their total consumption and overall inflation; (ii) rural net buyers – these buyers buy more than they sell, so the net-income effect may be negative; and (iii) rural net sellers – this population category benefits from the rise in food prices, but the effect size depends on the price elasticity of supply.

3.2 Effects of Increased Asian Demand on Food Prices

Considering strong growth demand as a long-term factor which can influence the increase in food prices, we analysed the effect of increased Asian demand on cereal prices. As presented in Fig. 3.4, Asian maize imports are not translated into a larger maize bill. India's maize imports have no influence on maize prices, which is to be expected since India is ranked as the 60th biggest importer of maize on the global level. Indonesia, as the 17th biggest maize importer in the world (in 2016), had some influence on maize prices in certain years, 1998–2003, while China ranked the highest in Asia and the 13th biggest importer globally, also influenced prices in both real and nominal terms only in certain years, for example in 2011. Asian wheat imports did not have any effect on wheat prices (Fig. 3.5), except in sporadic years. Indonesia, ranked as the second biggest wheat importer in 2016, influenced

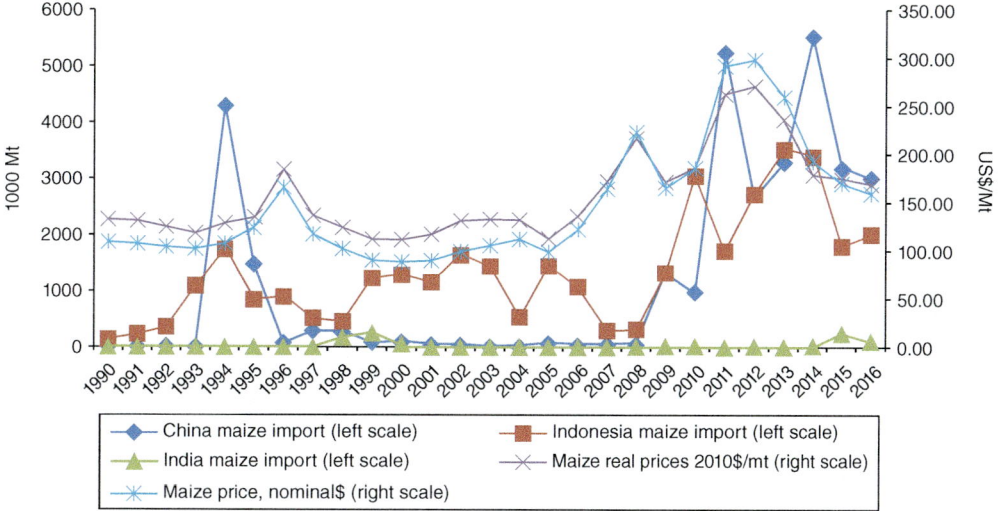

Fig. 3.4. Asian maize imports and international prices. (Adapted from World Bank, 2017.)

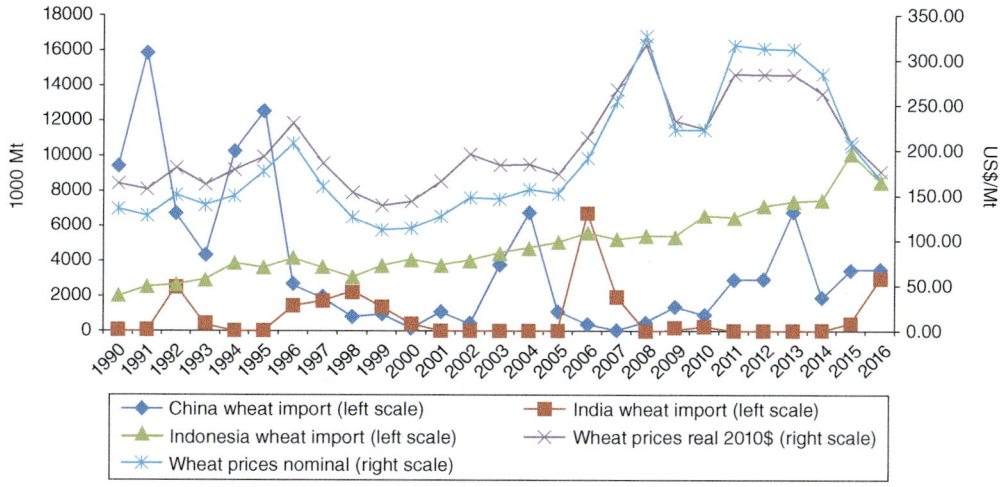

Fig. 3.5. Asian wheat imports and international prices. (Adapted from World Bank, 2017.)

prices in 1996 and 2006. However, this influence cannot be described as constant since, in 2007, when Indonesia significantly decreased its imports, wheat prices significantly increased, and in 2015 when Indonesia increased its imports, wheat prices decreased. No particular impact of China and India on wheat pricing is noticed because they are not top importing countries. China was ranked sixteenth and India twentieth in 2016.

Despite negligible impact on maize and wheat prices, China had a significant impact on crude oil and soybean prices as presented in Figs 3.6 and 3.7. China's decision to move away from soybean production and rely on imports has already been discussed in Chapter 2. From being self-sufficient in soybean production, China moved to a steady rise in soybean imports, accounting for 7.7% of global imports in 1990 to nearly 65% in

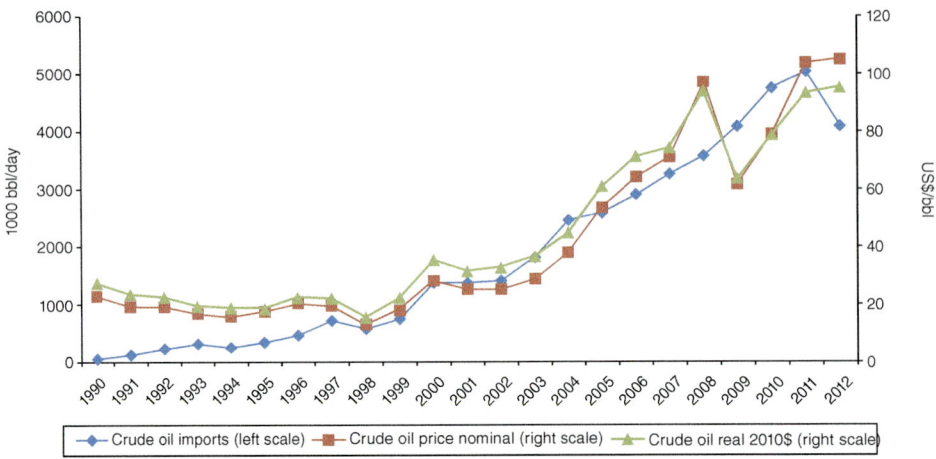

Fig. 3.6. China: Crude oil imports and international oil prices. (Adapted from IndexMundi (https://www. indexmundi.com/).)

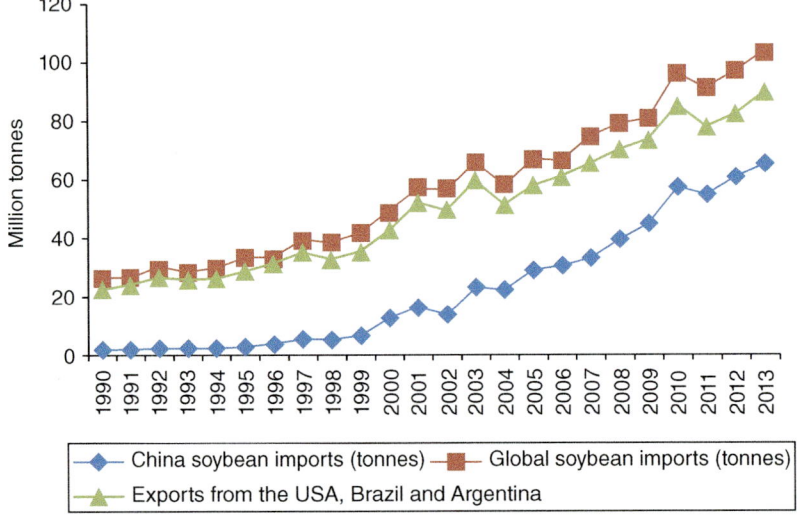

Fig. 3.7. Global imports, Chinese imports and exports from the USA, Brazil and Argentina: soybean, by years, tonnes. (Adapted from FAOSTAT (http://www.fao.org/faostat/en).)

recent years. In parallel with the increase in Chinese demand, the area under soybean cultivation in the USA, Brazil and Argentina also increased (Fig. 3.8). In other words, the biggest producers of transgenic soybeans, the USA, Brazil, and Argentina, increased their area under transgenic crops in accordance with Chinese demand. In this way, the Chinese demand for soy contributed to the increase of area under transgenic crops. In addition, Chinese demand indirectly led to a rise in food prices because the USA, Brazilian and Argentinian farmers shifted large amounts of land from grains to soybeans. All in all, 'although the direct impact of Chinese growth on food prices is small…the indirect impact is likely to have been large. This indirect impact comes largely through increasing the sensitivity of agricultural prices to demand shocks by decreasing supply responsiveness' (Gilbert, 2010, p. 3). For example, between the periods 1996 and 2016, consumption of meat in China increased significantly. Consumption of beef and veal increased from 2 to 4 kg/capita/year; sheep consumption increased from 1.3 to 3.1 kg; pork consumption from 19.8 to 30.8 kg; while poultry meat consumption increased

from 6.2 to 12.1 kg (OECD, 2017). This issue is certainly heavily related to GMOs.

The foregoing is evident from a comparison between harvested area under soybeans as the main important transgenic crop (and feeds) and wheat, a non-GMO crop but a staple food for millions (Fig. 3.8).

In all the selected countries there is a significant shift in the increase of soybean production after 1996, the year of the first introduction of transgenic soy. The area under soybeans in the USA increased by more than 10.5 million ha during 1990–2014. In the same period, areas under wheat decreased by 9.2 million ha. A similar process took place in Argentina – the area under soybeans increased by 6.8 million ha, while the area under wheat decreased by 1.4 million ha. Brazil is an exception since 'soybean exports were largely fueled by expansion of total agricultural area, such that the impact on production of other crops was not strong' (Headey and Fan, 2010, p. 17). The area under both soybeans and wheat increased in Brazil, the soybean area by almost 20 million ha and the wheat area only slightly. However, the foregoing is given just as a hypothetical example of comparison between areas under

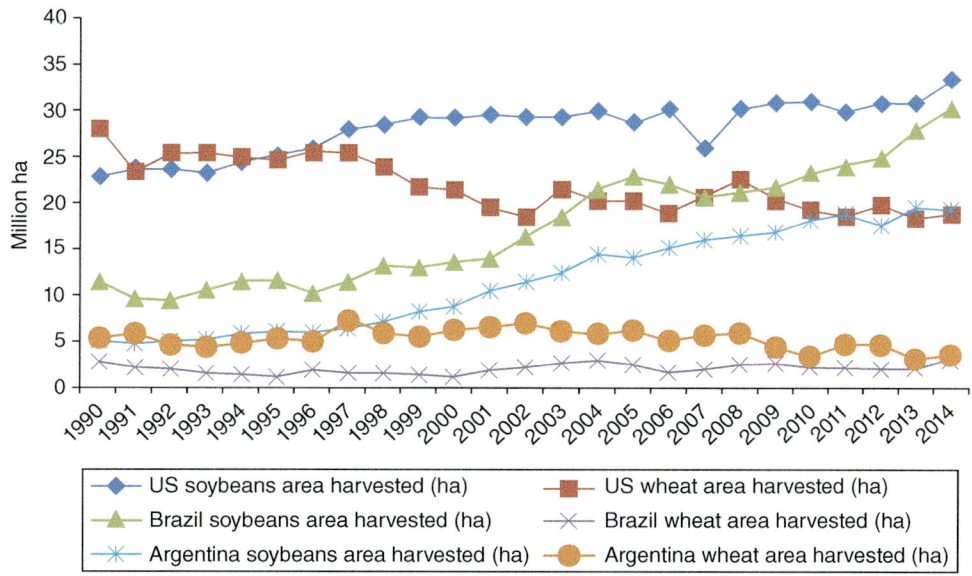

Fig. 3.8. Harvested area under soybeans and wheat in the USA, Brazil, and Argentina. (Adapted from FAOSTAT (http://www.fao.org/faostat/en).)

transgenic and non-transgenic crops, in most cases real competition for land exists between maize and soybean as well as between wheat and sunflower. Essentially, there are two issues here: competition between food and fuels, and whether we should be concerned about it.

3.3 US Ethanol and Maize Prices

For most agro-food products, price growth and volatility are the result of the composition of demand and/or supply shocks, reducing the product inventories. The interdependence between supply and demand changes, inventory status and the effect on price has been thoroughly analysed in numerous papers (e.g. Carter *et al.*, 2011; Hochman *et al.*, 2011; Wright, 2011). However, in addition to the above, many other factors can effect dramatic price changes. Economic analyses usually include factors such as: macroeconomic phenomena manifested by changes in nominal and real interest and exchange rates, speculations, cross-product linkages (through competition for land allocation), i.e. the effect of general equilibrium through substitution of inputs and costs, and state policies. Moreover, various export–import restrictions and barriers also affect price volatility, and have a backward effect on home markets. Public policy responses most often intensify the consequences of these internal and external market forces (Rausser and de Gorter, 2015). Therefore, the focus in this section is on causal mechanisms that emerged in the form of bioenergy policy of the USA, which induced the interdependence of prices of cereals and biofuels on the one hand, and the prices of cereals and fossil fuels on the other. The US bioenergy policy certainly has a decisive impact on the formation of grain prices, given that the USA participates with around 45% in the global biofuel production, while Brazil accounts with 25% and the EU with about 15%.

Given that the agribusiness sector has realized new sources of demand, which have higher price elasticity than agro-foods, biofuels have become increasingly attractive products. Therefore, it should not be surprising that US

agrarian legislation included bioenergy in 2008. However, although the Renewable Fuel Standard (RFS) was not implemented until 2005, it had been the subject of endless debates in the US Congress since 1978. The debates were strongly supported by the agrarian lobby, and resulted in the RFS for maize ethanol being doubled in 2007, together with the specified minimum production quantities for 2007–2022, with the RFS for 2022 set at four times that of 2007. This policy of the USA has led to a sevenfold increase in the production of ethanol in just 10 years (2003–2012). In the same period, the biodiesel production in the EU was also increased seven times, and the production of ethanol in Brazil was increased three times. Of course, the growth of biofuel production has increased the demand for maize, wheat and oilseeds, and therefore their prices as well.

Although there is disagreement among policymakers and analysts about the measurement of costs and benefits of biofuels, most of them agree that biofuels are drivers of food prices because biofuel production increases demand for maize, soybeans and edible oils. A number of studies by respected intergovernmental organizations have examined the link between the increase in food commodity prices and expanding biofuel production. FAO estimated that biofuels contributed 10–15% of the increase in food prices. The International Food Policy Research Institute (IFPRI) estimated the contribution was 25–30% (Henry, 2010). 'IMF estimates suggest that increased demand for biofuels accounts for 70 percent of the increase in corn [maize] prices and 40 percent of the increase in soybean prices' (Lipsky, 2008). A mathematical simulation shows 'that about 60 percent of the increase in maize prices from 2006 to 2008 may have been due to the increase in maize used in ethanol... a general equilibrium model calculated the long-term impact on weighted cereal prices of the acceleration in biofuel production from 2000 to 2007 to be 30 percent in real terms. Maize prices were estimated to have increased 39 percent in real terms, wheat prices increased 22 percent and rice prices increased 21 percent. During this period, the US CPI [consumer price index] increased by 20.4 percent, which would imply nominal price increases of 47, 26, and 25 [percent], respectively, for maize, wheat and rice prices. Differences in the estimates of the impact of biofuels on the price index of all food depend largely on how broadly the food basket is defined and what is assumed about the interaction between prices of maize and vegetable oils (directly influenced by demand for biofuels) to prices of other crops such as rice through substitution on the supply or demand side'. For example one study 'estimated that retail food prices increased only about 3 percent over the past 12 months due to ethanol production, in part because they only considered the impact of maize prices, directly and indirectly, on retail prices... The large increases in biofuel production in the US and EU were supported by subsidies, mandates, and tariffs on imports. Without these policies, biofuel production would have been lower and food commodity price increases would have been smaller. Biofuel production from sugar cane in Brazil is lower-cost than biofuel production in the US or EU and has not raised sugar prices significantly because sugar cane production has grown fast enough to meet both the demand for sugar and ethanol. Removing tariffs on ethanol imports in the US and EU would allow more efficient producers such as Brazil and other developing countries,

including many African countries, to produce ethanol profitably for export to meet the mandates in the US and EU' (Mitchell, 2008, pp. 4–17).

From our point of view, US ethanol production caused the rise in food prices, which is indirectly connected with transgenic crop production. 'Ethanol, made mostly from corn [maize] starch from kernels, is by far the most significant biofuel in the US, accounting for 94 percent of all biofuel production in 2012. Most of the remainder is biodiesel, made from vegetable oils (chiefly soy oil) as well as animal fats, waste oils, and greases' (USDA ERS, 2017a). Over time, ethanol share in the total maize use in the USA increased from 4.9% to 38.1%, in some years reaching almost 42% (Fig. 3.9). This is associated with a decreased share in US maize exports in the total use by about 6%, which in part can be attributed to the problems of finding export markets for transgenic maize. In addition, increased soybean production in the USA has had a very important impact on maize production, since these two crops compete for land. Since 1996, the first year of commercialization of transgenic crops, areas under soybeans increased by 10.5 million ha while maize areas increased significantly less, by 4.2 million ha. How US soybeans and maize compete with each other for land and how such competition influences food prices

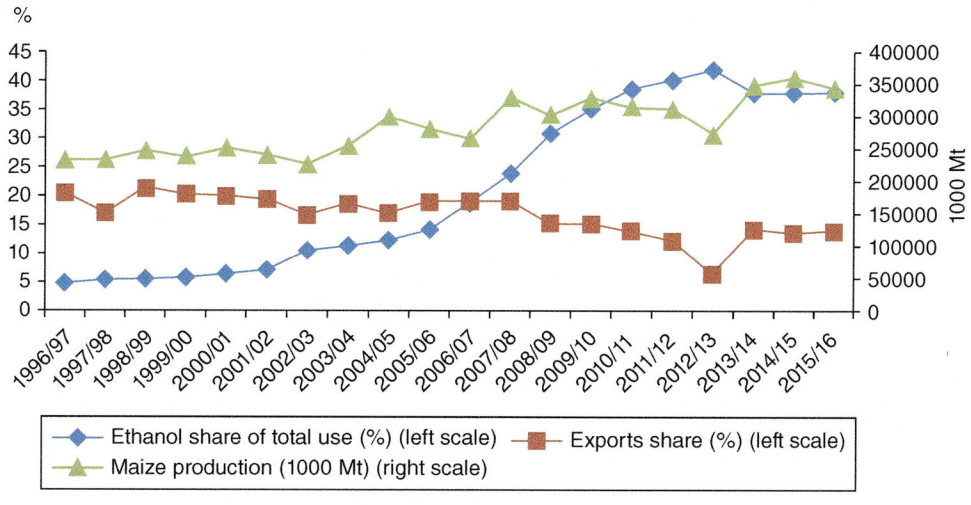

Fig. 3.9. Total US maize used for ethanol. (Adapted from USDA ERS, 2017a.)

is obviously visible in some years. When maize area in 2007 expanded by 22.4% compared with 2006, the area under soybeans dropped by 14% and that 'exchange' contributed to the increase of soybean prices in real terms of 70.2% in the period 2006–2008 (Fig. 3.2). When, in 2008, US soybean area increased by 16.4% and maize area dropped by 9.2% the real price of soybeans decreased by 11.5% in the period 2008–2010. Regardless of the 'exchange' the USA significantly increased maize production, mainly due to yield increase and to a lesser extent due to area expansion, but without any contribution to the relief of world hunger. Out of 5.7 billion Mt of accumulated maize production in the period 1996–2015, about 23% was dedicated to ethanol production. Our calculation shows that all the accumulated increase in maize production of about 1.3 billion Mt since 1996 was used for ethanol production (Table 3.2).

Such huge biofuel production has caused a decline in world grain stocks and reserves. The decline in grain stocks is not a new issue, it occurred some years before both

Table 3.2. US maize usage by ethanol segment. (Adapted from USDA ERS, 2017a.)

Fuel ethanol use (million Mt)	Increase in comparison to 1996 (million Mt)	Marketing year
10.9	0	1996/97
12.4	–0.18	1997/98
13.1	12.9	1998/99
14.4	18.4	1999/2000
16.0	24.1	2000/01
17.9	26.1	2001/02
25.3	17.8	2002/03
29.7	36.6	2003/04
33.6	47.5	2004/05
40.7	63.0	2005/06
53.8	61.4	2006/07
77.4	100.3	2007/08
94.2	83.0	2008/09
116.6	146.0	2009/10
127.5	107.8	2010/11
127.0	93.8	2011/12
117.9	58.3	2012/13
130.1	118.5	2013/14
132.1	125.9	2014/15
132.2	123.8	2015/16

crises. Considering FAO's recommendation 'that countries need to keep stocks of around 17–18% of total consumption or use level' (Headey and Fan, 2010, p. 31), it is noticeable that: maize stocks were below the recommended level before both crises; rice stocks were below the desirable level before the first one and at the border level before the second one; and that wheat stocks declined well before both crises but were always higher than the recommended level (above 20%) (Fig. 3.10). Figures 3.10 and 3.11 show that 'when stocks are high, prices are generally low and stable' (Headey and Fan, 2010, p. 31). When, in 2014, maize stocks increased above 21% of consumption, wheat stocks above 31%, and rice stocks above 24%, the average real price index of grain declined below 0.8. This logically leads to the question of why grain reserves are not always maintained at the desired level and, in answer, returns to the production of ethanol in the USA (Fig. 3.9) and transgenic technology. In the period 1996–2016, the USA increased its maize production by 64.9% and yet all the obtained gain was directed to ethanol production, and none at all to increasing reserves or to feeding the world.

Several very important motives influenced the introduction of renewable energy sources legislation in the USA. The first one was the effort to increase the income of cereal and oilseed producers. The second was the widespread political endeavour to reduce expenditure on subsidies of agriculture (i.e. deficit payments). The final was an effort to reduce dependence on fossil fuels (primarily oil) and alleviate the growing concern over its falling supply, and rising, unstable oil prices. In order to achieve these goals, the USA has implemented a number of measures and instruments including the following (Rausser and de Gorter, 2015): (i) biofuel consumption subsidies, such as the tax credits that expired in 2011; (ii) formal ethanol mandates such as the Renewable Fuel Standard; (iii) de facto mandates that ethanol be used in fuel to comply with Clean Air Act regulations on environmental protection; (iv) production subsidies for both biofuels and feedstocks, as well as

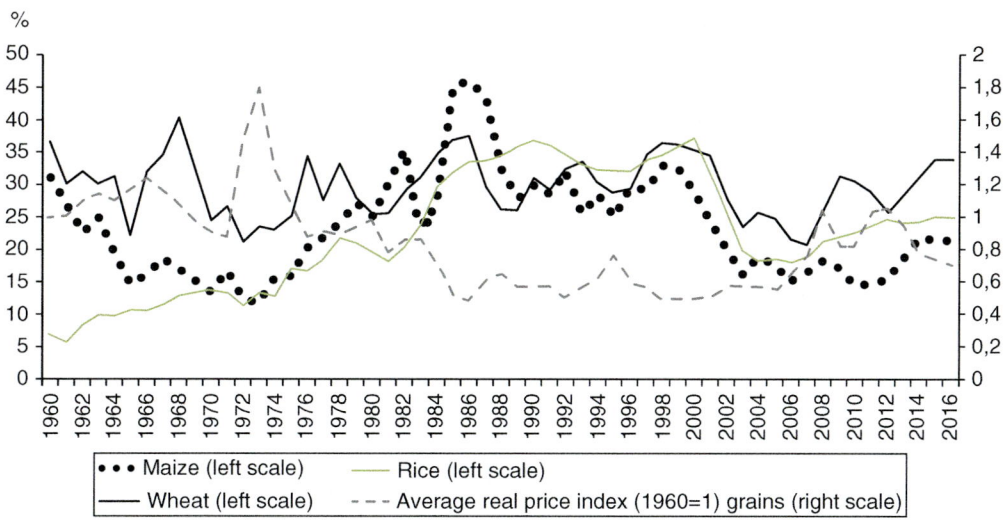

Fig. 3.10. Global trends in stock relative to consumption, 1960–2017. (Adapted from USDA FAS, 2017.)

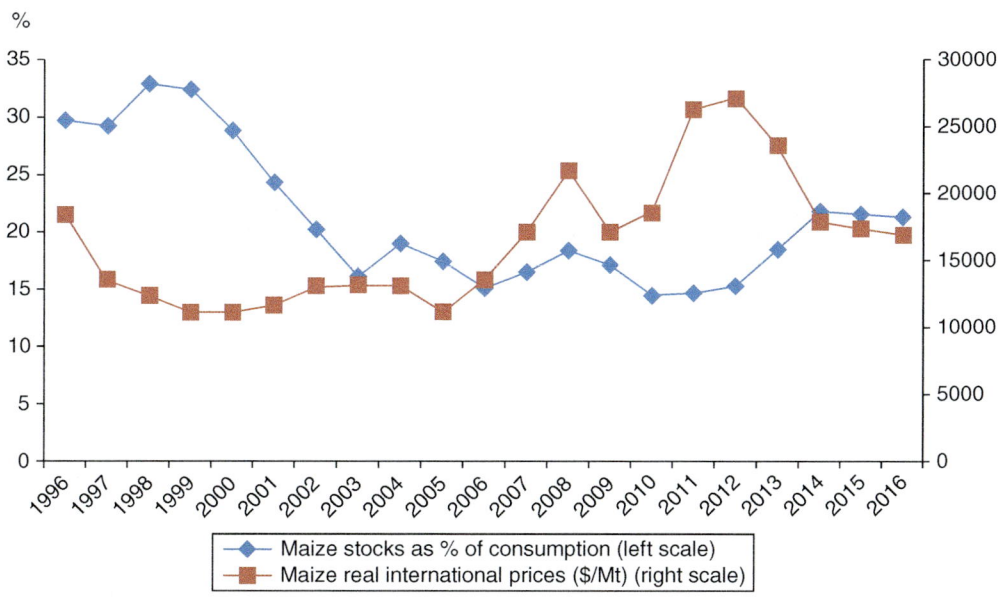

Fig. 3.11. Maize prices and stocks. (Adapted from USDA FAS, 2017.)

subsidies for alternative transport vehicles and gas stations; (v) import tariffs and tariff rate quotas that expired in 2011; and (vi) the standard defining that 1 gallon of corn ethanol must reduce greenhouse gas emissions by 20%. Many of these federal incentives, subsidies, and mandates were complementary to similar measures implemented by states.

The key to understanding how these biofuel policies affect the commodity prices is the interdependence between ethanol prices and maize prices. Ethanol prices are indirectly dependent on gasoline prices, while gasoline prices are directly dependent on oil prices. Of course, there is an indirect link between maize prices and oil prices through input costs, since maize production requires energy-intensive inputs. Therefore, the key factor for maize prices is the ethanol price transmission elasticity.[i]

The previous analysis has convincingly demonstrated that the US biofuel policies represent a new episode of the distorting effects of agrarian policies, which not only cause the instability of global prices but also have a negative impact on the welfare of consumers in food-importing countries. It has also been mentioned that the agrarian lobby, which advocates various forms of support for the ethanol production, is very aggressive. Major lobby groups promoting the ethanol production in the USA include numerous associations of cereal and oilseed producers, transporters, biofuel producers, transportation vehicle manufacturers, environmental associations, and the energy security community. However, the complexity of interdependence between biofuels policy, farmers' incomes and rural development has also produced novel effects. Higher maize prices due to the production of ethanol and higher oil prices due to biodiesel production represent a kind of implicit taxation for producers of meat products, thus reducing their income. In addition, measures that support the production of ethanol in a certain sense restrict economic growth in rural areas. Specifically, since the production of ethanol is capital intensive, its primary effect on rural development is to increase the land price, which is beneficial only for a small number of farmers. These novel effects led to the organization of producers into lobby groups that oppose the policy of supporting ethanol production. Accordingly, the organized interest groups that had shaped US agricultural policy in the past changed because of the support for ethanol production. In addition to the agrarian lobby, these groups also include energy and environmental groups, and they all participate in the throes for the potential transformation of agrarian, energy and environmental policy in the USA.

3.4 Transgenic Crops and Food Prices

Ethanol production gives one explanation for the rise of maize prices. The second specific reason for rising maize prices can be found in the varieties of transgenic maize produced. As Fig. 3.12 shows, the rise of maize prices and biofuel production, the decline in stock/use ratio, and the increase of areas under herbicide tolerant (HT) and Bt maize in the USA all happened at about the same time.

The story about the increase in adoption of transgenic varieties is much more understandable if we analyze trends in maize production costs. Figure 3.13 shows costs per ha divided by yield per ha in order to estimate costs per kilogram of produced maize. The total cost of produced maize significantly increased after 1996, the year of the first commercialization of transgenic corn. More precisely, total costs of maize production increased by 167% from 1990 to 2015. Accumulated costs of seed and chemicals in 1990 were US¢1.45, in 2000 US¢1.74, and US¢3.05 in 2015. Almost all the increase was due to an increase in seed prices. The largest seed price increase was recorded in a continuous trend in the period 2005–2012, with an average annual growth rate of 14.1%. In the same period, real international prices of maize significantly increased (Fig. 3.12).

The connection between HT soybean adoption in the USA and soy prices is obvious from Fig. 3.14, although the price trend best fits the logarithmic line (Eqn 3.1) while the HT soybean adoption trend best fits the polynomial line (Eqn 3.2).

$$y = 30.802\ln(x) + 9.265, R^2 = 0.94 \quad (3.1)$$

$$y = -0.0034x^4 - 0.1155x^3 + 6.6703x^2 - 57.675x + 405.31, R^2 = 0.86$$

$$(3.2)$$

The summation of seeds and chemical cost of US¢3.3 per kg of produced soybeans was almost the same in 1980 and 1990. After the introduction of transgenic soybean varieties the aggregated costs of seeds and

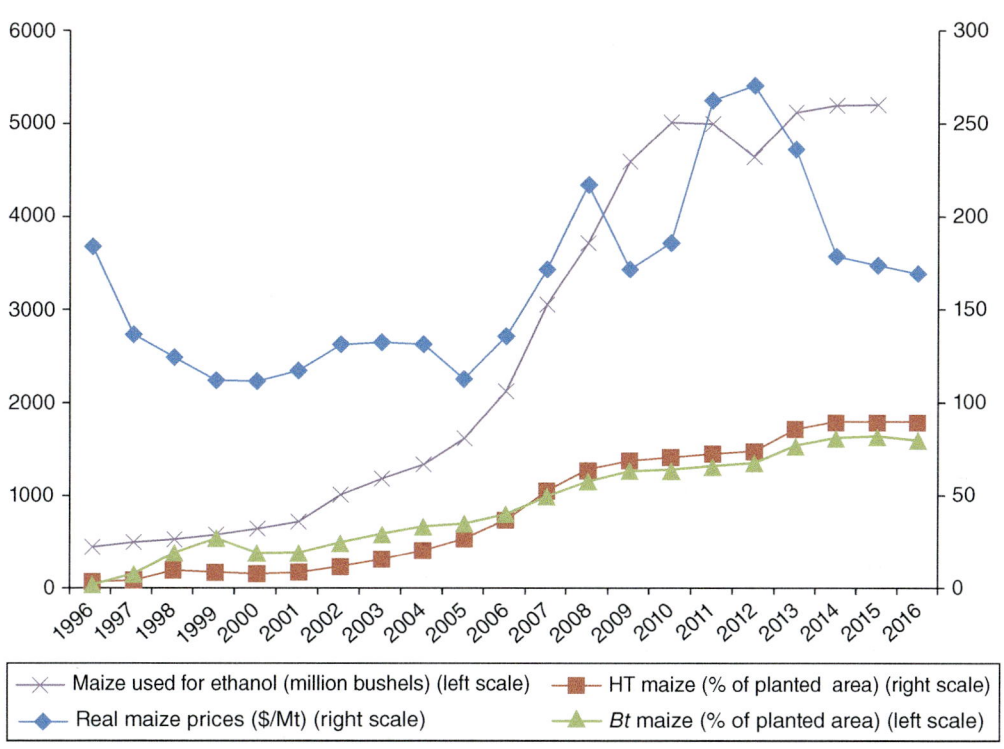

Fig. 3.12. Trends in prices, biofuel production, and transgenic variety adoption: US maize. (Adapted from USDA ERS, 2017b.)

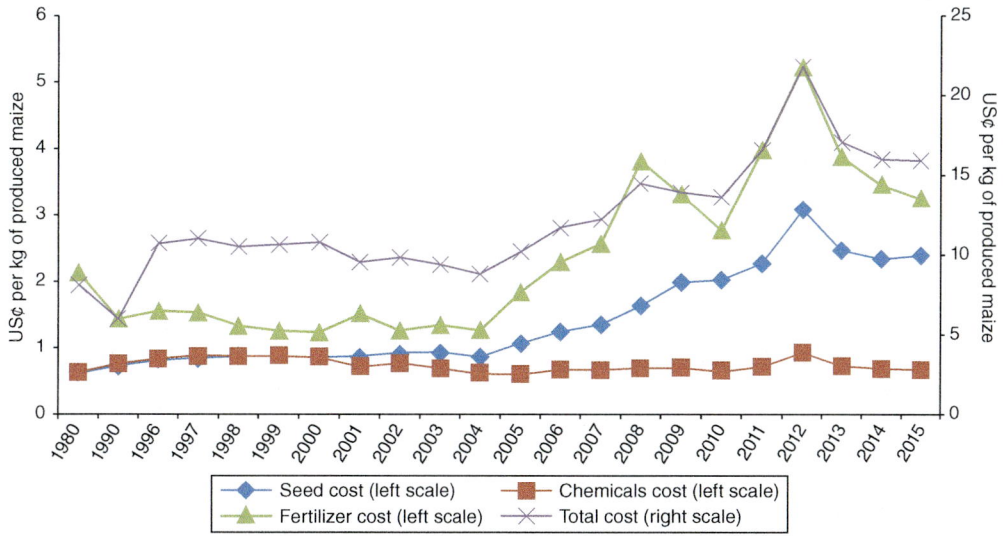

Fig. 3.13. US maize production costs. (Adapted from USDA ERS, 2017b.)

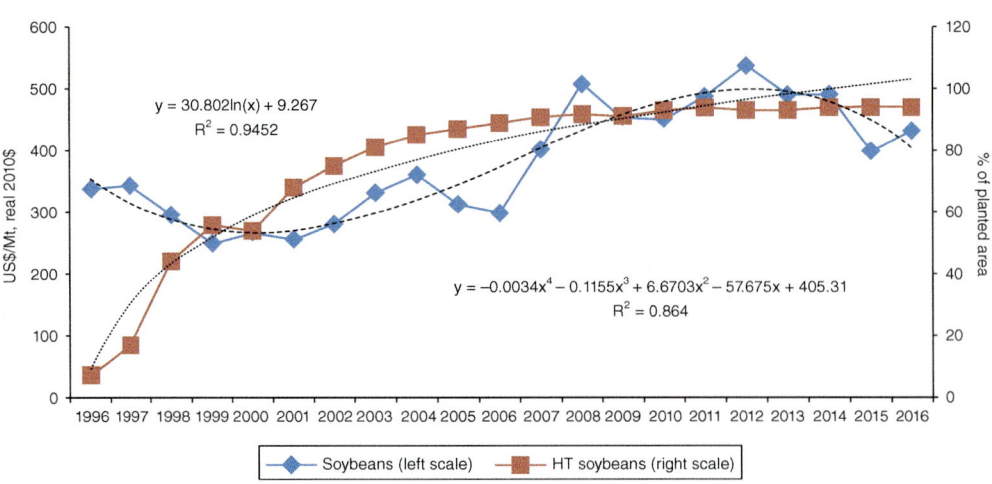

Fig. 3.14. Trends in HT soybean adoption in the USA and international soy prices. (Adapted from USDA ERS, 2017b.)

chemicals rose, from US¢3.9 in 1997 to US¢6.7 per kg of produced soy in 2015 (Fig. 3.15). On average seed cost grew by 5.5% each year from 1997 to 2015 and the share of seed costs in the total costs increased from 8.0% in 1997 to 12.8% in 2015 reaching a maximum share of 16.3% in 2010. Costs of chemicals, however, often varied: in the period 1998–2002 they decreased by 7.7% on average; after increasing by 10% in 2003 the costs decreased by 17.5% on average in the period 2003–2005; and finally, after increasing by 5.1% in the period 2005–2009 and significantly jumping by 68.6% in just 1 year (2012) they slightly decreased in the next period. Nevertheless, the total cost of soybean production per kilogram is significantly higher in the present than in the past, in 2015 it was 73.4% higher than in 1990.

Both maize seed and soybean seed as a percentage of operating cost per acre significantly increased after the commercialization of transgenic crops, and almost tripled in comparison to 1980, reaching more than 30% and 35% in 2015, respectively (Fig. 3.16). Different results come from analyzing seed costs as a percentage of crop revenue calculated as yield times price (Fig. 3.17), as the observed linear trend is less pronounced than in the previous case of seed costs per kilogram produced. However, the trend exists and on its basis seed costs as a percentage of

maize revenue increased from less than 10% in 1996 to 14% in 2015. Seed costs as a percentage of soybean revenue increased less so from below 10% in 1996 to above 12% in 2015. In relation to 1980, seed costs as a percentage of revenue increased more for soybeans than maize, by 279.5% and 225.5%, respectively.

Finally, Fig. 3.18 shows the ratio of seed price to harvest price per kilogram of produced maize and soybean. The observed linear trend is more pronounced for soybean (Eqn 3.3) than maize (Eqn 3.4).

$$y = 0.0024x + 0.045, R^2 = 0.7278 \quad (3.3)$$

$$y = 0.0022x + 0.605, R^2 = 0.6288 \quad (3.4)$$

After the adoption of transgenic technology, the ratio between maize seed price and harvest price increased by 127.4%, and by 155.2% for soybeans.

Overall, seed costs have increased on a per kilogram produced basis, as a percentage of operating cost and as a percentage of revenue basis. Increases in seed costs per kilogram of soybeans and maize produced indicates that yield increase did not keep pace with seed cost increase. If it is known that 'the $70 per bag price set for [Roundup Ready] RR 2 soybeans in 2010 is twice the cost of conventional seed' or that 'in 2009, the

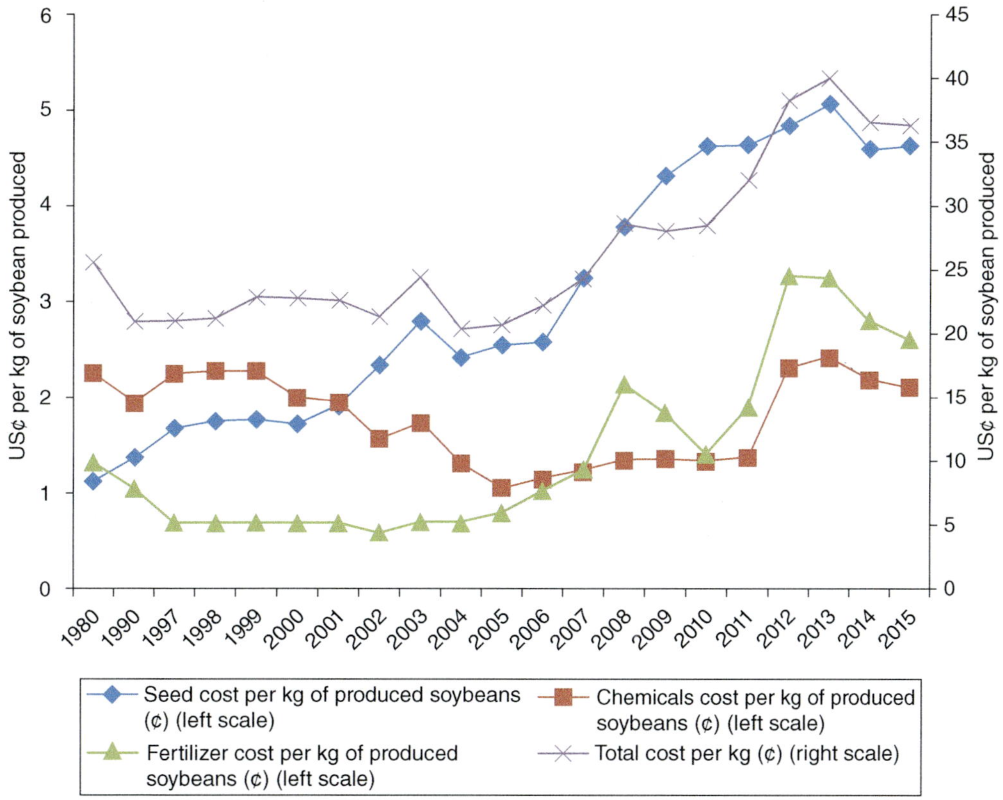

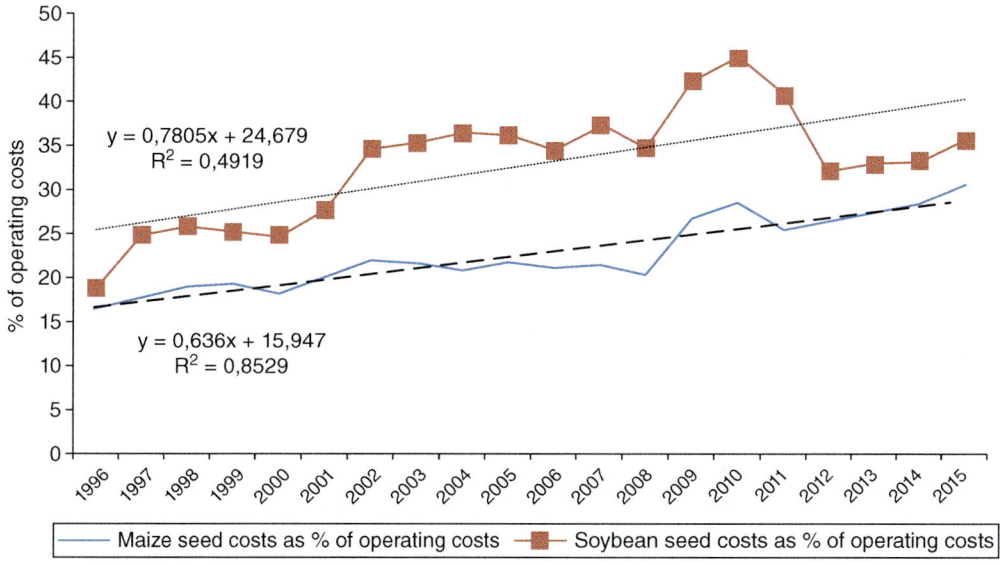

Fig. 3.15. US soybean production costs. (Adapted from USDA ERS 2017b.)

Fig. 3.16. Soybean and maize seed cost as a percentage of operating cost. (Adapted from USDA ERS, 2017b.)

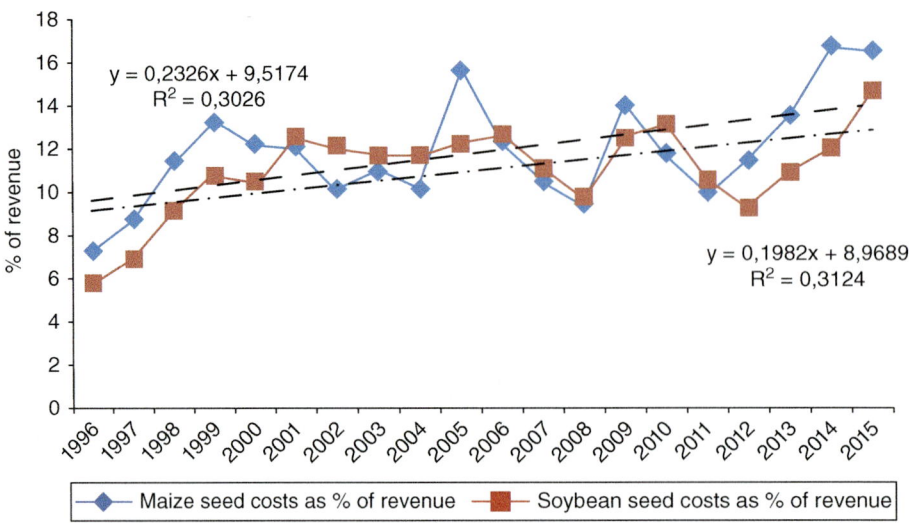

Fig. 3.17. Soybean and maize seed cost as a percentage of revenue. (Adapted from USDA ERS, 2017b.)

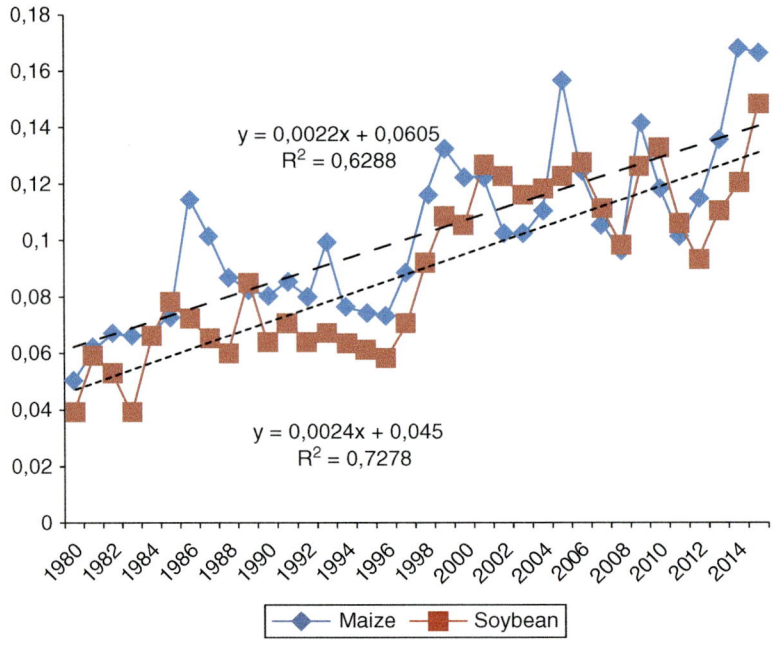

Fig. 3.18. Ratio between seed price and harvest price: US maize and soybean. (Adapted from USDA ERS, 2017a.)

[genetically engineered] GE corn [maize] to conventional corn [maize] seed premium was 69%, with GE seeds costing $235 per unit' (Benbrook, 2009, pp. 4–5) it seems reasonable to conclude that to the greatest extent, significant increases in maize and soybean seed prices occurred due to the introduction of transgenic varieties into production. Comparisons between seed prices of conventional and transgenic maize and soybean varieties leads to a similar conclusion. For comparison's sake, conventional maize seed price per bag (80,000 seeds) cost US$85.30 in 2001, while transgenic seeds cost US$110 per bag. The difference between these two prices increased in 2010, i.e. conventional maize per bag cost US$152.90, while transgenic maize cost US$270.25. A similar difference could be seen from prices of conventional and transgenic soybean seed per bushel. In 2001, the cost of conventional soybean seed per bushel was US$17.90, while transgenic seed cost was US$23.90. Nine years later, the prices of conventional and transgenic seeds per bushel were US$36.90 and US$59.50, respectively. In other words, transgenic seed premium increased over time as presented in Fig. 3.19. Transgenic maize seed premium increased significantly over time, and a clear linear trend can be observed (Eqn 3.5). On a trend basis, maize seed premium rose from 20% to more than 70%. Soybean seed premium fluctuated but a polynomial trend can be determined anyway (Eqn 3.6), and on its basis soybean seed premium increased from about 40% to 60%.

$$y = 5.5758x + 16.013, R^2 = 0.9271 \quad (3.5)$$

$$y = 0.0434x^4 - 0.6328x^3 + 0.5505x^2 + 18.122x + 22.642, R^2 = 0.4084$$
$$(3.6)$$

Unlike seed cost, although chemical costs used in soybean and maize production decreased in some years, in general they remained the same as before GMOs commercialization, which leads to the conclusion that the Gene Revolution did contribute to lowering chemicals costs, as was promised. Finally, fertilizer costs significantly increased in both maize and soybean production, by 126.6% and 149.5% from 1990 to 2015, respectively (Figs 3.13 and 3.15). However, these prices are directly affected by crude oil prices (Chen *et al.*, 2012) and by growing demand for grain crops, which keeps fertilizers prices high (Ruder and Bennion, 2013).

Finally, in order to analyse the impact of transgenic technology on food prices we compared producer and export prices between the three largest exporters of transgenic maize and soybeans: the USA, Brazil and

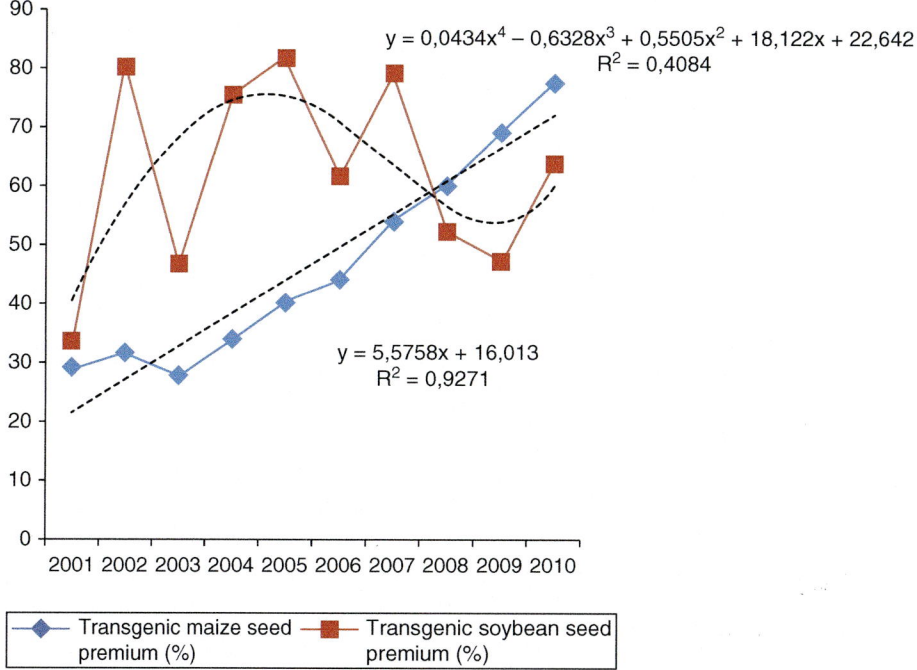

Fig. 3.19. Transgenic maize and soybean seed premium. (Adapted from Organic Center, 2017)

Argentina, and the world's largest exporter of non-transgenic maize and soybeans, Ukraine. Average producer prices of non-transgenic maize in the last 20 years in Ukraine (US$113.90/tonne) have been lower than average producer prices of transgenic maize in the USA (US$132.30/tonne), Brazil (US$144.70/tonne) and Argentina (US$117/tonne) (Fig. 3.20). Also, Ukrainian export maize prices have been lower than average world export prices since 2001 (Fig. 3.21) and since 2009 they have been continuously significantly lower than Argentinean and Brazilian prices. Comparison between the prices in Ukraine and the USA clearly shows that in 12 out of 20 observed years non-GMO maize export prices were lower than GMO maize export prices. It must be noted that the difference in price increase from year to year goes in favour of Ukraine: in 2008 Ukrainian maize export prices were 7.7% lower than US ones, while in 2013 the difference reached more than 24%. Thanks to non-transgenic production and favourable prices Ukraine is actively supplying maize both to traditional and new markets. With the exception of Ukrainian maize exports to the EU market, Ukraine is actively

supplying maize to Iran, the Republic of Korea, the Syrian Arabic Republic, Tunisia, Turkey, Libya, Egypt, Azerbaijan, Malaysia, Cuba, China, and many other countries. Another advantage that Ukraine has, is that there are almost no strong competitors in non-GMO maize. The two biggest Ukrainian competitors, Russia and Serbia, are facing some limitations. In 2016 Russia became the world's fifth-largest maize exporter, displacing the EU, but still needs some time to establish itself as a powerhouse in that product. In line with the prescribed doctrine, it is most probable that a large part of the future increase in maize production will be used on the domestic market due to high demand from the livestock industry. On the other hand, Serbia, the seventh largest maize exporting country, is limited by area size and has no possibility for further expansion of maize areas. Distribution of power on the world grain market would be quite different if there were no Russian–Ukrainian conflict. Russia, Ukraine, and Kazakhstan had a plan to form a Black Sea grain pool to establish common trading rules and implement a coordinated grain export policy. Should the plan ever be carried

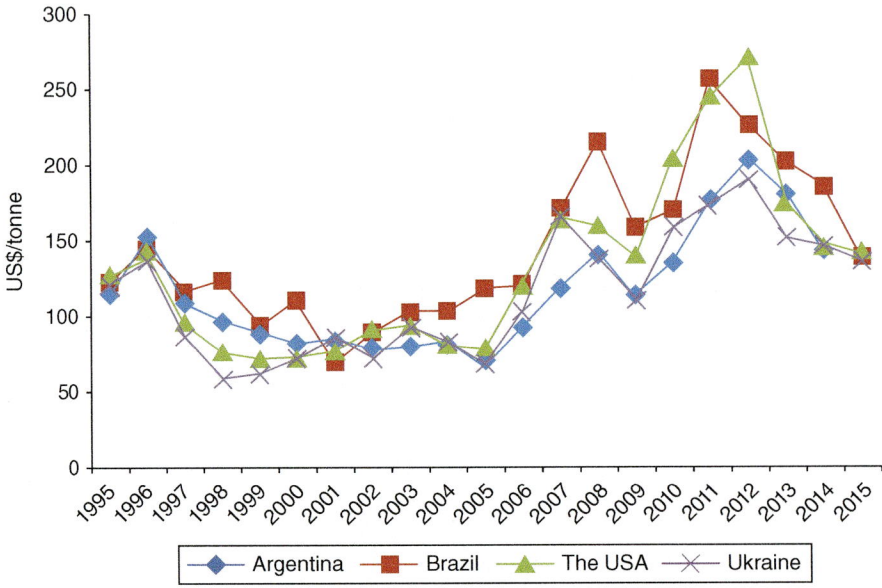

Fig. 3.20. Maize production price, different countries. (Adapted from FAOSTAT (http://www.fao.org/faostat/).)

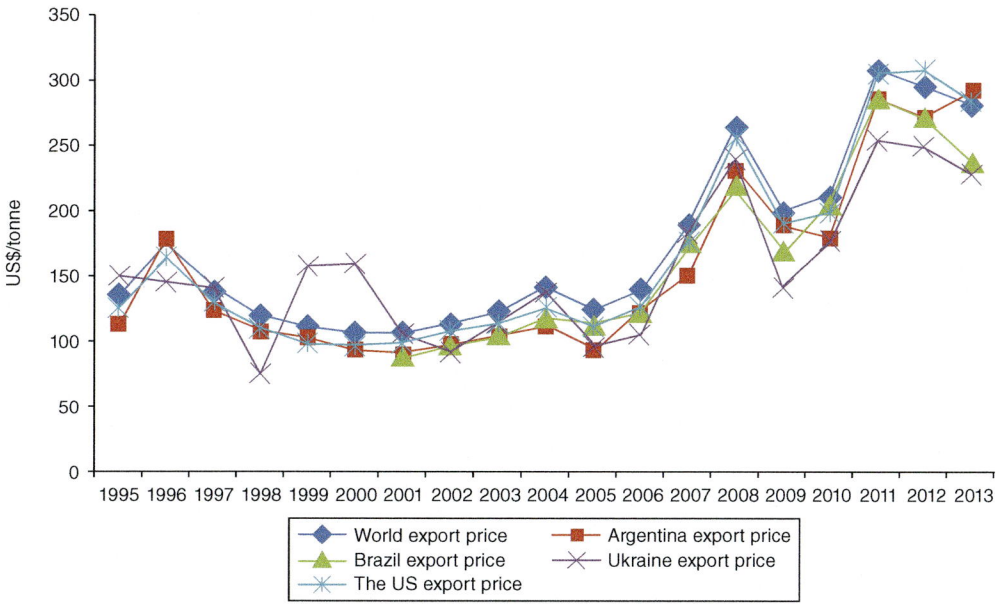

Fig. 3.21. Maize export price, different countries.

out, it can be assumed with high probability that the Black Sea grain pool would eclipse transgenic maize exporting countries for some decades. Suspended negotiations on the creation of the pool have left transgenic maize producing countries fighting among themselves for export domination in markets flexible to the importation of transgenic maize, such as: China (which does not allow seed import, but allows maize used as feed and for processing), Egypt, Mexico, Malaysia, Indonesia, Japan (relies on maize imports almost 100%), Vietnam, etc.

Ukraine is less competitive in soybean production than in maize production. Historically, soybeans have not been a typical crop for this country, but the situation has been gradually changing in the second half of the 2010s. As ShareUAPotential reported on November 1, 2016 during the last 10 years sown area under soybeans in Ukraine increased fourfold, average yield during the last 5 years was about 2.0 t/ha, while in the mid-2000s it made up only 1.5 t/ha, but was still behind Brazilian and the USA yield (ShareUAPotential, 2016). However, Ukraine, with its export of 2.3 million Mt of soybean oilseeds, is ranked the seventh export country

in the world. Ukrainian soybean production prices in most years were lower than Brazilian and American prices, but always higher than Argentinian prices (Fig. 3.22). On the other hand, since 2008 Ukrainian soybean export prices were lower than all other countries mentioned (Fig. 3.23). For example, in 2013 Ukrainian soybean export prices were lower: 5.6% than Argentinian; 7.2% than Brazilian; 10.3% than American; and 8.5% lower than average world export price. When we subtract the export soybean price from the producer price we obtain the smallest value for Ukrainian non-transgenic soybean.

The results presented illustrate that production and export prices for non-transgenic maize and soybean can be lower than transgenic ones. Significant differences exist between the Ukrainian export prices of maize and soybeans and the equivalent products from America, Brazil, and Argentina; undoubtedly this indicates that transgenic technology has contributed to rising food prices. Unfortunately, we cannot expect any major changes on the global export market because transgenic soybean and maize have already deeply strengthened their position. When we subtract all transgenic maize producing countries

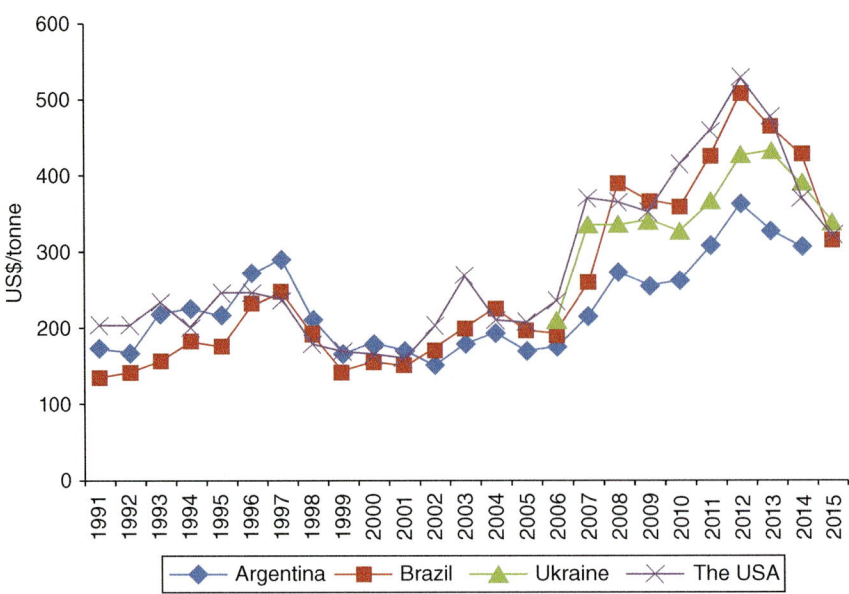

Fig. 3.22. Soybean production price, different countries. (Adapted from FAOSTAT (http://www.fao.org/faostat/).)

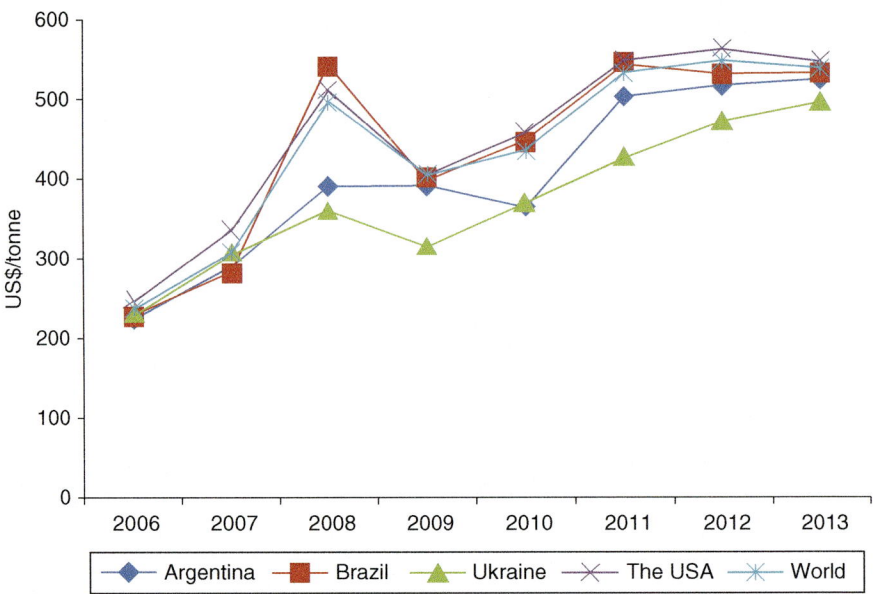

Fig. 3.23. Soybean export price, different countries. (Adapted from FAOSTAT (http://www.fao.org/faostat/).)

(the USA, Brazil, Argentina, Canada, Paraguay, South Africa, Uruguay, the Philippines, the EU, Colombia, Honduras, Chile, and Vietnam) and all soybean producing countries (the USA, Brazil, Argentina, Canada, Paraguay, South Africa, Uruguay, Bolivia, Mexico, Chile, and Costa Rica) we find that just 21.4% of maize and 2.5% of soybeans come from non-transgenic producing countries. Certainly, participation of conventional soybeans and maize in the world export market is somewhat bigger because transgenic producing countries also produce conventional crops to some extent, but such precise data cannot be found. Besides, it appears that the too often repeated story about cheaper transgenic soy and maize is purely and simply a myth. Conventional soybeans and maize can be very price competitive compared to transgenic soybeans and maize if appropriate cultivation practices and proper varieties are chosen. To confirm this theory we can observe Serbia, a small country that exports about eight times less maize than Ukraine, but that had lower export prices than the USA, Argentina, and Brazil, by 9.1%, 2.4%, and 2.5%, respectively, in 2011 and 2012.

The results illustrate that in addition to various long-, medium- and short-term factors, transgenic technology also contributed to rising food prices. The effect of was driven primarily through rising seed prices. It is well known that seed represents costs to farmers, but seed expenditure creates revenues for seed companies, meaning that farmers and end consumers end up paying more while companies reap benefits. Transgenic production jeopardized the traditional farmer's production practices such as 'brown

bagging' seed – a way of saving seeds (mostly soybean) from 1 year's harvest for cleaning and planting the next year. This practice has held seed prices at a lower level because 'every third or fourth year, farmers would purchase some new soybean seed, particularly if a promising new variety had been recently released, to grow on a portion of their land. If the variety performed well, the farmer would save some or all of the harvest for seed the next year, or purchase additional seed to plant the new variety on all fields... each acre devoted to soybean seed production will plant about 30 acres the next year' (Benbrook, 2009, p. 8).

3.5 Effects of Transgenic Technology on Food Prices in the Least-Developed Countries

'How individual countries will be affected by higher prices will depend on whether they are net agricultural commodity importers or net exporters' (FAO, 2008, p. 72). Rising commodity prices have caused a widening agricultural trade deficit in the LDCs over almost the last three decades (Fig. 3.24).

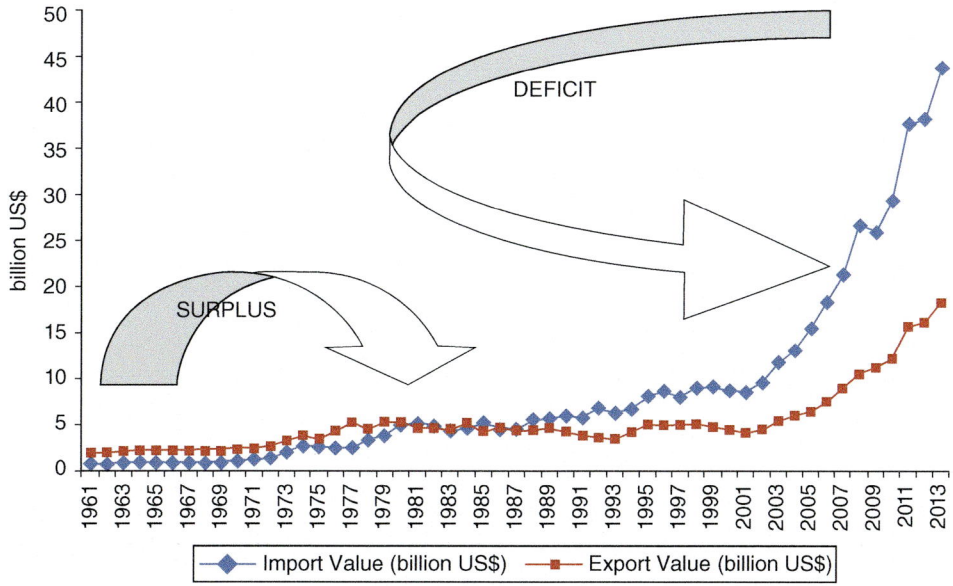

Fig. 3.24. LDCs: Agriculture trade balance. (Adapted from FAOSTAT (http://www.fao.org/faostat/).)

Rising commodity prices have pushed up the cost of imports with food import bills reaching record highs. Total food import bills (i.e. value of food commercially imported) on the global level increased by about 210%, and by about 537% in LDCs in the last three decades (Table 3.3). In the same period of time participation of LDCs in global expenditure on imported total agricultural products has doubled from 2% to 4%. LDCs' expenditure on imported cereals in 2013 rose by about 332% above the record of 1996. In addition, the import bill for meat, dairy and eggs, and animal and vegetable oil significantly increased by 1000%, 300%, and 425%, respectively.

In LDCs the share of the total agricultural products import bill in GDP terms has significantly increased over the past four decades from about 2% in the 1970s to about 5% in the 2010s (Fig. 3.25). This in effect means that the pressure on their already scarce resources grew further because, on average, the rate of growth in food import bills has been greater than that of the GDP. Although populations in LDCs almost trebled in the last four decades, from 309 million in the 1970s to about 910 million in the first years of the 2010s, rising agricultural product import bills cannot just be attributed to the undoubted population growth because food import bills per capita increased in both nominal and real terms (Fig. 3.25).

Table 3.3. Import bills of food commodities in the world and LDCs. (From FAOSTAT (http://www.fao.org/faostat/).)

Commodity	WORLD (billion USD)					LDCs (billion USD)				
	1980	1996	2000	2010	2013	1980	1996	2000	2010	2013
Cereals	44.8	55.3	40.8	96.9	138.8	1.9	2.8	2.6	7.9	12.1
Total meat	21.1	44.8	42.3	102.7	129.5	0.2	0.2	0.3	1.4	2.2
Dairy products and eggs	14.1	29.5	26.6	65.6	87.0	0.4	0.5	0.5	1.4	2.0
Animal and vegetable oil	11.5	25.1	21.3	77.3	100.9	0.4	1.2	1.4	4.7	6.3
Total agricultural products	254.5	480.2	433.1	1107.4	1429.3	5.0	8.7	8.7	29.5	44.0

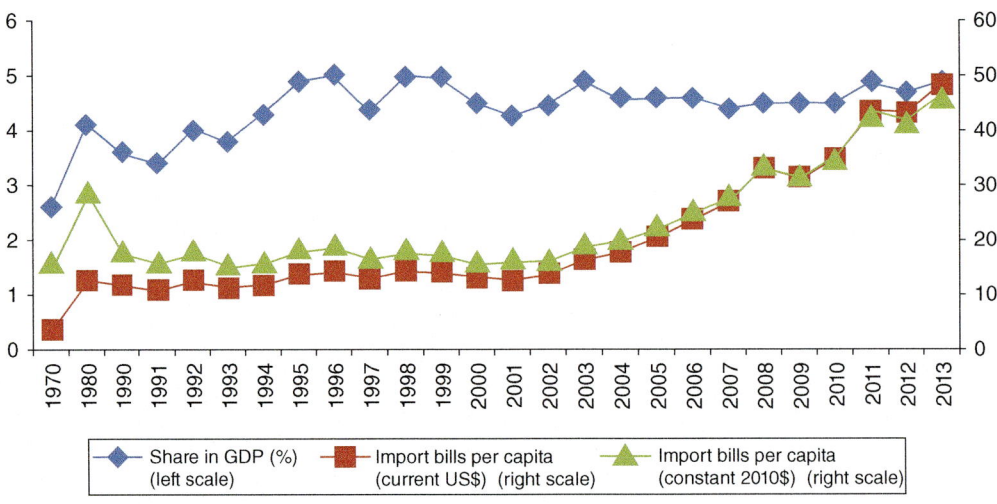

Fig. 3.25. Agricultural products import bill of LDCs: share in GDP and imports value per capita. (Adapted from FAOSTAT (http://www.fao.org/faostat/).)

Agricultural products import expenditure per capita rose more than 4 times in current US dollars and about 2.5 times in 2010 from the early 1990s to the early 2010s.

Just for comparison's sake, wholesale maize prices in Burundi, Ethiopia, and Rwanda are significantly higher than in other countries independently of whether or not they produce GMOs (Fig. 3.26). Nominal prices of maize in February 2017 were lower in Argentina and Brazil by about 450% than in Burundi. Also, in February 2017 Italy, which does not produce transgenic maize, had prices 381% lower than Burundi, 157% lower than Rwanda, and 21% lower than Ethiopia. Understandably, retail prices are consequently higher in LDCs than in the rest of the world. In January 2017, the retail price of soybean oil in China was 310% lower than in Angola (Fig. 3.27).

However, this very unfavourable situation cannot be attributed to transgenic technology. LDCs are neither significant producers nor significant importers of transgenic crops. Out of 48 LDCs, just four – Bangladesh, Burkina Faso, Myanmar, and Sudan – are very small GMO producers. Bangladesh, the most populous LDC with its 164 million people is also the only country in this group that produces an edible crop, *Bt* brinjal (aubergine) on 4,500 ha. Burkina Faso, Myanmar, and Sudan produce transgenic cotton. Burkina Faso, a leading transgenic country in Francophone West Africa produces cotton on about 0.4 million ha, Myanmar on about 0.3 million ha and Sudan on about 0.1 million ha. There is no official data about other transgenic crop production. In addition, LDCs are not among the world's top importing countries of maize or soybean. Yemen is

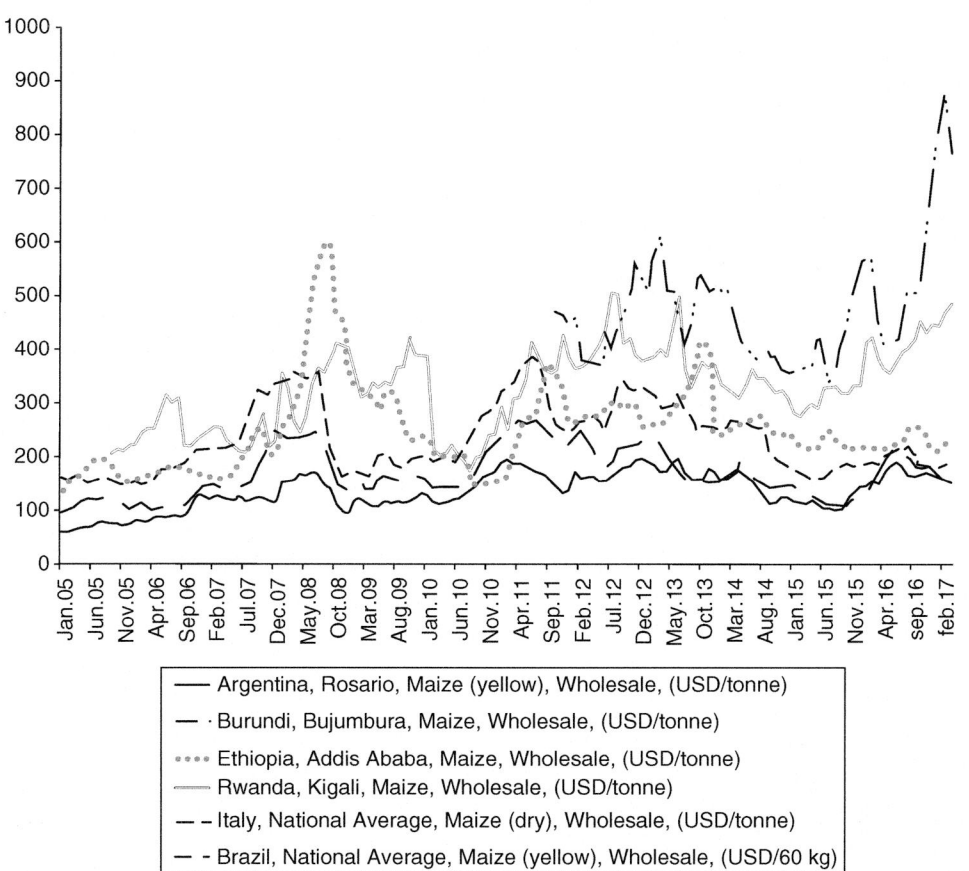

Fig. 3.26. Wholesale maize price in LDCs: Comparison with prices in Brazil, Argentina and Italy. (Adapted from FAO GIEWS, 2017.)

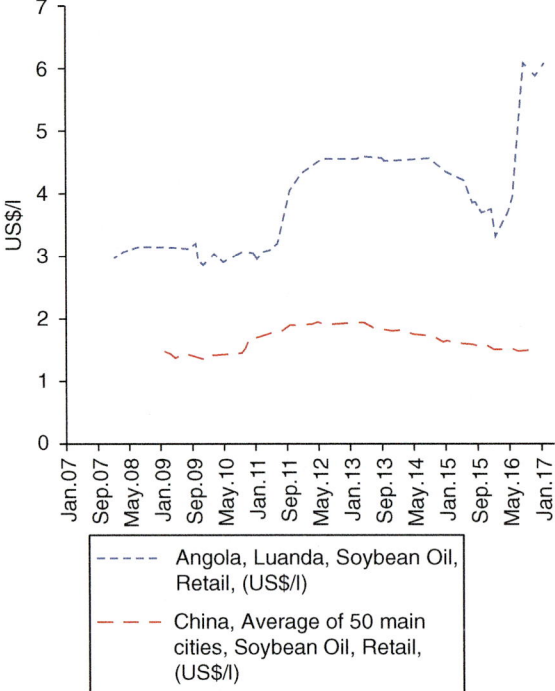

Fig. 3.27. Soybean oil retail price: Angola and China. (Adapted from FAO GIEWS, 2017.)

ranked 35th, Nepal 41st and Senegal 49th in global maize imports, while Bangladesh is the world's 15th largest soybean oilseed importing country. After all, LDCs mostly import the majority of the necessary quantities of maize and soybeans from neighbouring countries, and just a small amount from Argentina, Brazil, and the USA. Therefore it can be stated that the new technology had no impact on food prices in LDCs. Just as the Green Revolution has failed in sub-Saharan Africa, so the new Gene Revolution has bypassed in a wide arc the poorest countries in the world. However, the reasons for the difficult situations found in these countries should be sought elsewhere; it is outside the scope of this book.

Note

[i] The literature abounds in analyses of the ethanol price transmission elasticity at the price of maize. The maize ethanol price transmission coefficient was estimated (Drabik, 2011) to be 3.85 in 2010, which means that 45% of the maize price increase can be attributed to the price of ethanol.

References

Abbott, P., Hurt, C. and Tyner, E. (2011) *What's Driving Food Prices in 2011?* Farm Foundation, NFP, Oak Brook, Illinois.

Benbrook, C. (2009) *The Magnitude and Impacts of the Biotech and Organic Seed Price Premium.* Critical Issue Report: The Seed Price Premium, The Organic Center, Washington, DC, 17pp. Available at: http://www.organic-center.org/reportfiles/Seeds_Final_11-30-09.pdf (accessed 2 June 2018).

Carter, C., Rausser, C.G. and Smith, A. (2011) Commodity booms and busts. *Annual Review of Resource Economics* 3, 87–118.

Chen, P.Y., Chang, C.L., Chen, C.C. and McAleer, M. (2012) Modeling the effects of oil prices on global fertilizer prices and volatility. *Journal of Risk and Financial Management* 5, 78–114.

Da Silva, J.G., Nwanze, K.F. and Cousin, E. (2012) Tackling the Root Causes Of High Food Prices And Hunger. Available at: https://www.wfp.org/news/news-release/tackling-root-causes-high-food-prices-and-hunger (accessed 2 June 2018).

Drabik, D. (2011) *The Theory of Biofuel Policy and Food Grain Prices.* Working Paper No. 126615, Cornell Univesity, Ithaca, New York.

FAO (2008) Biofuels: Prospects, Risks, and Opportunities. The State of Food and Agriculture. FAO, Rome, Italy.

FAO (2009) High Food Prices and the Food Crisis – Experiences and Lessons Learned. FAO, Rome, Italy.

FAO GIEWS (2017) Food Price Monitoring and Analysis Tool. Available at: http://www.fao.org/giews/pricetool/ (accessed 2 June 2017).

Gilbert, C.L. (2010) How to understand high food prices. *Journal of Agricultural Economics* 61, 398–425.

Gilbert, L.C. and Morgan, C.W. (2010) Review: Food price volatilty. *Philosophical Transactions of the Royal Society* 365, 3023–3034.

Gilbert, L.C. and Morgan, C.W. (2011) Review: Food price volatilty. In: Piot-Lepetit, I. and Barek, M. (eds) *Methods to Analyse Agricultural Commodity Price Volatility*. Springer, New York, pp. 45–61.

Headey, D. and Fan, S. (2010) *Reflection on the Global Food Crisis: How Did it Happen? How Has It Hurt? And How Can We Prevent the Next One?* IFPRI, Washington, DC.

Henry, R. (2010) *Plant Resources to Food, Fuel and Conservation*. Earthscan, London, United Kingdom.

Hochman, G., Rajagopal, D., Timilsina, G. and Zilberman, D. (2011) *The Role of Inventory Adjustments in Quantifying Factors Changing Food Price Inflation*. Policy Research Working Paper No. 5744, World Bank, Washington, DC.

Lipsky, J. (2008) Commodity Prices and Global Inflation, Remarks by John Lipsky, First Deputy Managing Director, International Monetary Fund. Available at: https://www.imf.org/en/News/Articles/2015/09/28/04/53/sp050808 (accessed 2 June 2018).

Mitchell, D. (2008) A Note on Rising Food Prices. The World Bank Development Prospect Group. Policy Research Working Paper No. 4682. World Bank, Washington, DC. Available at: http://documents.worldbank.org/curated/en/229961468140943023/pdf/WP4682.pdf (accessed 2 June 2018).

OECD (2017) Meat Consumption. Available at: http://data.oecd.org/agroutput/meat-consumption.htm (accessed 2 June 2018).

Organic Center (2017). The Seed Premium-Farm Income Database, Version 1.1, The Organic Center, Washington, DC. Available at: https://organic-center.org/reportfiles/SeedPricesDatabase.pdf (accessed 3 June 2018).

Rausser, C.G. and de Gorter, H. (2015) US Policy Contribution to Agricultural Commodity Price Fluctuations, 2006-2012. In: Pinstrup-Andersen, P. (ed.) *Food Price Policy in an Era of Market Instability: A Political Economy Analysis*. Oxford University Press, Oxford, UK, pp. 433–456.

Ruder, J. and Bennion, E. (2013) Growing Demand for Fertilizers Keeps Prices High. *Beyond the Numbers: Global Economy*. Available at: https://www.bls.gov/opub/btn/volume-2/growing-demand-for-fertilizer-keeps-prices-high.htm (accessed 3 June 2018).

ShareUAPotential (2016) Ukrainian soybeans market is ready for new records. Available at: http://shareuapotential.com/BE/Ukrainian_soybeans_2016.html (accessed 8 June 2018).

Shaw, J.D. (2007) *World Food Security: A History since 1945*. Palgrave Macmillan, Hampshire, UK.

Swinnen, J. (2011) The Right Price for Food. Development Policy Review. Available at: https://www.econstor.eu/bitstream/10419/74923/1/dp259.pdf (accessed 3 June 2018).

Tadesse, G., Algier, B., Kalkuhl, M. and von Braun, J. (2016) Drivers and triggers of international food price spikes and volatility. In: Kalkuhl, M., von Braun, J. and Torero, M. (eds) *Food Price Volatility and Its Implications for Food Security and Policy*, Springer, Berlin, Germany, pp. 59–82.

USDA ERS (2017a) U.S. Bioenergy Statistics. Available at: https://www.ers.usda.gov/data-products/us-bioenergy-statistics/ (accessed 3 June 2018).

USDA ERS (2017b) Commodity Costs and Returns. Available at: https://www.ers.usda.gov/data-products/commodity-costs-and-returns/ accessed 3 June 2018).

USDA FAS (2017) Data & Analysis. Available at: https://www.fas.usda.gov/data (accessed 3 June 2018).

Von Braun, J. and Tadesse, G. (2012) *Global Food Price Volatility and Spikes: An Overview of Costs, Causes, and Solutions*. ZEF-Discussion Papers on Development Policy No. 161, University of Bonn, Bonn, Germany.

Wright, B.D. (2011) The economics of grain price volatility. *Applied Economics Perspectives and Policy* 33, 32–58.

World Bank (2017). DataBank. Available at: http://databank.worldbank.org/data/home# (accessed 4 June 2018).

4 Food Security and GMOs

Despite a growing area under genetically modified crops around the world, many nations have continued to fight the introduction of transgenic food. Resistance exists on all continents. After the abolition of the moratorium on the cultivation of GMO, Europe adopted strict labelling and traceability regulation for all food derived from GMOs on the grounds of human and animal health, and environmental protection. The EU has left its members a choice whether or not to grow GMOs, even if one GM maize MON810, is already authorized to be grown within the Union. Scotland, Wales and Northern Ireland in the UK, Austria, the Wallonia region in Belgium, Bulgaria, Croatia, Cyprus, Denmark, France, Germany, Greece, Hungary, Italy, Latvia, Lithuania, Luxembourg, Malta, the Netherlands, Poland, and Slovenia, have taken the 'opt-out' clause of a European Commission rule to abstain from growing GMO crops. 'A controversy over GM food arose in 2000 when it was discovered that some food aid donations contained GMOs and grew increasingly in 2002, when several Southern African countries refused GM food aid during a food crisis' (Brankov and Lovre, 2013, p. 100). The USA blamed Europe of being responsible for the African rejection, and therefore stopped feeding the famine. The GMO issue again became extremely relevant to the debate on food aid via the Doha Round talks in 2003–2004. Despite pressure from other countries, the US rejected replacing in-kind food aid with cash. This deepened the suspicion that the US was insisting on in-kind food aid because of the intention to spread transgenic crops more broadly. In addition, controversy surrounding numerous international organizations that had influenced the diffusion of genetically modified food (Brankov and Lovre, 2012) and possible anticompetitive practices in the seed industry (Brankov and Lovre, 2010) heated up the question: Do GMOs help to solve world hunger or just contribute to the enrichment of multinational companies?

4.1 Food Security Dimensions

The FAO defines hunger as an uncomfortable or painful sensation caused by insufficient food energy consumption. Scientifically, hunger is referred to as:

> ...food deprivation or undernourishment, as the consumption of food that is not sufficient to provide the minimum amount of dietary energy that each individual requires for living a healthy and productive life, given his or her sex, age, stature and physical activity level ... Simply put, all hungry people are food insecure, but not all food insecure people are hungry, as there are other causes of food

insecurity, including those due to poor intake of micronutrients. Malnutrition results from deficiencies, excesses or imbalances in the consumption of macro- and/or micronutrients. Malnutrition may be an outcome of food insecurity, or it may relate to non-food factors, such as: inadequate care practices for children, insufficient health services and an unhealthy environment. While poverty is undoubtedly a cause of hunger, lack of adequate and proper nutrition itself is an underlying cause of poverty. A current and widely used definition of poverty is: Poverty encompasses different dimensions of deprivation that relate to human capabilities including consumption and food security, health, education, rights, voice, security, dignity and decent work. It is argued that a strategy for attacking poverty in conjunction with policies to ensure food security offers the best hope of swiftly reducing mass poverty and hunger. However, recent studies show that economic growth alone will not take care of the problem of food security. What is needed is a combination of: income growth supported by direct nutrition interventions and investment in health, water and education.

(FAO, 2017)

The FAO stated that food security exists when all people, at all times, have physical and economic access to sufficient, safe, and nutritious food that meets their dietary needs and food preferences for an active and healthy life. Concepts of food security have evolved over time, from the 1970s when it was understood as food supply or basic foods availability and price stability, through the 1980s when focus was on balance between demand and supply or on food access, to the multidimensionality concept in the 1990s (availability, access, food use and stability). More recently, the FAO 'introduced a suite of food security indicators, which measures separately the four dimensions of food security to allow a more nuanced assessment of food insecurity' (FAO, 2015a, p. 48).

The first dimension of food security, food availability, represents availability of sufficient quantities of food of appropriate quality, supplied through domestic production or imports (including food aid) and is analysed through five indicators such as average value of food production and average protein supply. Food access – access by individuals to adequate resources for acquiring appropriate foods for a nutritious diet – is analysed through nine indicators, among which are road density, domestic food price index and GDP in PPP. Food stability refers to both the availability and access dimensions of food security, since a population, household, or individual must have access to adequate food at all times regardless of any seasonal shocks or economic crises. Seven indicators belong to this food security dimension: (i) cereal import dependency ratio; (ii) percentage of arable land equipped for irrigation; (iii) value of food imports over total goods exports; (iv) political stability and absence of violence/terrorism; (v) domestic food price volatility; (vi) per capita food production variability; and (vi) per capita food supply variability. The last food security dimension represents utilization of food through adequate diet, clean water, sanitation, and healthcare to reach a state of nutritional well-being whereby all physiological needs are met. Food utilization that brings out the importance of non-food inputs in food security issues is described through ten indicators (Table 4.1).

The three drivers that determine outcomes of food availability are: production, distribution, and exchange. Three components – affordability, allocation, and preference – drive food accessibility. Nutritional value, social value, and food safety are three elements that contribute to food utilization (Ericksen, 2007).

4.1.1 Food security at the global level

'Since the early 1990s, the number of hungry people has declined by 216 million globally, a reduction of 21.4 percent, notwithstanding a 1.9 billion increase in the world's population. This is mainly due to changes in highly populated countries like China and India, where rapid progress was achieved during the 1990s. Marked differences in progress occur not only among individual countries, but also across regions and sub region' (FAO, 2015a, pp. 2–3). For example, the prevalence of hunger reduced rapidly in the Caucasus and Central Asia from 14.1%

Table 4.1. Suite of food security indicators. (From FAO, 2015a.)

Food security indicators	Dimensions
Average dietary energy supply adequacy	Availability
Average value of food production	
Share of dietary energy supply derived from cereals, roots, and tubers	
Average protein supply	
Average supply of protein of animal origin	
Percentage of paved roads over total roads	Access
Road density	
Rail line density	
Gross domestic products (in purchasing power parity)	
Domestic food price index	
Prevalence of undernourishment	
Share of food expenditure of the poor	
Depth of the food deficit	
Prevalence of food inadequacy	
Cereal import dependency ratio	Stability
Percent of arable land equipped for irrigation	
Value of food imports over total goods exports	
Political stability and absence of violence/terrorism	
Domestic food price volatility	
Per capita food production variability	
Per capita food supply variability	
Access to improved water sources	Utilization
Access to improved sanitation facilities	
Percentage of children under 5 years of age affected by wasting	
Percentage of children under 5 years of age who are stunted	
Percentage of children under 5 years of age who are underweight	
Percentage of adults who are underweight	
Prevalence of anaemia among pregnant women	
Prevalence of anaemia among children under 5 years of age	
Prevalence of vitamin A deficiency in the population	
Prevalence of iodine deficiency in the population	

to 7% of the population; in Eastern Asia from 23.2% to 9.6%; in Southern Asia from 23.9% to 15.7%; in Latin America from 13.9% to less than 5%; and in South America from 15.1% to less than 5% (FAO, 2015b, pp. 44–47). But, if we exclude China, the proportion of undernourished people in the total population in Eastern Asia increased from 9.6% to 14.6%, because of significant increases in the Democratic People's Republic of Korea and Mongolia. One out of four people or 23.2% of the population are hungry in sub-Saharan Africa. The highest burden of hunger occurs in Southern Asia: 281 million people.

When expressed as the IFPRI Global Hunger Index (GHI), a tool for tracking hunger and malnutrition (IFPRI, 2017), serious problems persist in the developing world. If we use a revised formula that states: values less than 9.9 reflect low hunger, values between 10.0 and 19.9 reflect moderate hunger, values between 20 and 34.9 indicate a serious problem, values between 35 and 49.9 are alarming, and values exceeding 50 are extremely alarming, then we can see that the developing world as a whole, Africa south of the Sahara, and South Asia, with scores of 21.3, 30.1, and 29 points in 2016, respectively, are seriously hungry. East and Southeast Asia with a score point of 12.8 and Near East and North Africa with 11.7 points are moderately hungry. Just two regions tracked by IFPRI reflect low hunger, Eastern Europe and CIS (8.3) and Latin America and the Caribbean (7.8). Obvious areas of concern are countries where hunger is still persisting in over 20% of the total population (Fig. 4.1). Due to insufficient data, 2016 GHI scores could not be calculated for a number of countries which are also a cause for significant concern: Burundi, the Comoros, the Democratic Republic of the Congo, Eritrea, Libya, Papua New Guinea, Somalia, South Sudan, Sudan, and the Syrian Arab Republic.

In any case, global food security nowadays has improved compared with the past. About 795 million people, or one person out of nine in the world, are undernourished, a number that has changed since 25 years ago when it was about 1 billion.

The year 2015 is a milestone, marking the end of the Millennium Development Goal (MDG) monitoring period. For the developing regions as a whole, the target to reduce the proportion of the world's hungry by 50% by 2015 was missed by a small margin. Some regions, such as Latin America, the eastern and south-eastern regions of Asia, the Caucasus and Central Asia, and the northern and western regions of Africa have reached the target, as they made fast progress in reducing undernourishment. As many as 72 developing countries out of 129 have reached the MDG hunger target. Most of these enjoyed stable political conditions and economic growth, along with sound social protection policies aimed at assisting the most vulnerable

(FAO, 2015a, p. 1).

In the period 1990–2016, with the increase of GDP per capita, the depth of the food deficit decreased (Fig. 4.2). Food availability at the global level increased, whereby average dietary energy supply adequacy, average value of food production, average protein supply, and average supply of protein of animal origin increased, while share of dietary energy supply derived from cereals, roots, and tubers decreased (Fig. 4.3). In general, access to improved water resources and sanitation facilities were upgraded (Fig. 4.4). The percentage of arable land equipped for irrigation increased, while the value of food imports over total goods exports decreased. Per capita food production variability, that corresponds to the net food production value in constant 2004/06 prices divided by the population number, has increased since 1990 (Fig. 4.5).

4.1.2 Food security by special groups and regions

Despite some important improvements on a global level, a wide difference in progress occurs at regional level and among individual countries. Over a 23-year period, domestic food production in monetary terms has doubled with respect to developing countries and almost quadrupled with respect to production in the LDCs (Fig. 4.6). This indicates that developing countries and especially the

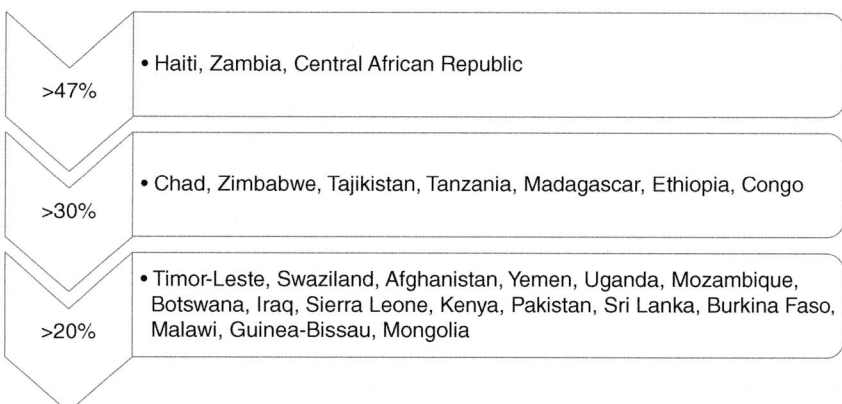

Fig. 4.1. Most hungry nations, percentage of undernourished in the population. (Adapted from IFPRI, 2017.)

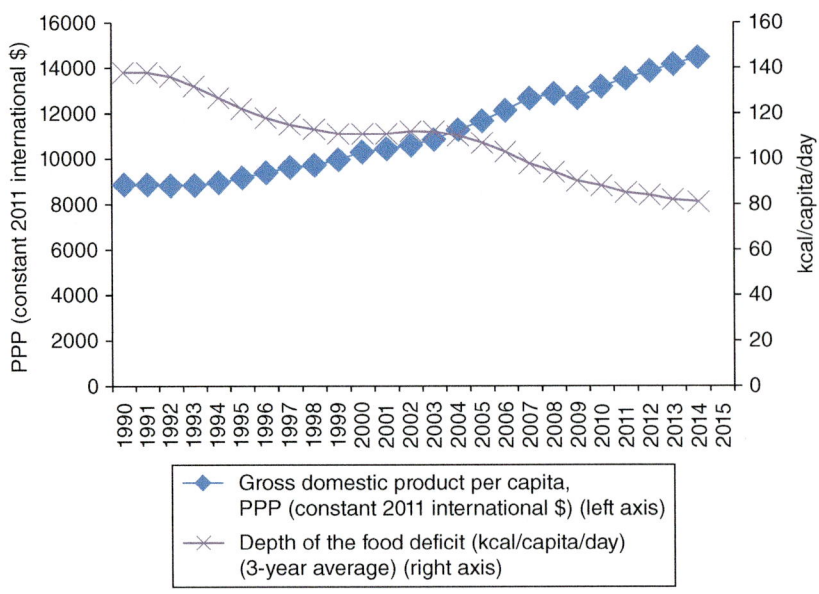

Fig. 4.2. Access to food in the world, by year. (Adapted from FAOSTAT (http://www.fao.org/faostat/).)

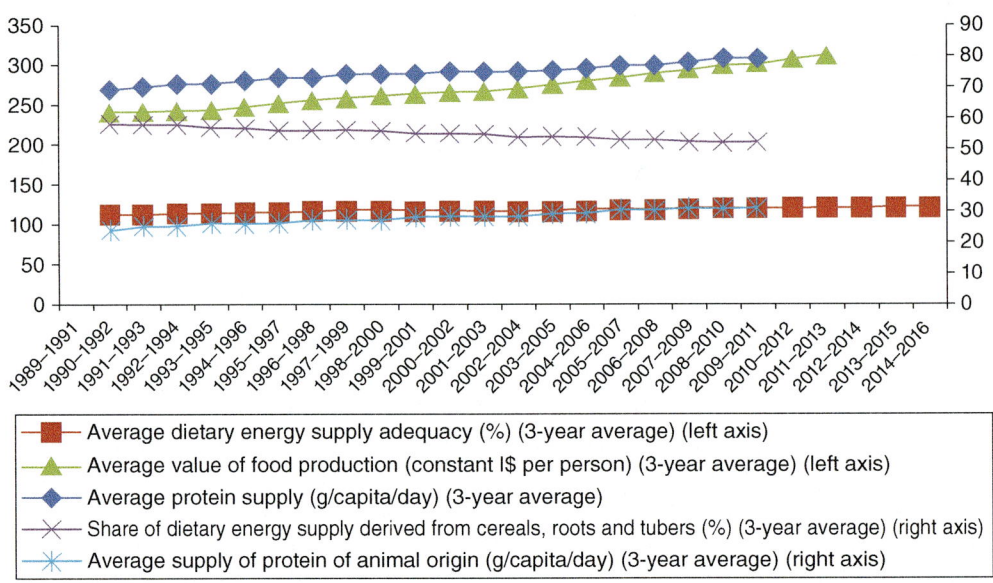

Fig. 4.3. Food availability in the world, by year. (Adapted from FAOSTAT (http://www.fao.org/faostat/).)

LDCs have serious problems in continually providing sufficient quantities of food.

Relative prices of food, expressed as a relative level of domestic food prices compared to those in the USA (the USA is equal to 1 in the index), is 4.7 times higher in the LDCs than in developed countries. Relative food prices are 2.4 times higher in developing countries than in developed ones (Fig. 4.7). Data from FAOSTAT shows that developed countries have about five and two times higher road density per 100 sq. km of land area than the LDCs

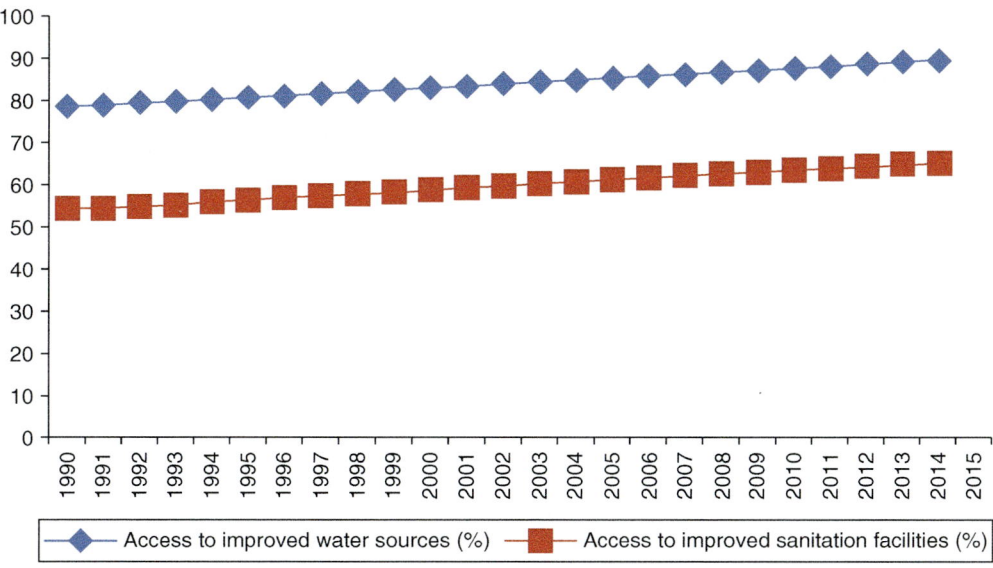

Fig. 4.4. Access to water sources and sanitation facilities, by year. (Adapted from FAOSTAT (http://www. fao.org/faostat/).)

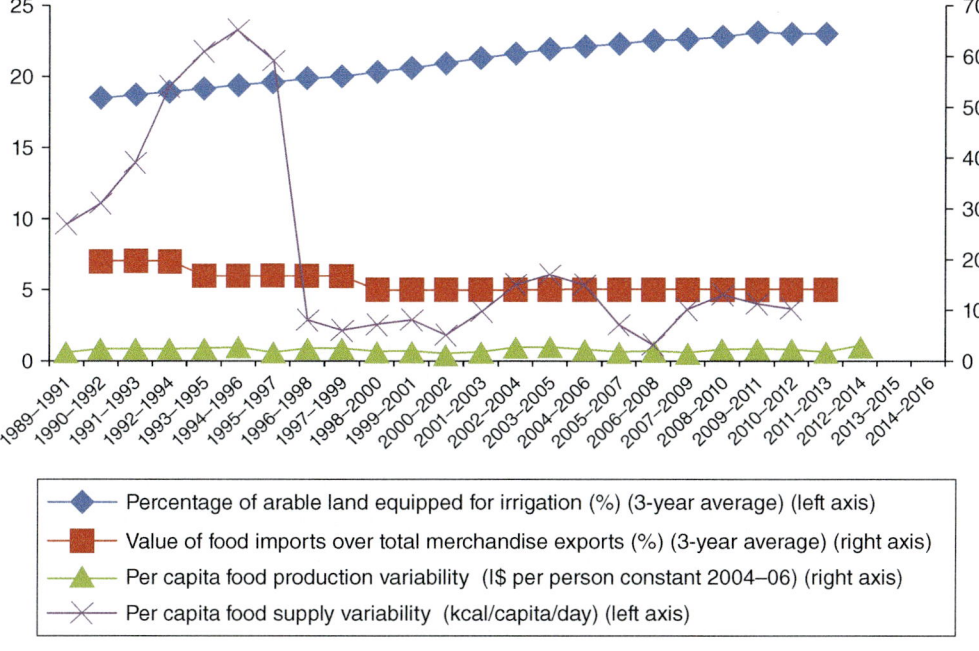

Fig. 4.5. Food stability in the world, by year. (Adapted from FAOSTAT (http://www.fao.org/faostat/).)

and developing countries, respectively. In addition, the GDP per capita converted into international dollars using the PPP rates is 18 times higher in developed countries than in

the LDCs. Developing countries have four times lower GDP PPP than developed ones. Income, prices, and infrastructure comparison clearly shows that developing countries

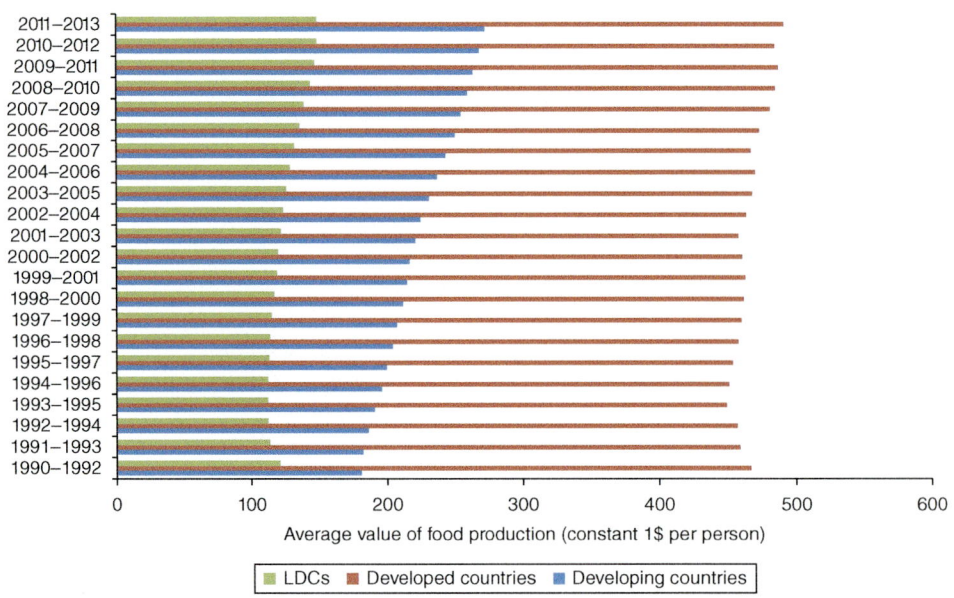

Fig. 4.6. Average value of food production (3-year average), by special group. (Adapted from FAOSTAT (http://www.fao.org/faostat/).)

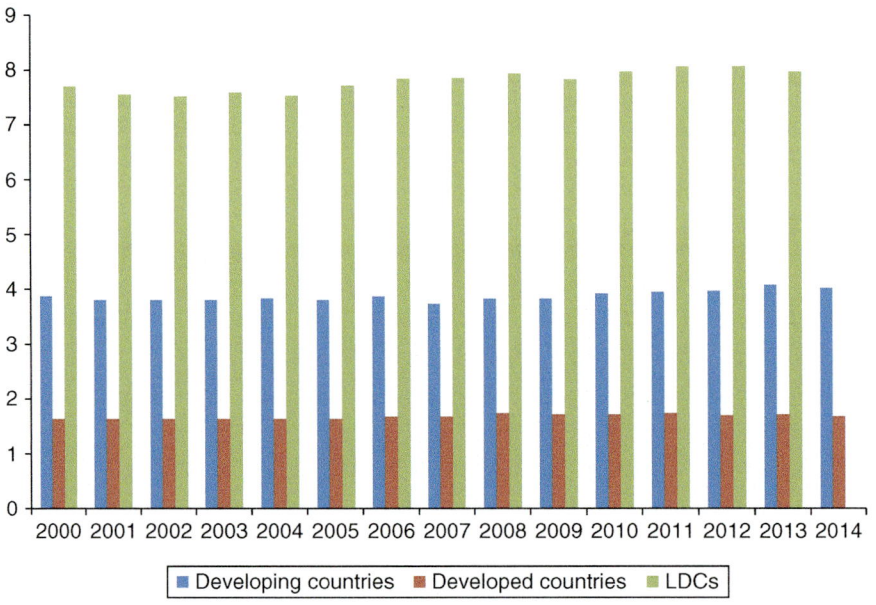

Fig. 4.7. Domestic food price index, by special group. (Adapted from FAOSTAT (http://www.fao.org/faostat/).)

and the LDCs have serious problems with both economic and physical access to agricultural and food markets. Developed countries have low domestic relative prices of food, high road density, and high GDP PPP. Thus, developed countries have high access to food.

Apart from higher relative prices in developing countries and LDCs, variations in domestic food prices are much more distinct than in developed countries (Fig. 4.8). Cereal import dependency ratio [(cereal imports − cereal exports)/(cereal production + cereal imports − cereal exports) *100] significantly increased in developing countries (from 9.5% in 1995–1997 to 50.7% in 2011–2013) and the LDCs (from 11.9% in 1994–1996 to 46.2% in 2011–2013), while it decreased in developed countries (from −16.6% in 1995–1997 to −18.2% in 2009–2011). High cereal import dependence ratio, as well as other indicators that belong to supply stability dimensions, demonstrate problems with self-sufficiency in developing countries and the LDCs. Recent research shows that 'low and middle income developing countries had the highest variability in per capita food supply, ranging from 63 to 101 calories per person per day. On the other hand, the vast majority of countries with low per capita food supply variability (between 6 and 11 kcal/capita/day) were high-income developed countries. On the whole it can be concluded that low-income developing countries lack food stability in general, when measured by food trade, price volatility and food supply variability indicators. On the other hand, high income developed

countries have stable food supplies' (Jambor and Babu, 2016, p. 91).

Domestic food production in monetary terms is most vulnerable in Africa. For comparison it is about four times lower than in the USA (Fig. 4.9). The biggest improvement of about 60% is recorded in South America and Asia. Domestic food prices are the lowest in South America and the highest in Africa. Africa has a domestic food price index that is more than 5 times higher than the USA (Fig. 4.10). All regions face similar problems with domestic food price volatility (Fig. 4.11). The cereal import dependency ratio in Africa is 42%, 93.3% in Asia, 49.1% in Central America, −12.7% in South America, 51.8% in the Caribbean, and 95.4% in Oceania. This means that South America is the only region in the group that is a net exporter of cereals.

4.2 Effects of GMOs on World Hunger

Since Quintus Septimius Florens Tertulianus (3rd century CE), through Thomas Robert Malthus (18th century), to Paul Erlich (20th century), numerous authors have made their

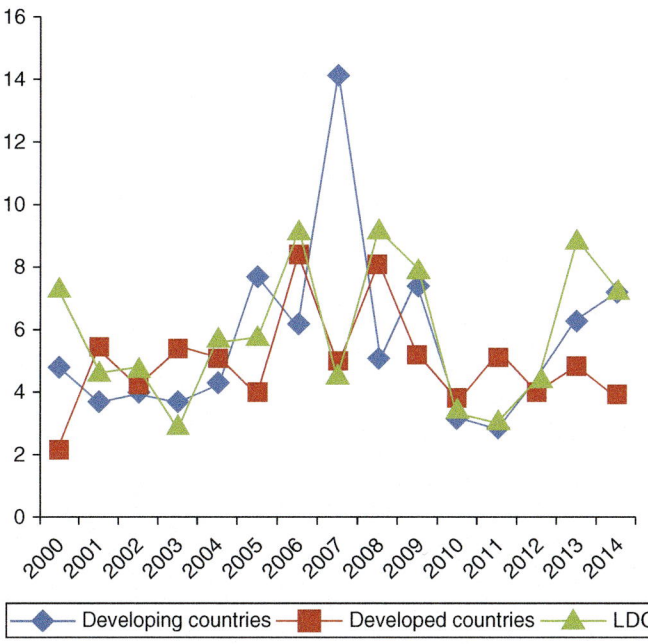

Fig. 4.8. Domestic food price volatility (index), by special group. (Adapted from FAOSTAT (http://www.fao.org/faostat/).)

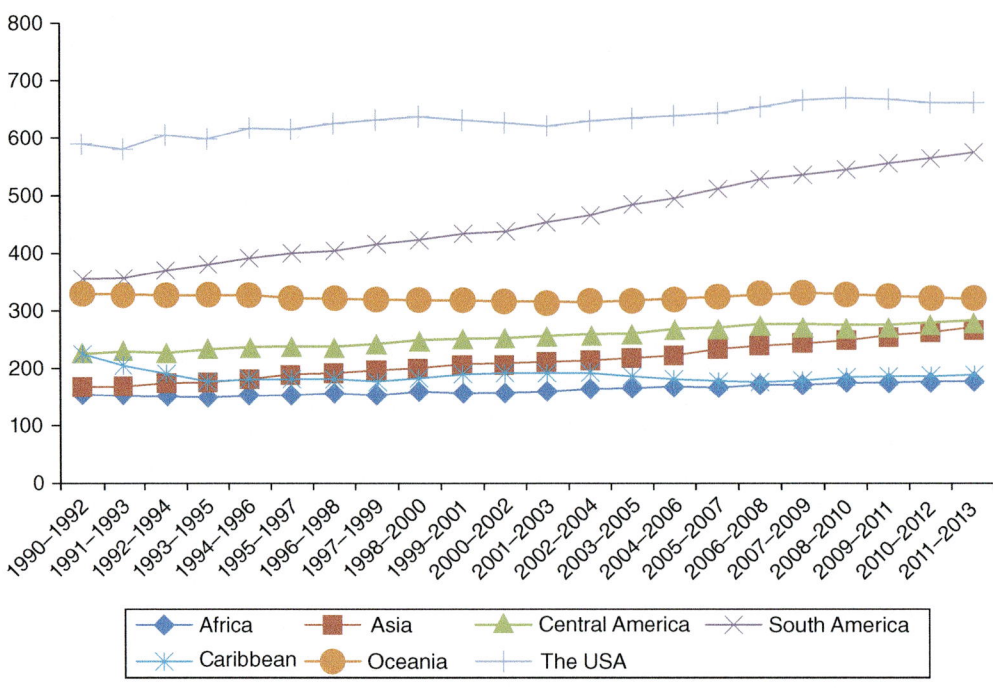

Fig. 4.9. Average value of food production (constant International $ per person) (3-year average), by region (Adapted from FAOSTAT (http://www.fao.org/faostat/).)

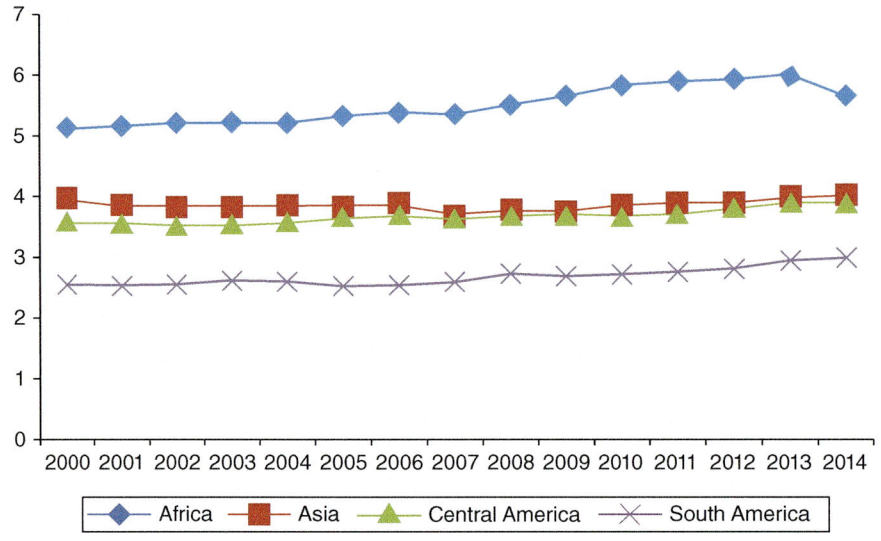

Fig. 4.10. Domestic food price index, by region. (Adapted from FAOSTAT (http://www.fao.org/faostat/).)

grim predictions of an overpopulated planet and mass hunger, believing in the impossibility of achieving food security for the growing population. However, science, technological advance, and innovation have since negated the pessimistic predictions. During the past decades, the food production rate managed to surpass the population growth rate

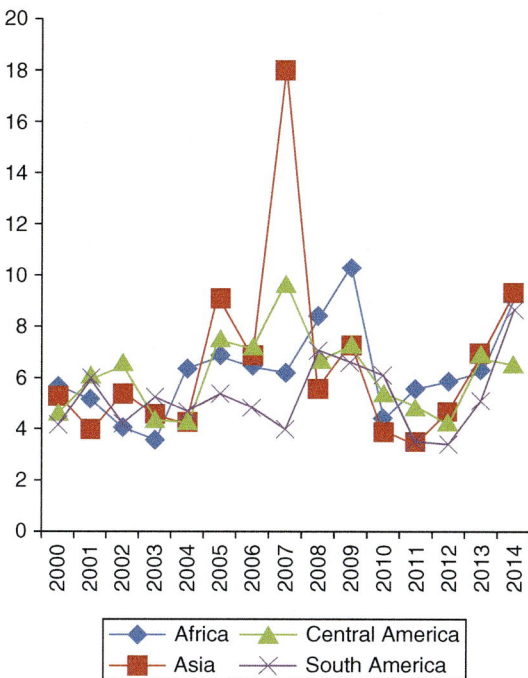

Fig. 4.11. Domestic food price volatility (index), by region. (Adapted from FAOSTAT (http://www. fao.org/faostat/).)

(Holt-Giménez *et al.*, 2012). The Green Revolution made a great 'breakthrough' in agricultural production by combining high-yield grain cultivars, artificial fertilizers, pesticides, and irrigation (Tilman, 1998). Selected genetic traits increased yield, yield stability, and wide-scale adaptability of certain varieties. Essentially, the Green Revolution industrialized agriculture production, allowed for greater control over the growing conditions, and reduced the interaction of genotype and environment. At the same time, the Revolution caused numerous negative consequences (Matson *et al.*, 1997), of which the consequences of intensive use of artificial fertilizers and pesticides have been particularly extensively analyzed (Phipps and Park, 2002; Mickaël *et al.*, 2012; Pingali, 2012; Sun *et al.*, 2012; Whitehorn *et al.*, 2012; Liu *et al.*, 2015).

The problems of achieving food security have become relevant again, because of a predicted alarming growth rate in population and consequently the need to increase global food production by about 70% by 2050 (UN, 2015). One of the solutions offered to solve global food security was transgenic technology: 'The

world has the technology that is either available or well advanced in the research pipeline to feed a population of 10 billion people...' (Borlaug, 2000, p 490); transgenic technology has the potential 'for improving the living conditions of the poor and underprivileged in developing countries' (Potrykus, 2001,' p. 1161). But, after 20 years of implementation and after occupying about 13% of arable land (CBAN, 2015) or about 180 million ha (James, 2015), the technology is still the subject of a wide debate that is not winding down, and the world is divided into opponents and advocates of GMOs.

In order to participate in the global debate, hereafter we analyse some changes in world food production after the commercialization of GMOs, considering the following facts: (i) transgenic technology by itself cannot solve world hunger, which is obvious by observing the previously outlined complexity of food security; (ii) only four crops – soybean, maize, cotton, and canola – are grown in significant quantities, of which the share of soybeans in total transgenic areas is 51.2%, maize about 30%, cotton 13.3%, and canola 4.7%; (iii) there are no commercialized transgenic varieties of wheat, rice, barley, millet, sorghum, cassava, and other key staple crops; and (iv) transgenic fruit and vegetables are grown on 1% of global transgenic areas (Brankov *et al.*, 2016) but we focused only on the production of cereals and oilseeds. Our research made a clear distinction between two periods, before and after the year 1996 encompassing a period of five decades. For the analysis of historical trends in grain production in the period 1958–2012, we used the database of the Earth Policy Institute, while the data on global oilseed crops production (1961–2010) and arable land were acquired from the FAOSTAT database. A database of total global population was obtained from the UN Population Department website. The data on GMO areas were provided by the annual reports of the International Service for the Acquisition of Agri-biotech Applications (ISAAA). For the statistical analysis of the data, we used Microsoft Office Excel and R Studio software. For determining the trends in the observed variables, the gathered data was compared by linear regression methods, using the factor time as an independent variable. The t-test for independent samples was used for testing the significance of differences in grain and oilseed crop production per capita in two periods.

We found that grain production on the global level increased 2.8 times, and the world population 2.4 times in the period between 1958 and 2012. Although both parameters have a linear growth trend (population: Eqn 4.1; grain: Eqn 4.2) the growth rate of cereal production is more pronounced than the population growth rate.

$$y = 78.87x + 2702.6; R^2 = 0.998 \qquad (4.1)$$

$$y = 27.26x + 762.86; R^2 = 0.9774 \qquad (4.2)$$

Per capita world cereal production increased by 16%. The total grain production reached 2.2×10^9 tonnes in 2012 (Fig. 4.12). The t-test for independent samples showed no statistically significant difference in grain production between the periods of 1958–1995 and 1996–2012, i.e. that there are no important differences in grain production per capita in the periods before and after the GMOs introduction into the market (t= –1.481; df=53; p =0.145) (Fig. 4.13).

Quite the opposite phenomenon happens with oilseeds (Fig. 4.14). In the observed period (1961–2010) global production of oil crops increased threefold (Eqn 4.3) from 8 to 25 kg per capita.

$$y = 0.316x + 5.889, R^2 = 0.920 \qquad (4.3)$$

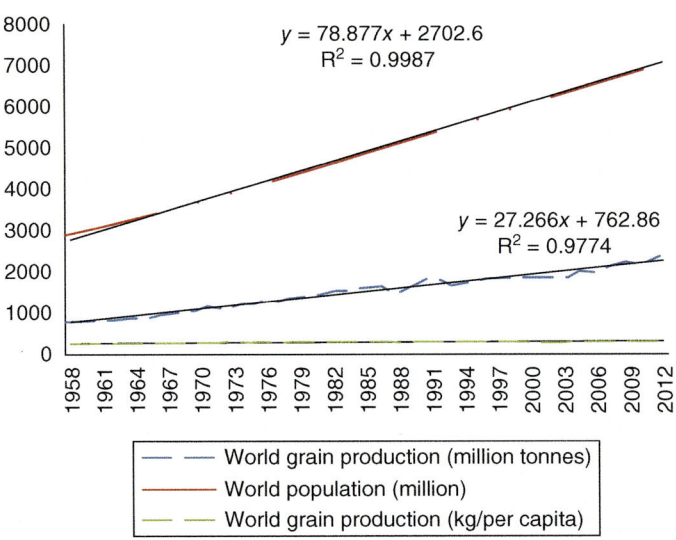

Fig. 4.12. World grain production, 1958–2012. (Adapted from Earth Policy Institute, 2017.)

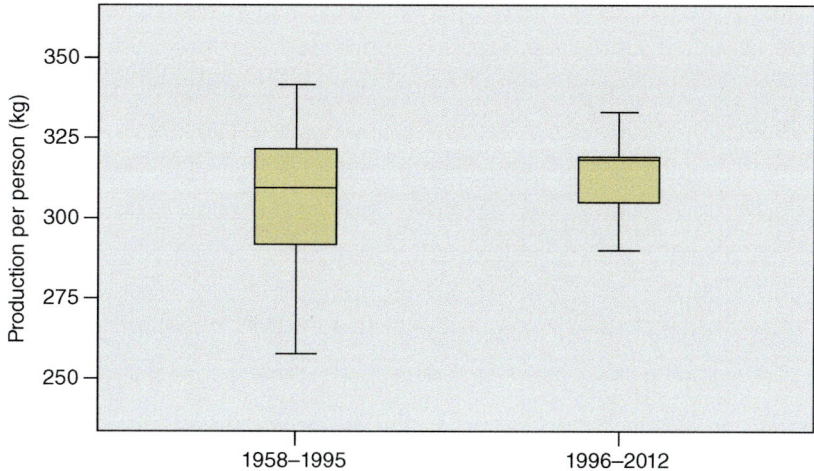

Fig. 4.13. World grain production per capita, the period before and after GMOs. (Authors' calculation.)

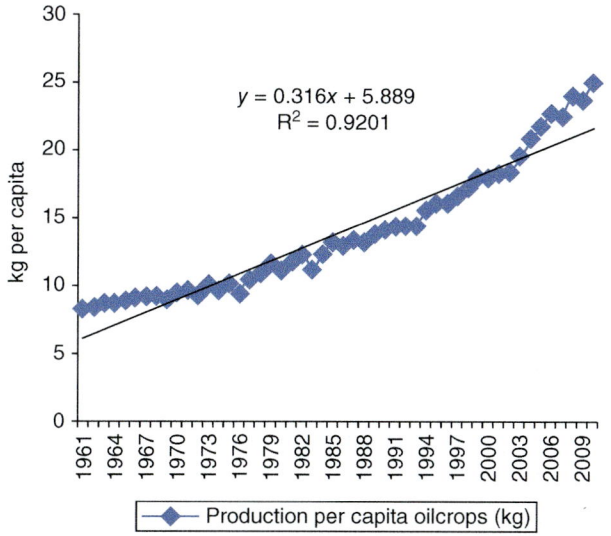

$$y = 0.316x + 5.889$$
$$R^2 = 0.9201$$

Fig. 4.14. World oilseed crop production per capita, 1961–2010. (Adapted from FAOSTAT (http://www.fao.org/faostat/).)

Also, statistically significant differences in oilseed production have been detected between the periods of 1961–1995 and 1996–2010 (t=−11,600; $p < 0,001$) (Fig. 4.15).

These results do not dispute the effects of the Green Revolution. In the period 1950–1995 the world population grew from 2.532 billion to 5.726 billion, and the Green Revolution, with its crop genotype improvements and greater use of different inputs was very successful in meeting the growing demands. In the period 1996–2012 the rate of population

growth slowed (from 5.807 to 7.052 billion) and, consequently, demand reduced as well. The annual growth rate of world demand for cereals declined from 2.5% a year in the 1970s and 1.9% a year in the 1980s to only 1% a year in the 1990s while between 2000 and 2003, growth was almost zero (FAO, 2013). Also, over time growth rate of crop yields has slowed down, so the current rate is insufficient to feed the global population in the coming decades (Ray et al., 2013). Out of all cereals, only transgenic maize is grown in significant

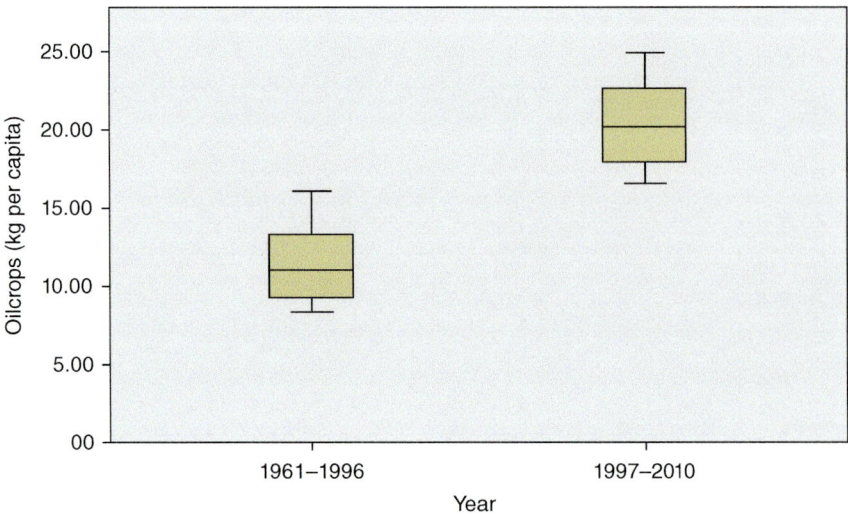

Fig. 4.15. Oilseed crop production per capita, the period before and after GMOs. (Authors' calculation)

quantities (30% of total GM area, with 30% global adoption rate) (James, 2015); there is no commercially important cultivation of other grain types. Considering maize accounts for about 38% of worldwide grain production the absence of effect of transgenic crops on overall grain supply is not so surprising.

In the period 1961–2010 oil crop production increased threefold and significantly after transgenic technology diffusion. In general, this could be seen as part of a dramatic shift in global diet forced by income growth, urbanization and retail globalization. New technologies, inter alia, contributed to the worldwide increase in use of inexpensive vegetable oil (soybean, sunflower, rapeseed, palm and groundnut oil) (Popkin and Gordon-Larsen, 2004). For example, individual intake of the vegetable oils increased between threefold and sixfold throughout the developing world. Furthermore, increase in per capita income, together with population growth and progressive urbanization is highly correlated with the 'livestock revolution' (Pica-Ciamarra and Otte, 2011) as well as consumption of more processed food (Popkin *et al.*, 2012). It is well known that soybean meal is the most important protein source used to feed farm animals, and soybeans are widely used in processed food products. Apart from this, oil crops are a

base for biodiesel production, an industry poised for growth. Out of four major types of oilseeds produced throughout the world (soybeans, rapeseed, cottonseed, and peanuts) three are principal GM crops, accounting on average for about 80% of world oilseed production. Given that soybean accounts for more than 50% of total transgenic area (with an 82% global adoption rate), cotton for 14% (with 68% of adoption), and canola for 5% (with 25% adoption rate) (James, 2014) transgenic technology, without any doubt, had a great impact on the increase of oil crop production per capita. Essentially, the above means that the multinationals directed their research and investments toward the production of those crops that will bring greater profit. The needs of the poorest, whose diet is based on grains, has been completely ignored. Although production of oilseed feedstuffs significantly increased after commercialization of transgenic crops, prices of meat and animal products also increased as already elaborated in Chapter 3. This could be one reason why developing countries are lagging behind developed countries regarding the average supply of protein of animal origin, a food quality indicator. Still, developing countries are 2.4 times less supplied and the LDCs are 5 times less supplied with animal proteins than developed ones (Fig. 4.16).

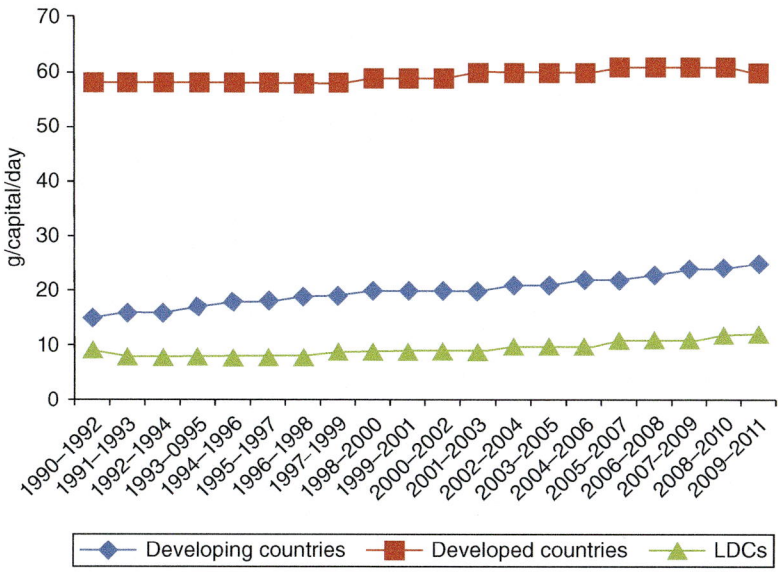

Fig. 4.16. Average supply of protein of animal origin (3-year average). (Adapted from FAOSTAT (http://www.fao.org/faostat/).)

Literature review identifies the linkages between increases in agricultural productivity and poverty reduction. The evidence suggests that there are multiple pathways through which increases in agricultural productivity can reduce poverty, including 'real income changes, employment generation, rural non-farm multiplier effects, and food prices effects' (Schneider and Gugerty, 2011, p. 56). In order to analyse agricultural productivity in GMO producing countries, we compare agriculture value added per worker among countries that have more than 100,000 ha under GMOs (Table 4.2, Fig. 4.17 and Fig. 4.18).

Since 1997 agriculture value added per worker grew with a CAGR of: 4.74% in the USA, 3.84% in China, 2.49% in Mexico, 2.67% in Argentina, 2% in Australia, 3.99% in South Africa, and in Canada 3.5% (data available since 2007, and so not entirely reliable because Canada was an early adopter of GMOs). Uruguay planted transgenic crops for the first time in 2000, and since then agriculture value added per worker grew with a CAGR of 1.96%. India and Columbia, after GMO introduction in 2002, increased this parameter by 2.14% and 2.67% CAGR, respectively. Relatively late GMOs adopters Brazil, Philippines,

and Spain (an early adopter of GMOs but due to an EU moratorium production broke up for some years) since approval of GMOs in 2003 increased agriculture value added per worker with CAGR of 4.86%, 1.53%, and 3.60%, respectively. More countries have since become transgenic-producing, such as Paraguay (2004), Burkina Faso, and Bolivia (2008), Pakistan and Myanmar (2010), Sudan (2012) and after this their agriculture value added per worker grew (or decreased) with a CAGR of 4.05%, −1.82%, 1.07%, 0.77%, 1.04%, and −0.48%, respectively.

As can be seen from the results presented, agriculture value added per worker significantly differs among GMO producing countries. Two producing countries, Burkina Faso and Sudan have experienced decreases in agriculture value added per worker since their introduction of transgenic technology. Another 17 analysed countries recorded growth rate increases, the biggest ones being Brazil and the USA. However, before GMO cultivation Brazil increased its agriculture value added per worker with a CAGR of 5.08% in the period 1997–2002, i.e. even more than after transgenic crop cultivation. Crucially, the gap in agriculture value added per worker has increased over time between

Table 4.2. Global area of transgenic crops in 2015 by country. (From James, 2015.)

Rank	Country	Area (million ha)	Transgenic crops
1	USA	70.9	Maize, soybean, cotton, canola, sugar beet, alfalfa, papaya, squash, potato
2	Brazil	44.2	Soybean, maize, cotton
3	Argentina	24.5	Soybean, maize, cotton
4	India	11.6	Cotton
5	Canada	11.0	Canola, maize, soybean, sugar beet
6	China	3.7	Cotton, papaya, poplar
7	Paraguay	3.6	Soybean, maize, cotton
8	Pakistan	2.9	Cotton
9	South Africa	2.3	Maize, soybean, cotton
10	Uruguay	1.4	Soybean, maize
11	Bolivia	1.1	Soybean
12	Philippines	0.7	Maize
13	Australia	0.7	Cotton, canola
14	Burkina Faso	0.4	Cotton
15	Myanmar	0.3	Cotton
16	Mexico	0.1	Cotton, soybean
17	Spain	0.1	Maize
18	Colombia	0.1	Cotton, maize
19	Sudan	0.1	Cotton
20	Honduras	<0.1	Maize
21	Chile	<0.1	Maize, soybean, canola
22	Portugal	<0.1	Maize
23	Vietnam	<0.1	Maize
24	Czech Republic	<0.1	Maize
25	Slovakia	<0.1	Maize
26	Costa Rica	<0.1	Cotton, soybean
27	Bangladesh	<0.1	Brinjal/aubergine
28	Romania	<0.1	Maize
	Total	179.7	

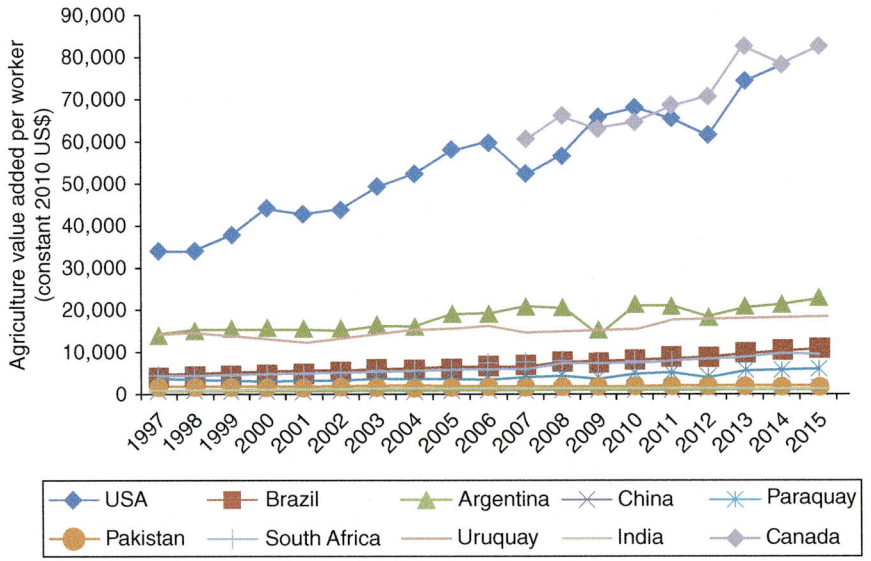

Fig. 4.17. Agriculture value added per worker in top ten GMO producing countries. (Adapted from World Bank (http://databank.worldbank.org/data/home#).)

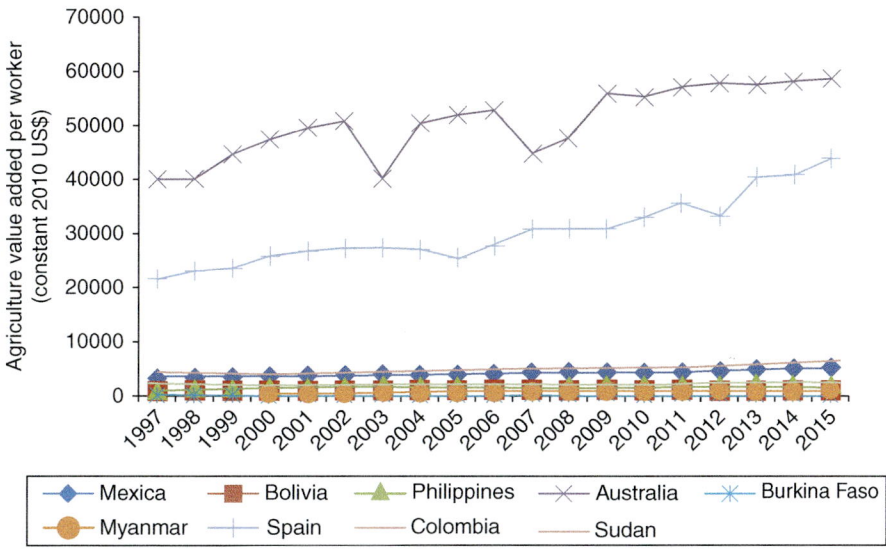

Fig. 4.18. Agriculture value added per worker in selected countries. (Adapted from World Bank (http://databank.worldbank.org/data/home#).)

the USA and other early adopters. The USA agriculture value added per worker was 2.4 times higher than in Argentina in 1997, and 3.6 times in 2014. Similarly, the USA had a 47.4 times higher value than China in 1997; in 2014 the difference increased and reached 56.0 times. Also, the difference between Canada and India grew over time, from 59.6 times in 2007 to 71.9 times in 2015. Roughly speaking, the above may be interpreted as the influence of transgenic technology on deepening the gap between developed and developing countries. Because of this we compared agriculture productivity in high-income, upper middle-income, lower middle-income and low-income countries after transgenic technology entered food production. Once again, we estimated a widening gap between the rich and the poor (Fig. 4.19). Just before the start of the diffusion of technology, in 1995, high-income countries had 43 times higher agriculture value added per worker than low-income countries. Two decades after transgenic crop commercialization the difference reached 77 times. High-income countries had 14.1 times higher agriculture value added per worker than upper middle-income countries and 24.4 times higher than lower middle-income countries in 1995. In 2014, the difference had increased by an additional 3.5 and 7.8 times, respectively.

Since the introduction of transgenic technology, high-income countries have increased agriculture productivity (measured by agriculture value added per worker) by CAGR of 4.35%, upper middle-income countries by 3.21%, lower middle-income countries by 2.35%, and low-income countries by just 1.36%.

The approach presented of measuring agriculture productivity had some limitations. First of all, it is very difficult to assess to what degree transgenic technology influenced agriculture value added per worker. Secondly, the indicator of agriculture value added per worker itself as a measure of agricultural productivity sometimes leads to crude approximations that can differ from the true values due to differences in national statistics methods and classifications, climate conditions or farming techniques. On the other hand, transgenic technology diffusion and changes in agriculture productivity in the observed countries move together in parallel. Thus, the results presented should not be ignored. The gap in agricultural productivity between the rich and the poor has deepened in the last 20 years. No doubt, without reducing the gap it is not possible to solve global hunger. The gap between the rich and the poor cannot be reduced without the support of small farmers. Unlike the

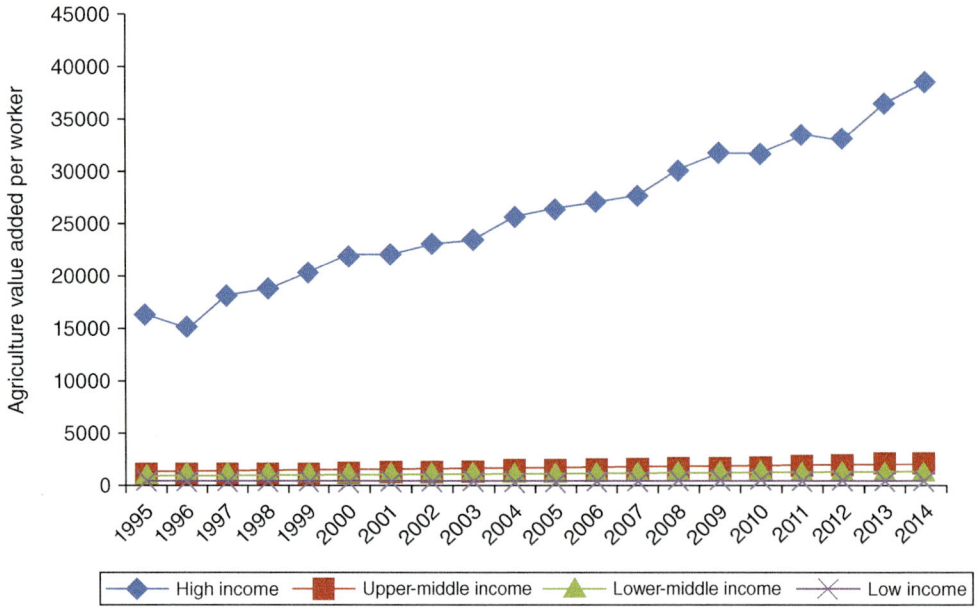

Fig. 4.19. Agriculture value added per worker by income country level. (Adapted from World Bank (https://data.worldbank.org/).)

Green Revolution that was designed to help small farmers in developing countries, but was also suitable for large-scale farms, it seems that the Gene Revolution led by multinational companies will never reach small farmers. The world is still hungry.

References

Borlaug, N.E. (2000) Ending world hunger. The promise of biotechnology and the threat of antiscience zealotry. *Plant Physiology* 124, 487–490.

Brankov, T. and Lovre, K. (2010) Implications of global economic crisis on biotechnology industry. *Economics of Agriculture* [Ekonomika poljoprivrede] 3, 369–376.

Brankov, T. and Lovre, K. (2012) The role of international organizations in the spread of genetically modified food. *Proceedings of Matica Srpska for Social Sciences* [Zbornik Matice srpske za drustvene nauke] 138, 29–38.

Brankov, T. and Lovre, K. (2013) WTO law and genetically modified products. In: Tomic, D., Sevarlic, M., Lovre, K., Zekic, S. (eds) *Challenges for the Global Agricultural Trade Regime After Doha.* Serbian Association of Agricultural Economists and Faculty of Economics, Belgrade and Subotica, Serbia, pp. 99–113.

Brankov, T., Lovre, K., Popovic, B. and Bozovic, V. (2016) Gene Revolution in Agriculture: 20 years of Controversy. In: Jamal, F. (ed.) *Genetic Engineering: An Insight into the Strategies and Spplications.* InTech, Rijeka, Croatia, pp.1–22.

CBAN (2015) *Where in the World are GM Crops and Foods?* Report 1. Canadian Biotechnology Network Action, Ontario, Canada.

Earth Policy Institute (2017) Data Center. Available at: http://www.earth-policy.org/data_center/ (accessed 3 June 2018).

Ericksen, P.J. (2007) Conceptualizing food systems for global environmental change research. *Global Environmental Change* 18, 1–11.

FAO (2013) *FAO Statistical Yearbook 2013. World Food and Agriculture.* FAO, Rome, Italy.

FAO (2015a) *State of the Food Insecurity in the World.* Available at: http://www.fao.org/3/a-i4646e.pdf (accessed 8 June 2018).

FAO (2015b) *The State of Food Insecurity in the World.* FAO, Rome, Italy.

FAO (2017) *An Introduction to the Basic Concepts of Food Security.* FAO, Rome, Italy. Available at: http://www.fao.org/docrep/013/al936e/al936e00.pdf (accessed 3 June 2018).

Holt-Giménez, E., Shattuck, A., Altieri, M., Herren, H. and Gliessman, S. (2012) We already grow enough food for 10 billion people… and still can't end hunger. *Journal of Sustainable Agriculture* 36, 595–598.

IFPRI (2017) Global Hunger Index. Available at: http://ghi.ifpri.org/about/ (accessed 23 June 2018).

Jambor, A. and Babu, S. (2016) Competitiveness of Global Agriculture: Policy Lessons for Food Security. Springer, Cham, Switzerland.

James, C. (2014) *Global Status of Commercialized Biotech/GM Crops: 2014*. ISAAA Brief No. 49, ISAAA, Ithaca, New York.

James, C. (2015) 20th Anniversary (1996 to 2015) of the Global Commercialization of Biotech Crops and Biotech Crop Highlights in 2015. ISAAA Brief No. 51, ISAAA, Ithaca, New York.

Liu, Y., Pan, X. and Li, J. (2015) A 1961–2010 record of fertilizer use, pesticide application and cereal yields: a review. *Agronomy for Sustainable Development* 35, 83–93.

Matson, P.A, Parton, W.J., Power, A.G. and Swift. M.J. (1997) Agricultural intensification and ecosystem properties. *Science* 277, 504–509.

Mickaël, H., Beguin, M., Requier, F., Rollin, O., Odoux, J.F. *et al.* (2012) A common pesticide decreases foraging success and survival in honey bees. *Science* 336, 348–350.

Phipps, R.H. and Park, J.R. (2002) Environmental benefits of genetically modified crops: Global and European perspectives on their ability to reduce pesticide use. *Journal of Animal and Feed Sciences* 11, 1–18.

Pica-Ciamarra, U. and Otte, J. (2011) The 'Livestock Revolution': Rhetoric and reality. *Outlook on Agriculture* 40, 7–19.

Pingali, P. (2012) Green Revolution: Impacts, limits, and the path ahead. *Proceedings of the National Academy of Sciences* 109, 12302–12308.

Popkin, B.M. and Gordon-Larsen, P. (2004) The nutrition transition: worldwide obesity dynamics and their determinants. *International Journal of Obesity* 28, S2–S9.

Popkin, B.M., Adair, L.S. and Ng. S.W. (2012) Global nutrition transition and the pandemic of obesity in developing countries. *Nutrition Reviews* 70, 3–21.

Potrykus, I. (2001) Golden rice and beyond. *Plant Physiology* 125, 1157–1161.

Ray, D.K, Mueller, N.D., West, P.C. and Foley. J.A. (2013) Yield trends are insufficient to double global crop production by 2050. *PLOS ONE* 8, e66428.

Schneider, K. and Gugerty, M.K. (2011). Agricultural productivity and poverty reduction: Linkages and pathways. *The Evans School Review* 1, 56–74.

Sun, B., Linxiu, Z., Linzhang, Y., Fusuo, Z., David, N. and Zhaoliang, Z. (2012) Agricultural non-point source pollution in China: Causes and mitigation measures. *Ambio* 41, 370–379.

Tilman, D. (1998) The greening of the green revolution. *Nature* 396, 211–212.

UN (2015) World Population Prospects: The 2015 Revision, Key Findings and Advance Tables. ESA/P/WP.241. UN Department of Economic and Social Affairs, New York.

Whitehorn, P.R., O'Connor, S., Wackers, F.L. and Goulson, D. (2012) Neonicotinoid pesticide reduces bumble bee colony growth and queen production. *Science* 336, 351–352.

5 Is GMO Farming an Eco-Friendly Choice?

As already mentioned in Chapter 4 the Green Revolution caused numerous negative consequences (Matson *et al.*, 1997), of which the consequences of intensive use of artificial fertilizers and pesticides have been particularly accented (Mickaël *et al.*, 2012; Pingali, 2012; Sun *et al.*, 2012; Whitehorn *et al.*, 2012; Liu *et al.*, 2015a). Transgenic technology had been promoted as a solution for environmental problems. The father of the Green Revolution, Norman E. Borlaug in his paper published in *Plant Physiology* put it:

> Transgenic varieties and hybrids of cotton, maize, and potatoes, containing genes from Bacillus thuringiensis [*Bt*] that effectively control a number of serious insect pests, are now being successfully introduced commercially in the United States [US]. The use of such varieties will greatly reduce the need for insecticides. Considerable progress also has been made in the development of transgenic plants of cotton, maize, oilseed rape, soybeans, sugar beet, and wheat, with tolerance to a number of herbicides. The development of these plants could lead to a reduction in overall herbicide use through more specific interventions and dosages. Not only will this development lower production costs; it also has important environmental advantages ... There are also hopeful signs that we will be able to improve fertilizer-use efficiency by genetically engineering wheat and other crops to have high levels of Glu dehydrogenase
>
> (Borlaug, 2000, p. 487)

In 2000, the USDA's Economic Research Service (ERS) report 'between average pesticide use of adopters and non-adopters, revealed that adopters of GE [genetically engineered] corn [maize], soybeans, and cotton combined, used 7.6 million fewer acre-treatments (2.5 percent) of pesticides than non-adopters in 1997 ... The difference rose to nearly 17 million fewer acre-treatments (4.4 percent) by adopters in 1998' (ERS, 2000). The ERS report about pesticide use in the USA published in 2014 stated: 'The total quantity of pesticides applied to the 21 crops analysed grew from 196 million pounds of pesticide active ingredients in 1960 to 632 million pounds in 1981. Improvements in the types and modes of action of active ingredients applied along with small annual fluctuations resulted in a slight downward trend in pesticide use to 516 million pounds in 2008. These changes were driven by economic factors that determined crop and input prices and were influenced by pest pressures, environmental and weather conditions, crop acreages, agricultural practices (including adoption of genetically engineered crops), access to land-grant extension personnel and crop consultants, the cost-effectiveness of pesticides and other practices in protecting crop yields and quality, technological innovations in pest management systems/practices, and environmental and health regulations' (Fernandez-Cornejo *et al.*, 2014). Besides, there are different opinions about this important issue. An utterly optimistic opinion is that *Bt* technology has the potential to be the 'best environment-friendly strategy' (Kumar *et al.*, 2008, p. 650). Moderate stances express mixed trends in HT crops: 'a decrease in herbicide use in the first few years, but later an increase in it' (Bonny, 2011, p. 1316). An extremely negative opinion stresses: 'There are a growing number of studies that have associated certain pesticides with increased cancer risk and with diseases such as Parkinson's and Alzheimer's, especially among those with high exposures. Unborn children are especially vulnerable because pesticides have also been associated with birth defects. The surge in genetically engineered crops in the past few decades is one of the main drivers of increased pesticide use and chemicals in agriculture. As a matter of fact, genetically engineered crops directly promote an industrial and chemical-intensive model of farming harmful to people, the environment, and wildlife' (Greenpeace, 2017). One of the most cited works states:

> Contrary to often-repeated claims that today's genetically-engineered crops have, and are reducing pesticide use, the spread of glyphosate-resistant weeds in herbicide-resistant weed management systems has brought about substantial increases in the number and volume of herbicides applied. If new genetically engineered forms of corn [maize] and soybeans tolerant of 2,4-D [2,4-Dichlorophenoxyacetic acid] are approved, the volume of 2,4-D sprayed could drive herbicide usage upward by another approximate 50%. The magnitude of increases in herbicide use on herbicide-resistant ha has dwarfed the reduction in insecticide use on *Bt* crops over the past 16 years, and will continue to do so for the foreseeable future ...
> Herbicide-resistant [HT] crop technology has led to a 239 million kilogram (527 million pound) increase in herbicide use in the United States [US] between 1996 and 2011, while *Bt* crops have reduced insecticide applications by 56 million kg (123 million pounds). Overall,

pesticide use increased by an estimated 183 million kg (404 million pounds), or about 7%. (Benbrook, 2012, p. 24).

5.1 Global Pesticide and Fertilizer Use

Considering that Benbrook in his work estimated pesticide use in the USA, not in the whole world, further in the text we explore the potential impact of transgenic technology on agrochemical use on a global scale. Of course, our estimation is rougher. Benbrook used a model 'to quantify by crop and year the impacts of six major transgenic pest-management traits on pesticide use in the US over the 16-year period, 1996–2011: herbicide-resistant corn [maize], soybeans, and cotton; *Bt* maize targeting the European corn borer; *Bt* corn [maize] for corn rootworms; and *Bt* cotton for Lepidopteran insects' (Benbrook, 2012, p. 24). We started with a presumption that two decades after GMOs were introduced, the predicted reduction of specific herbicides used in their production would have an impact on total (including non-GMO) pesticide usage.

As suggested by Liu *et al.* (2015b), since there is no available data on pesticide usage (particularly on the amount of glyphosate, glufosinate or delta-endotoxin based pesticides which are widely used in transgenic plant production), global import value of pesticides, including insecticides, herbicides, fungicides, and bactericides in the period 1996–2012 were downloaded from FAOSTAT (http://www.fao.org/faostat/en/) and used as a proxy. In addition, we analysed fertilizer usage before and after the introduction of GMOs. Data on the import value of nitrogen–phosphorus–potassium (NPK) fertilizers were also obtained from FAOSTAT, while the data on nitrogen, phosphorus, and potassium fertilizer consumption were provided by the Earth Policy Institute database (http://www.earth-policy.org/data_center/). For the purpose of calculating the real import value of pesticides and fertilizers per ha of arable land, the data provided from the aforementioned databases, expressed in current prices, were first transformed into constant prices, i.e. base indices. The global fertilizer and pesticide efficiency of cereal production was defined as annual global cereal production divided by annual global application of fertilizers and pesticides, respectively (Liu *et al.*, 2015b). For the statistical analysis of the data, we used Microsoft Office Excel and R Studio software. For determining the trends in the observed variables, the gathered data were compared by the linear regression method, using the factor time as an independent variable. For determining the relation of areas under transgenic crops on the one hand and the import value of pesticides or fertilizers on the other, we used the Spearman correlation. The starting years of the research differ according to the available data, but on the whole, a period spanning five decades was analyzed. The research made a clear distinction between two periods, before and after the year 1996.

Base indices (1991=100) in constant prices (CPI 2010) of the global pesticide import value have increased approximately 15-fold, from 15.3 to 229.2 in the period from 1961 to 2012 (Fig. 5.1).

Considering the growth in arable land of 9.1%, the calculations have shown that the pesticide import value per ha of arable land has increased 13.7 times. The polynomial trend line of pesticide import is expressed in Eqn 5.1, while the pesticide import value per ha of arable land is expressed in the potential trend line of Eqn 5.2.

$$y = 0.0308x^2 + 1.703x + 19.82,$$
$$R^2 = 0.907 \qquad (5.1)$$

$$y = 0.790x^{0.729}, R^2 = 0.9103 \qquad (5.2)$$

The import value indices for fertilizers varied, especially for phosphorus fertilizers which had their first maximum in 1975, when the value of the index had increased 15 times compared to 1962, and their second maximum in 2008 with an increase of 17 times. Import value indices for nitrogen fertilizers increased approximately fivefold in 2012 compared to 1962, while the same index in potassium fertilizers increased more than 330 times (Fig. 5.2). The import value of NPK fertilizers per ha of arable land increased 7.5 times in the observed period (Fig. 5.3).

Great import variations did not allow for defining a reliable trend line, so we defined a clear linear growth trend in global consumption of fertilizer expressed in quantitative indices (Eqn 5.3) and the polynomial growth trend in fertilizer use per ha of arable land (Eqn 5.4) as follows:

$$y = 2.5127x + 51.09, R^2 = 0.9069 \qquad (5.3)$$

$$y = -0.0268x^2 + 3.1447x + 26.56,$$
$$R^2 = 0.9329 \qquad (5.4)$$

In the period 1962 to 2013, the global use of fertilizers and the use of fertilizers per ha

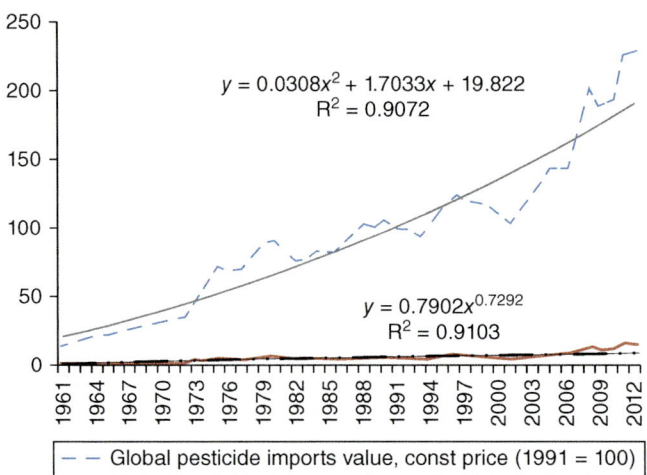

Fig. 5.1. Global pesticide import value. (Adapted from FAOSTAT (http://www.fao.org/faostat/en/).)

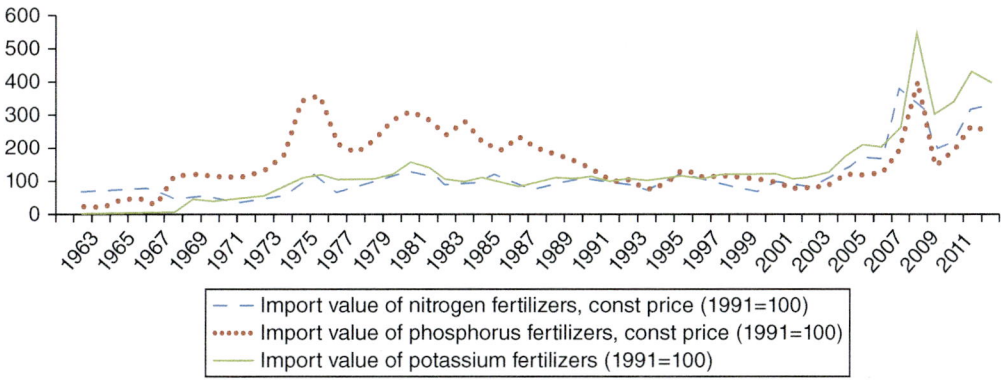

Fig. 5.2. Global nitrogen, phosphorus, and potassium fertilizer import value. (Adapted from FAOSTAT (http://www.fao.org/faostat/en/).)

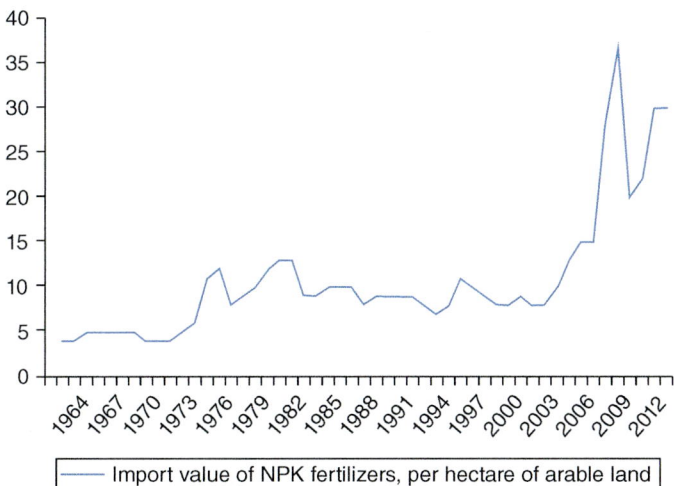

Import value of NPK fertilizers, per hectare of arable land

Fig. 5.3. NPK fertilizer import value. (Adapted from FAOSTAT (http://www.fao.org/faostat/en/).)

of arable land increased approximately five-fold, from 34 to 181 million tonnes, and from 26 to 128 kg/ha, respectively (Fig. 5.4).

A strong, positive correlation between transgenic crop cultivation and the import value of pesticides (rho=0.8627; n=17; p<0.001) (Fig. 5.5), as well as between GMOs cultivation and the import value of NPK fertilizers (rho=0.8259, n=17, p< 0.001) (Fig. 5.6) was found.

5.2 Efficiency of Fertilizer and Pesticide Use

In general, there is a trend of diminishing fertilizer efficiency in the period of 1962–2012,

with efficiency dropping faster in the period before introducing transgenic cultures into production (1962–1995) (Eqn 5.5), while after their introduction (1996–2012) the decrease in efficiency becomes more stable (Eqn 5.6):

$$y = 0.0231x^2 - 1.0842x + 24.651,$$
$$R^2 = 0.959 \tag{5.5}$$

$$y = 0.0091x^2 - 0.2032x + 14.094,$$
$$R^2 = 0.229 \tag{5.6}$$

Starting from about 13–14 tonnes of cereal produced per tonne of fertilizer in the years before transgenic technology introduction, there

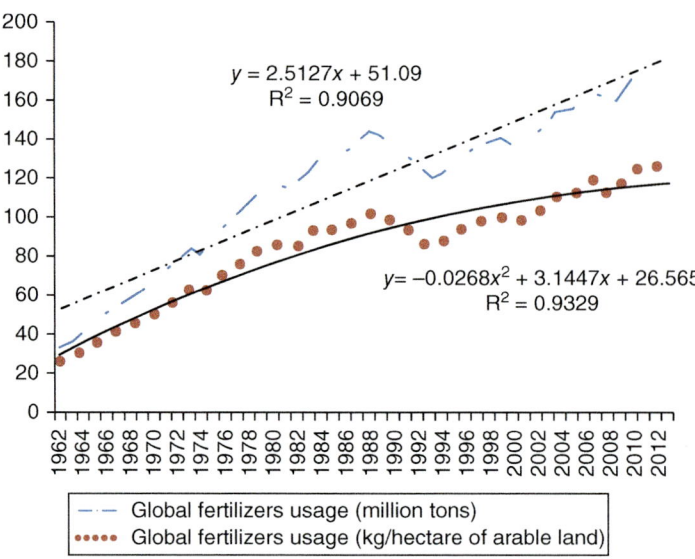

Fig. 5.4. Global consumption of fertilizers. (Adapted from Earth Policy Institute, 2017.)

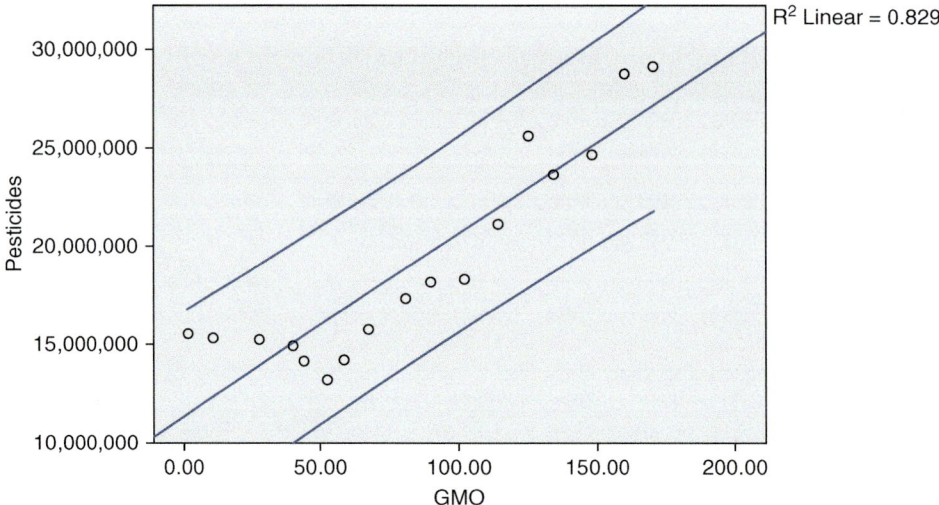

Fig. 5.5. Correlation between the production of transgenic crops and the import value of pesticides (constant price). (Authors' calculation.)

was just a slight decrease to below 13 tonnes observed in the 2010s (Fig. 5.7 and Fig. 5.8).

Similarly, the drop in pesticide efficiency also slowed down in the period after transgenic plant cultivation (Fig. 5.9 and Fig. 5.10).

In the period 1962–1995, pesticide efficiency is expressed in Eqn 5.7, and in the period 1996–2012 with Eqn 5.8:

$$y = 0.0004x^2 - 0.0215x + 0.4056,$$
$$R^2 = 0.897 \tag{5.7}$$

$$y = -0.0004x^2 + 0.003x + 0.0116,$$
$$R^2 = 0.912 \tag{5.8}$$

In nominal prices, pesticide efficiency decreased 40-fold over the past 50 years. In the

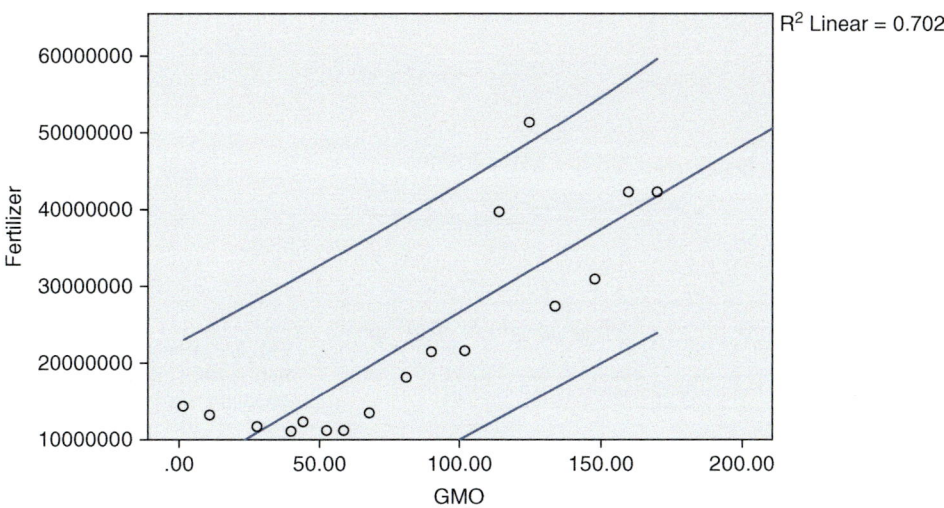

Fig. 5.6. Correlation between the production of transgenic crops and the import value of NPK fertilizers (constant price). (Authors' calculation.)

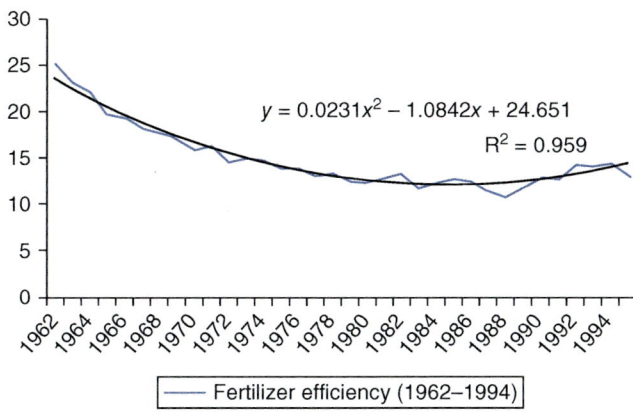

Fig. 5.7. Fertilizer efficiency before transgenic technology adoption. (Authors' calculation.)

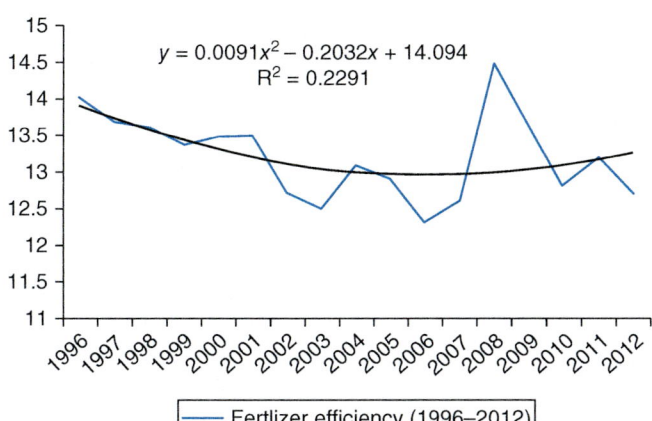

Fig. 5.8. Fertilizer efficiency after transgenic technology adoption. (Authors' calculation.)

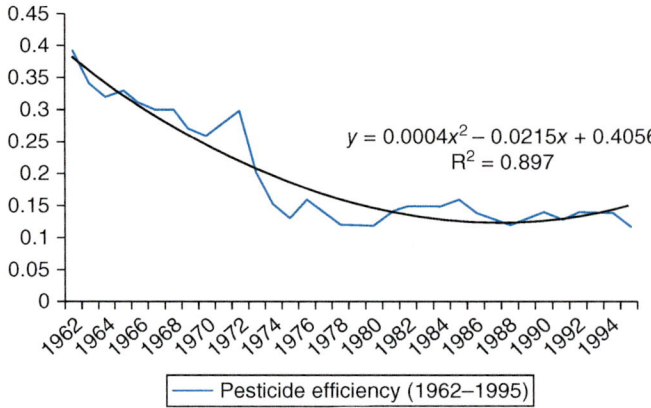

Fig. 5.9. Pesticide efficiency before transgenic technology adoption. (Authors' calculation.)

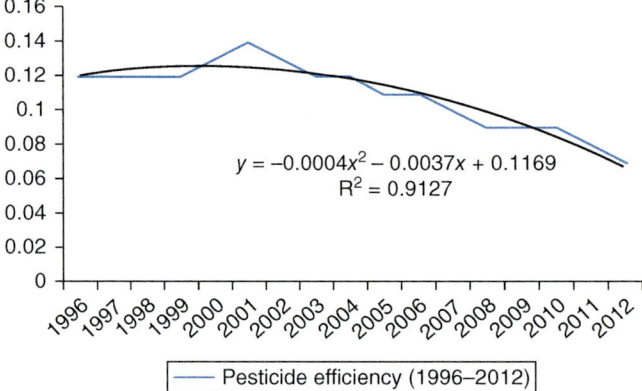

Fig. 5.10. Pesticide efficiency after transgenic technology adoption. (Authors' calculation.)

1960s for each US dollar spent on pesticide imports, 3.3 tonnes of cereal were produced. In the 2010s, for each US dollar spent on pesticide imports 0.1 tonnes of cereals were produced. In real prices pesticide efficiency decreased more than 5.5 times over the past 50 years. For each US dollar spent on pesticide imports, 0.45 tonnes of cereal were produced in the 1960s, and 0.08 tonnes in the 2010s. Just before transgenic technology diffusion, in 1995, for each nominal US dollar of pesticide imports, 1.84 tonnes of cereals was produced or 0.13 tonnes for each real US dollar. This means that after transgenic production pesticide efficiency decreased more than 20 times in nominal prices and about 1.6 times in real prices. In other words, pesticide efficiency in the period before transgenic production (1962–1995) decreased with a CAGR of 3.51%, after transgenic technology introduction (1996–2012) with a CAGR of 3.12%.

Import value of pesticide and NPK fertilizers per ha in constant prices has increased by 13.7 and 7.5 times, respectively in the observed period. If we neglect the limitation of this study (at the global level there is no available data about specific pesticides used in GMO production) these results together with the stated correlation between GMO cultivation and agricultural chemicals, indicates that transgenic technology did not contribute to a decrease in the use of agrichemicals in agricultural production; on the contrary, an increase in their use per ha of arable land was observed. Data show that the drop in pesticide efficiency as well as the drop in fertilizer efficiency became less pronounced after 1996. The slowdown in decreasing rate of pesticide and fertilizer efficiency after 1996 can be explained as a continuation of a trend that has lasted for more than 30 years (Liu *et al.*, 2015a).

5.3 Pesticide Consumption Trends: Comparison Between GMO and non-GMO Producing Countries

Before drawing conclusions, after considering global changes in pesticide consumption before and after transgenic crops production, we compared the usage of herbicides in GMO producing countries, and Ukraine, the world's largest exporter of non-GMO maize and soybean. First of all, using FAO data and speaking in terms of percentage of transgenic area in comparison with total permanent arable land (arable and permanent cropland area), we found that the leading countries are Paraguay, Argentina, Uruguay, and Brazil. Although the USA is the absolute leader in relation to number of ha (70.9 million) dedicated to GMO production, the transgenics area occupies less than half of its permanent arable land. Paraguay, an upper middle-income country is the leading country in percentage devoted to transgenic production, as much as 73.5% of its permanent arable land. Argentina, Uruguay, and Brazil also have more than half of their permanent arable land intended for transgenic production, 60.9%, 58.3%, and 51%, respectively. Uruguay is the only high-income country that has more than half of its arable land intended for GMO production. Excluding the USA, other high-income countries have significantly less percentage of arable land for this purpose: Canada (21.7%), Australia (1.5%), Spain (0.6%), Chile (0.6%), Portugal (0.5%), Czech Republic (0.06%), and Slovak Republic (<0.07%). Out of all the lower middle-income countries, Bolivia has more than one-fifth of its land (23.4%) for this kind of production. Two other lower middle-income countries, Pakistan and India, have a smaller part of land, 9.3%, and 6.8%, intended for transgenic crops despite the fact that they are well known for transgenic cotton production. Low-income countries have only one representative, Burkina Faso that has adopted the technology on about 6.5% of its arable land. All in all, if we add to Paraguay, Argentina, and Brazil, one more country, South Africa, with its 17.8% share of transgenic area in permanent arable land, it seems that upper middle-income

countries are the countries with the highest level of adoption of transgenic technology (Table 5.1).

In the next step, based on FAO data we calculated changes of active ingredient use of pesticides on permanent arable land over time in Uruguay, Paraguay, Argentina, Brazil, Bolivia, and Ukraine. Uruguay, which devoted more land to transgenic production than other high-income countries, was selected for analysis as a representative of high-income countries. We have excluded the USA because of the aforementioned specific Benbrook analysis. Paraguay was selected because of having the largest share under GMOs in total arable land in the group of upper middle-income countries. Brazil and Argentina, also upper middle-income countries, were analysed not only because of high transgenic land share, but also because of their huge importance in international trade. We chose Bolivia as a representative of lower middle-income countries, since it devoted more than three times higher share of land to GMOs than India. Finally, we calculated changes of active ingredient use of pesticides in permanent arable land over time in Ukraine, a lower middle-income non-GMO country, and compared it with the other listed countries.

Uruguay is the tenth largest producer of transgenic crops in the world. It produces transgenic maize and soybean on 1.4 million ha or on 58.3% of its permanent arable land. Since 2000 when it introduced transgenic varieties, it has increased consumption of total pesticides (hereafter the sum of total herbicides, insecticides, and fungicides, and bactericides) by more than three times (Fig. 5.11). Uruguay increased active ingredient use of insecticides by 6.6 times and herbicides by 3.7 times, while it decreased fungicide and bactericide consumption by 20% in the period 2000–2013. The country increased total consumption of pesticides with CAGR of 8.7%, of which insecticides were the most with CAGR of 14.4%. Uruguay spent significantly more on pesticides than Ukraine (Fig. 5.12). Unlike Uruguay, which increased arable land surface by about 70% since the 2000s, Ukraine had no increase in arable land surface; it remained stable at the level of about 33.4 million ha since the 2000s, but it is

Table 5.1. Genetically modified crop producing countries, by percent of transgenic area in total arable land. (Adapted from FAOSTAT (http://www.fao.org/faostat/en), World Bank (2017) and James (2015).)

Rank	Country	Transgenic area (million ha)	% of transgenic area in total permanent arable land	World Bank classification by income (as of March 2017)
1	USA	70.9	45.1	High income
2	Brazil	44.2	51	Upper-middle income
3	Argentina	24.5	60.9	Upper-middle income
4	India	11.6	6.8	Lower-middle income
5	Canada	11.0	21.7	High income
6	China	3.7	3	Upper-middle income
7	Paraguay	3.6	73.5	Upper-middle income
8	Pakistan	2.9	9.3	Lower-middle income
9	South Africa	2.3	17.8	Upper-middle income
10	Uruguay	1.4	58.3	High income
11	Bolivia	1.1	23.4	Lower-middle income
12	Philippines	0.7	6.4	Lower-middle income
13	Australia	0.7	1.5	High income
14	Burkina Faso	0.4	6.5	Low income
15	Myanmar	0.3	2.4	Lower-middle income
16	Mexico	0.1	0.4	Upper-middle income
17	Spain	0.1	0.6	High income
18	Colombia	0.1	2.8	Upper-middle income
19	Sudan	0.1	5	Lower-middle income
20	Honduras	0.03	2	Lower-middle income
21	Chile	0.01	0.6	High income
22	Portugal	0.009	0.5	High income
23	Vietnam	<0.1	<0.9	Lower-middle income
24	Czech Republic	0.002	0.06	High income
25	Slovak Republic	<0.001	<0.07	High income
26	Costa Rica	<0.001	<0.2	Upper-middle income
27	Bangladesh	<0.001	<0.01	Lower-middle income
28	Romania	<0.001	<0.01	Upper-middle income
	Total	179.7		

about 13 times larger than in Uruguay. After coming out of an 8-year recession Ukraine intensified its agriculture production in the 2000s. In the period 2003–2014 it increased total pesticide consumption by about six times, which is slightly above the level in the first years of the 1990s. Insecticide, fungicide, and bactericide consumption per ha of permanent arable land is below the 1990s level, but herbicide consumption is 44% higher than in the 1990s. As a whole, in 2013 Ukraine's total pesticide consumption was 2.41 tonnes of active ingredient per 1000 ha of permanent arable land. In the same year, Uruguay used 7.26 tonnes of active ingredients of pesticides per 1000 ha of permanent arable land. Ukraine used 1.57 tonnes of

active ingredients of herbicides per 1000 ha of permanent arable land or 282.2% less than Uruguay. Also, Ukraine used 95.4% less insecticide than Uruguay, while the consumption of fungicides and bactericides was at the same level in both countries. In 1992, 8 years before transgenic technology production started in Uruguay the situation was quite different. Ukraine used 17% more insecticides, 57% more herbicides, and 62.5% less fungicides and bactericides per 1000 ha of permanent arable land than Uruguay.

Paraguay, an upper middle-income country with the world's largest share of permanent arable land intended for GMO production is the world's fifth largest exporter of soybean. It has grown transgenic soybean for 11 years

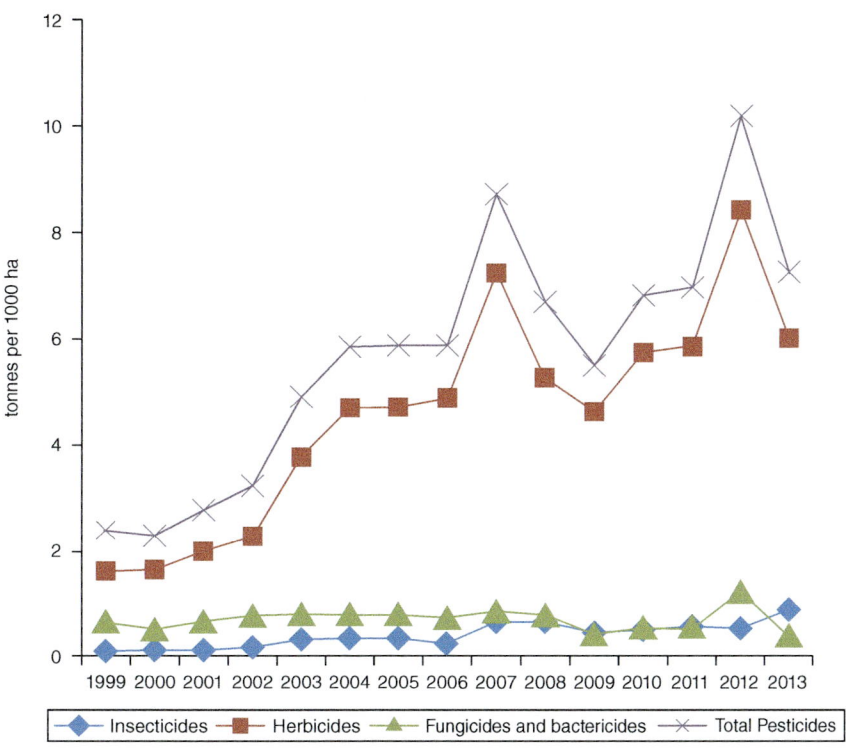

Fig. 5.11. Uruguay: Active ingredient use of pesticides in permanent arable land. (Adapted from FAOSTAT (http://www.fao.org/faostat/en).)

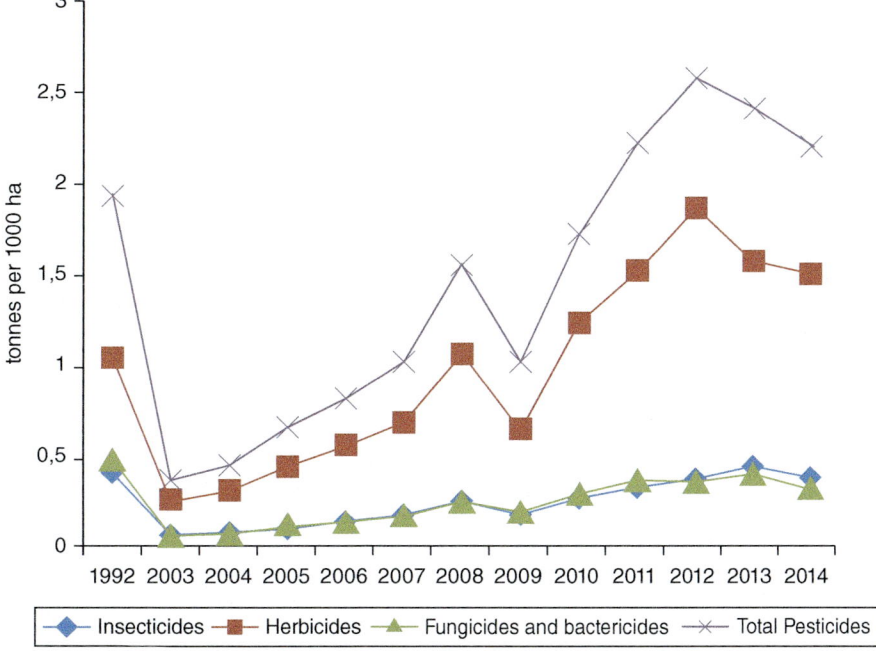

Fig. 5.12. Ukraine: Active ingredient use of pesticides in permanent arable land. (Adapted from FAOSTAT (http://www.fao.org/faostat/en).)

since 2004. In 2015, transgenic soybean occupied 3.3 million ha. In addition, Paraguay planted 305,000 ha of GMO maize and 12,000 ha of transgenic cotton (ISAAA, 2017). Data concerning pesticide consumption in Paraguay can be calculated only for 4 years (Fig. 5.13). Despite this and despite large fluctuations in insecticide and fungicide consumption, significant increases in herbicide usage are visible. Compared with the first available year, 2000, herbicide consumption in the last available year, 2013, had increased by more than five times. This consumption is 50.7% lower than in Uruguay and 60.5% higher than in Ukraine. In the same year Paraguay spent 41.2% and 70.2% more on insecticide active ingredients per 1000 ha of permanent arable land than Uruguay and Ukraine, respectively. Expressed in the same way, the country spent 17% less on fungicides and bactericides than the aforementioned two countries.

As Indexmundi listed on its site (https://www.indexmundi.com/), the 2014 soybean area harvested in Brazil was 32.1 million ha, maize area harvested was about 15.7 million ha, and cotton area harvested was 1.0 million ha. Considering adoption rates of transgenic maize (83%), soybean (93%), and cotton (67%) (GAIN, 2015) as a whole, Brazil's harvested area under GMOs is about 44 million ha. Knowing that Brazil has about 86.6 million ha of permanent arable land (based on FAO data), it means that the harvested area under GMOs occupied about 51% of permanent arable land. Since the 1990s the total pesticide usage has increased by five times, of which insecticide consumption increased by 165.6%, fungicide and bactericide by 253.3%, and herbicides by 622.5% (Fig. 5.14). The highest increase occurred in herbicide consumption from 0.4 tonnes of active ingredient per 1000 ha of permanent arable land in 1990 to 2.9 tonnes of active ingredient per 1000 ha in 2013. However, Brazil officially approved transgenic production for the first time in 2003, thus not all of these changes can be attributable to the

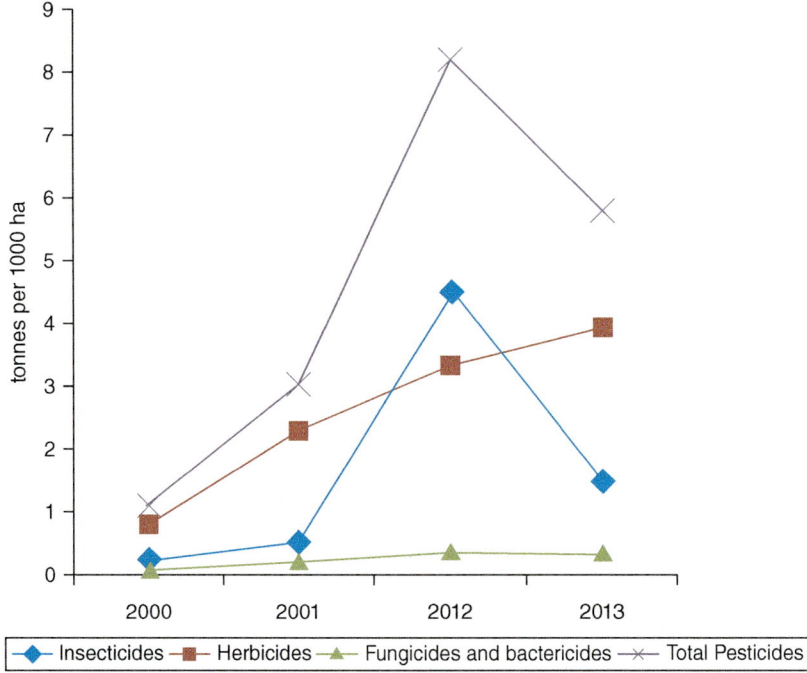

Fig. 5.13. Paraguay: Active ingredient use of pesticides in permanent arable land. (Adapted from FAOSTAT (http://www.fao.org/faostat/en).)

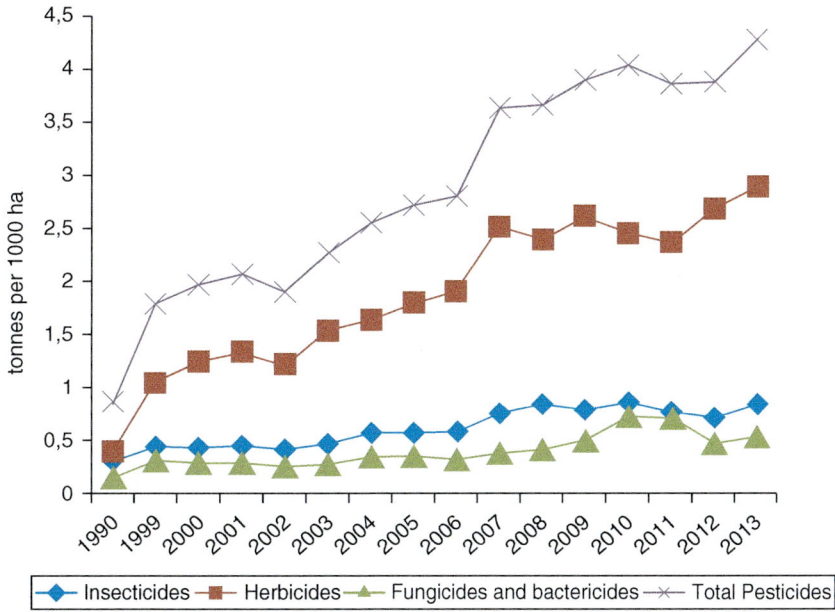

Fig. 5.14. Brazil: Active ingredient use of pesticides in permanent arable land. (Adapted from FAOSTAT (http://www.fao.org/faostat/en).)

transgenic technology. As a whole in the period 2003–2013 Brazil has doubled its consumption of pesticides. Overall consumption of pesticides in Brazil expressed as active ingredient use of pesticides in permanent arable land is lower than in Uruguay (by 70.0%) and Paraguay (by 36.0%) but higher than in Ukraine (by 43.5%). In 2013 Brazil used almost twice as much pesticide as Ukraine, 48.0% more insecticides, 46.0% more herbicides, and 24% more fungicides and bactericides.

Argentina with 14% of the world's total production of transgenic crops continues to be the third largest producer of biotech crops, after the USA and Brazil. Almost all area under soybean and cotton are planted with transgenic seed varieties, while 95% of the area under maize is planted with transgenic varieties (GAIN, 2016). In 2014, total maize area harvested was 3.5 million ha, soybean area harvested was 19.3 million ha, and cotton area harvested was 0.5 million ha. This means that about 23.1 million ha were under transgenic crops. Considering there are 40.2 million ha of permanent

arable land, we can estimate that 57.5% of permanent arable land in Argentina was dedicated to GMOs in 2014. In 2016 the percentage increased to about 60% because of large increases in maize plantation. Both countries, Brazil and Argentina, have significantly increased arable land surfaces since the 1990s, by more than 44%. Brazil has twice as large an area of arable land as Argentina, but the smaller part of it is intended for GMOs. Since there is no data for Argentina in 2013, we compared consumption of pesticides in 2011. In that year Argentina used 31% less insecticides, 55.4% more herbicides, and 42.0% less fungicides and bactericides expressed as active ingredient use of pesticides in permanent arable land than Brazil (Fig. 5.15). After Uruguay, Argentina is the second largest country by pesticides used. Between these two countries the differences are the smallest. Uruguay used 8.7% more herbicides than Argentina, while in insecticide, fungicide, and bactericide consumption there are almost no differences between the two countries. In 2011, Argentina consumed 43% more

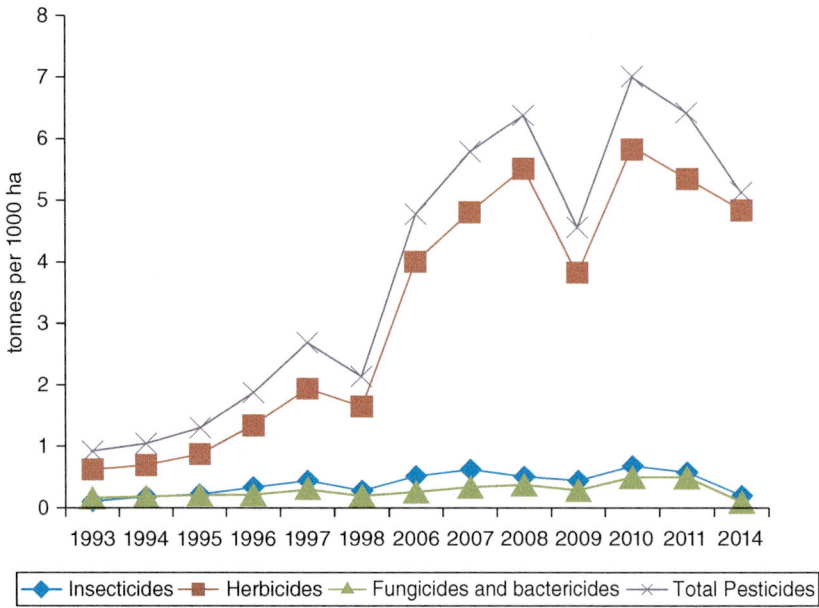

Fig. 5.15. Argentina: Active ingredient use of pesticides in permanent arable land. (Adapted from FAOSTAT (http://www.fao.org/faostat/en).)

insecticides, 71.5% more herbicides, and 26% more fungicides and bactericides than Ukraine.

Finally, analysis of Bolivia, a lower middle-income country, has shown that total pesticide consumption in this country is significantly above the level in all other analysed countries, as much as 23% above the 2013 Uruguay level. Since 1990 Bolivia has seen a more than fourfold increase of pesticide consumption, a tenfold increase of herbicides, fourfold increase in insecticides, and a 16.8% increase in fungicides and bactericides. Bolivia included transgenic plants (soybean) in its agriculture for the first time in 2008. Since that time, the country has increased pesticide usage expressed as active ingredient use of pesticides in permanent arable land as follows: 42% insecticides, 66% herbicides, and 58% fungicides and bactericides (Fig. 5.16). Like all other analysed countries Bolivia used significantly more pesticides than Ukraine, about four times more herbicides and insecticides and about three times more fungicides and bactericides.

The foregoing discussion shows that active ingredient use of pesticides on permanent arable land is significantly higher in GMO producing countries than in Ukraine, a non-GMO country. This conclusion applies to all countries, regardless of their income level. Average GMO producing countries used two to four times more active ingredient use of pesticides on arable land and permanent crops than Ukraine. The highest difference between Ukraine and GMO producing countries exists in herbicide consumption, while the smallest difference is observed in fungicide and bactericide consumption. Knowing that almost all commercialized transgenic crops are designed to be tolerant to certain herbicides or resistant to insecticide, the smaller difference in fungicide and bactericide consumption among countries is quite understandable.

Although GMO crops have never been created to require a smaller amount of fertilizer, it should be noticed that despite oscillations the largest GMO exporting countries increased fertilizer consumption in the period 2002–2014 (Fig. 5.17). This means that after transgenic technology was introduced the countries have continued to intensify their agricultural production.

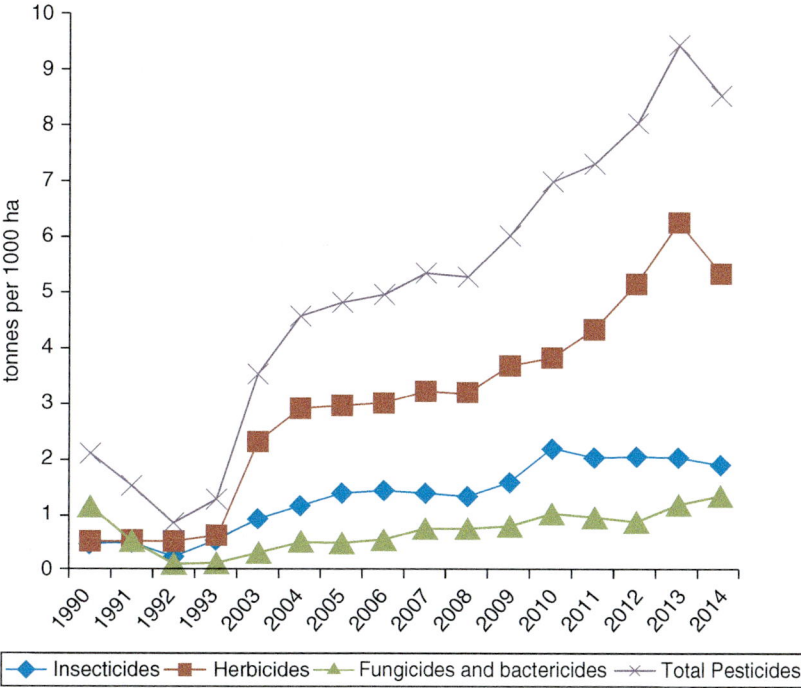

Fig. 5.16. Bolivia: Active ingredient use of pesticides in permanent arable land. (Adapted from FAOSTAT (http://www.fao.org/faostat/en).)

In addition to the conclusion from the previous chapter, that non-transgenic maize and soybean produced in Ukraine can be more price competitive than transgenic maize and cotton from the largest GMO producing countries, we must add that agriculture production in Ukraine seems to be more environmentally friendly than in GMO producing countries. Ukraine used significantly lower amounts of pesticides than the most important GMO producing countries and significantly lower amounts of fertilizers than Brazil and the USA (though somewhat above Argentina in the most recent years). However, at the same time we must take into account some opinions such as that of the Regional Strategic Programmes Coordinator for FAO in Europe and Central Asia, Raimund Jehle, who declared that 'Ukraine is one of the biggest markets for illegal pesticides' (FAO, 2017). Because of this we cannot simply claim that there is such a huge difference between Ukraine and other countries in terms of pesticide consumption in the range

that we get. However, it is certainly unlikely that the black or grey market can be so large as to make up for the differences obtained.

The results presented illustrate a potential crisis in food production. Bearing in mind the expected alarming population growth in the upcoming decades, the slowing growth rate for yields, and increased consumption of chemical substances, it is clear that future market demand for food can only be met by the *adoption of sustainable technological solutions*. The complexity of the problem indicates the need for the emergence and diffusion of a newer 'greener' revolution that would make use of the accumulated knowledge of structure and behaviour of cultivated plants (target selection), ecological processes, disease dynamics, soil and microbial processes and their use in ecologically sustainable food production. Our data indicate that transgenic technology most probably contributed to an increase of agrochemical use in agricultural production, but the technology did not impede the trend of a

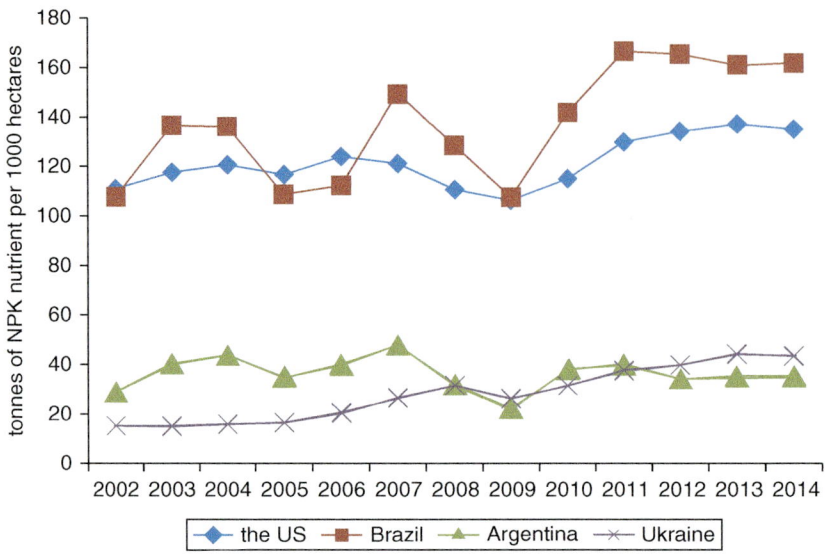

Fig. 5.17. Nutrient use in arable and permanent crop area. (Adapted from FAOSTAT (http://www.fao.org/faostat/en).)

slowdown in pesticide and fertilizer efficiency. Thus, transgenic technology can be interpreted rather as a logical continuation of intensification and industrialization of agricultural production than an environmentally friendly technology.

Acknowledgments

We are grateful to Bozidar Popovic and Vladimir Bozovic, Professors from University of Montenegro, and to Professor Miodrag Dimitrijevic from University of Novi Sad for useful comments on this chapter.

References

Benbrook, C.M. (2012) Impacts of genetically engineered crops on pesticide use in the US – the first sixteen years. *Environmental Sciences Europe* 24, 1–13.

Bonny, S. (2011) Herbicide-tolerant transgenic soybean over 15 years of cultivation: pesticide use, weed resistance, and some economic issues. The case of the USA. *Sustainability* 3, 1302–1322.

Borlaug, N.E. (2000) Ending world hunger. The promise of biotechnology and the threat of anti-science zealotry. *Plant Physiology* 124, 487–490.

Earth Policy Institute (2017) Data Center. Available at: http://www.earth-policy.org/data_center/ (accessed 3 June 2018).

ERS (2000) Genetically engineered crops: Has adoption reduced pesticide use? *Agricultural Outlook* August 2000, 13–17 Available at: https://wayback.archive-it.org/5923/201203 11013300/http://ers.usda.gov/publications/ agoutlook/aug2000/ao273f.pdf (accessed 3 June, 2018).

Fernandez-Cornejo, J., Nehring, R., Osteen, C., Wechsler, S., Martin, A. and Vialou, A. (2014) Pesticide Use in US Agriculture: 21 Selected Crops, 1960–2008. Available at: https://www. researchgate.net/publication/314459368_ Pesticide_Use_in_US_Agriculture_21_Selected_ Crops_1960-2008 (accessed 8 June 2018).

FAO (2017) Interview: Agriculture in Ukraine – what does the future hold? Available at: http://www. fao.org/europe/news/detail-news/en/c/447159/ (accessed 3 June 2018).

GAIN (2015) *Brazil Agricultural Biotechnology Annual*. Available at: https://gain.fas.usda.gov/Recent% 20GAIN%20Publications/Agricultural%20 Biotechnology%20Annual_Brasilia_Brazil_ 7-8-2015.pdf (accessed 3 June 2018).

GAIN (2016) *Argentina Agricultural Biotechnology Annual*. Available at: https://gain.fas.usda.gov/ Recent%20GAIN%20Publications/Agricultural %20Biotechnology%20Annual_Buenos%20 Aires_Argentina_12-27-2016.pdf (accessed 3 June 2018).

Greenpeace (2017) GMOs & Toxic Pesticides. Available at: http://www.greenpeace.org/usa/sustainable-agriculture/issues/gmos/ (accessed 3 June 2018).

ISAAA (2017) Biotech Country: Facts & Trends: Paraguay. Available at: https://www.isaaa.org/resources/publications/biotech_country_facts_and_trends/download/Facts%20and%20Trends%20-%20Paraguay.pdf (accessed 3 June 2018).

James, C. (2015) *20th Anniversary (1996 to 2015) of the Global Commercialization of Biotech Crops and Biotech Crop Highlights in 2015.* ISAAA Brief No. 51. ISAAA, Ithaca, New York.

Kumar, S., Chandra, A. and Pandey, K.C. (2008) *Bacillus thuringiensis (Bt)* transgenic crop: an environment-friendly insect-pest management strategy. *Journal of Environmental Biology* 29, 641–653.

Liu, Y., Pan, X. and Li, J. (2015a) A 1961–2010 record of fertilizer use, pesticide application and cereal yields: a review. *Agronomy for Sustainable Development* 35, 83–93.

Liu, Y., Pan, P. and Li, J. (2015b) Current agricultural practices threaten future global food production. *Journal of Agricultural and Environmental Ethics* 28, 203–216.

Matson, P.A, Parton, W.J., Power, A.G. and Swift, M.J. (1997) Agricultural intensification and ecosystem properties. *Science* 277, 504–509.

Mickaël, H., Beguin, M., Requier, F., Rollin, O., Odoux, J.F. *et al.* (2012) A common pesticide decreases foraging success and survival in honey bees. *Science* 336, 348–350.

Pingali, P. (2012) Green Revolution: Impacts, limits, and the path ahead. *Proceedings of the National Academy of Sciences* 109, 12302–12308.

Sun, B., Linxiu, Z., Linzhang, Y., Fusuo, Z., David, N. and Zhaoliang, Z. (2012) Agricultural non-point source pollution in China: causes and mitigation measures. *Ambio* 41, 370–379.

Whitehorn, P.R., O'Connor, S., Wackers, F.L. and Goulson, D. (2012) Neonicotinoid pesticide reduces bumble bee colony growth and queen production. *Science* 336, 351–352.

6 GMOs: What are We Eating?

Legislative frameworks of different countries and public attitudes towards transgenic crop cultivation have already been discussed in Chapter 2. Our analyses have shown that national GMO politics should be analysed not separately but as an integral part of agricultural policy. Each country accepts or refuses transgenic products depending on its overall agricultural policies. The countries whose main goal is to achieve self-sufficiency have adopted, with various degrees of success, strong regulatory oversights. The countries whose main goal is to expand exports have approved a weak regulatory approach to GMOs. In addition, in Table 5.1 we presented an overview of land area under transgenic crops of 28 transgenic crop producing countries. EU countries, Bangladesh, Costa Rica, Vietnam, Chile, and Mexico dedicated less than 1% of their arable land to transgenic crop production; China, Australia, Myanmar, and Colombia less than 3%; Pakistan, Philippines, Burkina Faso, Sudan, and India less than 10%; Canada, South Africa, and Bolivia about 20%. The top five GMO producing countries by percentage of available land intended for transgenic production are Paraguay, Argentina, Uruguay, Brazil, and the USA. For this purpose Paraguay invested 73.5% of its arable land, Argentina 60.9%, Uruguay 58.3%, Brazil 51% and the USA 45.1%. The absolute leader in relation to the number of ha (70.9 million) dedicated to transgenic crop production is the USA. It is followed by Brazil (44.2 million ha), Argentina (24.5 million ha), India (11.6 million ha), and Canada (11.0 million ha). Bolivia, Uruguay, South Africa, Pakistan, Paraguay, and China produce GMO crops on significant areas: 1.1, 1.4, 2.3, 2.9, 3.6, and 3.7 million ha, respectively. Philippines, Australia, Burkina Faso, Myanmar, Mexico, Spain, Colombia, Sudan, Honduras, Chile, Portugal, Vietnam, the Czech Republic, the Slovak Republic, Costa Rica, Bangladesh, and Romania produce GMO crops in significantly smaller areas than the aforementioned countries. The rest of the world's countries are not involved in transgenic production. However, a commitment by the state not to produce transgenic crops does not mean that citizens of that state do not consume GMOs. Also, if a country produces just a single transgenic crop, for example, maize, it does not mean that that single crop is the only GMO involved in the food chain.

GMOs can enter the food chain through transgenic foodstuff and feedstuff import or by contamination. The case of Russia has already been outlined (Chapter 2). Public attitude has helped the Russian government not just to adopt de jure prohibition of GMO cultivation but also to tolerate transgenic import. On the one hand, Russians have an extremely negative attitude towards GMO cultivation. On the other hand, they are totally indifferent regarding the import of transgenic ingredients. Because of that, in Russia, a country with one of the most restrictive GMO laws in the world (*de jure* prohibition of cultivation) citizens consume meat or milk that comes from livestock fed with GMOs. Moreover, in Russia one can most probably consume ice cream or burgers with GMO ingredients. Such an ambivalent situation in Russia is present because of the country's international obligations and because of its real needs. As a member of the WTO, Russia cannot completely prohibit the import and marketing of GMOs. On the other hand, Russia needs to feed a growing livestock sector and, to meet those needs, the country depends on the import of transgenic feedstuffs. Another option existed for those countries where transgenic production is fully equal to conventional production. The concept of substantial equivalence, developed by the OECD and further elaborated by FAO/WHO 'embodies the concept that if a new food or food component is found to be substantially equivalent to an existing food or food component, it can be treated in the same manner with respect to safety (i.e. the food or food component can be concluded to be as safe as the conventional food or food component)' (FAO/WHO, 1996). Such a situation exists in, among other countries, the USA where it has been estimated that upward of 75% of processed foods on supermarket shelves, from soda to soup, and from crackers to condiments contains GMO ingredients (Center for Food Safety, 2017). The third scenario exists in the countries that have been bastions of defence against GMOs. Such an example is Serbia, also already discussed in Chapter 2. Serbia is not a WTO member and its 2009 legislation strictly prohibits production, commercialization and importation of transgenic crops and products. But, GMOs found two ways to enter the food chain in Serbia as well as other countries with similar transgenic national policies. The first way is through contamination, accidental or intentional. Each year hundreds of ha of illegally planted transgenic crops are discovered. Transgenic soy has been found even in health food stores. To what extent and how the

state treats illegal plantations in Serbia is difficult to answer precisely. In fact, it is difficult to say what the true intentions of the government apparatus are, and whether some of the steps undertaken are only to calm the general public. The second way for GMO ingredients to enter the food chain is through processed food in which GMOs are present in extremely low amounts.

Transgenic technology issues are a matter of high politics. Sometimes the countries do not destroy illegal GMO plantations deliberately, sometimes because they are powerless. All the approaches shown for the penetration of GMOs into the food chain have caused a great expansion of transgenic components in the daily diet.

6.1 GMOs Approved For Use in Food

As of April 2017, at the global level there have been a total of 486 GMO events that have got some kind of approval (Fig. 6.1), of which almost half are maize events. Maize events together with cotton, potato, Argentinian canola, soybean, carnation, and tomato events account for 90% of the total GMO events approved to date. The other 10% are approvals for rice, alfalfa (lucerne), papaya, apple, chicory, sugar beet, sugarcane, melon, poplar, rose, squash, bean, eggplant (aubergine), creeping bentgrass, eucalyptus, flax, petunia, plum, Polish canola, tobacco, and wheat. 439 events have been for direct food use or additives, 353 as direct feed use or additives, and 339 for cultivation (direct or non-domestic use).

Many countries, whether GMO producers or not, have issued permission for the direct use of GMOs in food or as an additive (Table 6.1). This kind of permission encompasses 23 transgenic crop varieties: alfalfa (lucerne) (*Medicago sativa*), apple (*Malus* × *domestica*), Argentine canola (*Brassica napus*), bean (*Phaseolus vulgaris*), chicory (*Cichorium intybus*), cotton (*Gossypium hirsutum* L.), creeping bentgrass (*Agrostis stolonifera*), eggplant (aubergine) (*Solanum melongena*), eucalyptus (*Eucalyptus sp.*), flax (*Linum usitatissimum* L.), maize (*Zea mays* L.), melon (*Cucumis melo*), papaya (*Carica*

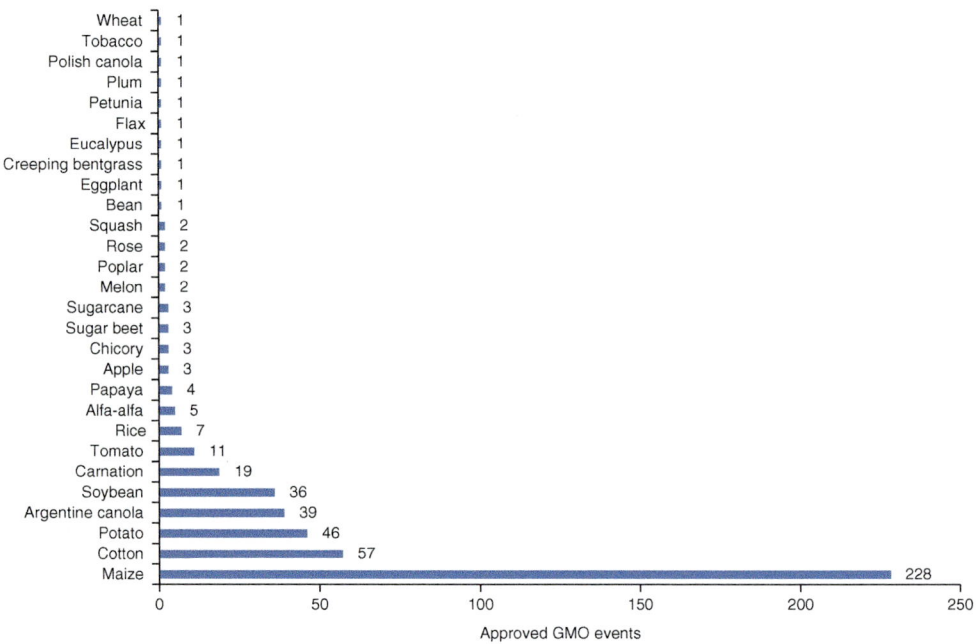

Fig. 6.1. GMO events approved around the world. (Adapted from ISAAA, 2005.)

papaya), plum (*Prunus domestica*), potato (*Solanum tuberosum* L.), rice (*Oryza sativa* L.), soybean (*Glycine max* L.), squash (*Cucurbita pepo*), sugarbeet (*Beta vulgaris*), sugarcane (*Saccharum sp.*), sweet pepper (*Capsicum annum*), tomato (*Lycopersicon esculentum*), and wheat (*Triticum aestivum*).

As can be seen from Table 6.1, Japan, although not a producer of transgenic food crops, is the country that has allowed the largest number of GMO events to be used in food, summarized as follows: 5 alfalfa, 20 Argentine canola, 36 cotton, 198 maize, 1 papaya, 8 potato, 26 soybean, and 3 sugarbeet. Thus, Japan is one of the world's largest per capita importers of foods and feeds that have been produced using transgenic technology. 'The group's list of products covers a wide variety of processed foods, including snacks, ice cream, soda, soy milk, vegetable oil, and ready-to-eat foods' (GAIN, 2013). The country with second most GMO events approved for use as food or additive is the USA. The US approved the most diverse transgenic varieties, in total 183 GMOs: 3 alfalfa, 3 apple, 19 Argentinian canola, 3 chicory, 27 cotton, 1 creeping bentgrass, 1 flax, 41 maize, 2 melon, 2 papaya, 1 plum,

43 potato, 3 rice, 20 soybean, 2 squash, 3 sugarbeet, 8 tomatoes, and 1 wheat. Mexico has issued permission for 153 GMOs including alfalfa (lucerne), Argentinian canola, cotton, maize, potato, rice, soybean, sugarbeet, and tomato. South Korea gave permission for 140 GMOs, Canada for 134, Australia for 105, New Zealand for 96, Philippines and the EU for 88, Brazil for 66, South Africa for 64, Argentina for 61, China for 54, Malaysia for 26, Singapore for 24, Vietnam for 22, Russia for 21, Indonesia for 18, Paraguay for 17, Thailand for 15, Uruguay for 11, India for 5, Switzerland for 4, and Honduras for 3 GMO varieties. Four countries allowed just one single GMO event in food: Bangladesh – aubergine, Bolivia – soybean, Iran – rice, and Panama – maize. Theoretically, it means that any product that contains an ingredient of crops and which have an approved transgenic version, can contain GMOs. Depending on the national laws concerning labelling, people may or may not be aware of and informed about GMO consumption.

However, this does not mean that all events with permission are really available in a particular country's market. For some events the

Table 6.1. Number of GMO events approved for direct food use or additives, by country. (Adapted from ISAAA, 2005.)

	Alfalfa (Lucerne)	Apple	Argentinian canola	Bean	Chicory	Cotton	Creeping bentgrass	Eggplant (Aubergine)	Eucalyptus	Flax	Maize	Total
AR						6					43	61
AU	3		21			22					27	105
BD								1				1
BO												1
BR				1		15			1		38	66
BF						1						1
CA	3	2	18		22					1	36	134
CN			12			8					16	54
CO			12			11					45	71
EU						11					48	88
HN											2	3
IN												5
ID											9	18
IR												1
JP	5		20			36					198	297
MY			1			4					14	26
MX	5		13			30					68	158
NZ	3		14			21					27	96
PA											1	1
PY						3					12	17
PH	2		2			8					52	88
RU											11	21
SG			1			4					10	24
ZA			5			6					41	64
KR	4		13			27					66	140
CH											3	4
TW			6			22					71	124
TH											12	15
US	3	3	19		3	27	1			1	41	183
UY											9	11
VN											14	22

Continued

Table 6.1. Continued.

	Melon	Papaya	Plum	Potato	Rice	Soybean	Sugarcane	Squash	Sugarbeet	Sweet pepper	Tomato	Wheat	Total
AR						12							61
AU				11	1	17			2			1	105
BD													1
BO						1							1
BR						11							66
BF													1
CA		1		24	1	19		1	2		4		134
CN					1	12			1	1	3		54
CO					2	11			1			1	71
EU				1		15			1				88
HN					1								3
IN						5							5
ID						6	3						18
IR					1								1
JP		1		8		26			3				297
MY						7							26
MX				13	1	22			1		5	1	158
NZ				10	1	17			2				96
PA													1
PY						2							17
PH				8	1	14			1				88
RU				2	1	6			1				21
SG						8			1				24
ZA					1	11							64
KR				8		21			1				140
CH						1							4
TW						24			1				124
TH						3							15
US	2	2		43	3	20		2	3		8	1	183
UY			1			2							11
VN						8							22

Note: AR-Argentina; AU-Australia; BD-Bangladesh; BO-Bolivia; BR-Brazil; BF-Burkina Faso; CA-Canada; CN-China; CO- Colombia; EU-European Union; HN-Honduras; IN-India; ID-Indonesia; IR-Iran; JP-Japan; MY-Malaysia; MX-Mexico; NZ-New Zealand; PA-Panama; PY-Paraguay; PH-Philippines; RU-Russia; SG-Singapore; ZA-South Africa; KR-South Korea; CH-Switzerland; TW-Taiwan; TH-Thailand; US-United States of America; UY-Uruguay; VN-Vietnam.

producer has given up production or withdrawn it even though it had a deregulated or a newly regulated status, or the country stopped imports for a certain period of time due to other reasons. Some such cases are, for example, wheat, tomato, and rice. There is no transgenic wheat officially grown anywhere in the world. 'In 2004, Monsanto withdrew requests for government approval of its GM [genetically modified] herbicide-tolerant [HT] Roundup Ready wheat in Canada and the USA because of widespread farmer and consumer opposition in both countries, as well as in major international wheat markets' (CBAN, 2015, p. 20). Delayed-ripening 'Flavr Savr' tomato developed by the company Calgene was available for a short time in some grocery markets in the 1990s. 'By the summer of 1995, Calgene was close to bankruptcy and was bought out by Monsanto in 1996' (CBAN, 2015, p. 19). There are five GMO rice events approved for food use; one of them was developed by Agricultural Biotech Institute Iran, one was created by Huazhong Agricultural University in China, and three were patented by Bayer Crop Science. 'Bt rice, officially released in Iran in 2004, was grown on approximately 4000 ha in 2005 by several hundred farmers who initiated commercialization of biotech rice in Iran and produced supplies of seed for full commercialization in 2006' (ISAAA, 2005). Very soon, though, the government decided to drop GMO commercialization. China gave safety approval for its transgenic rice in 2009, but has never grown it officially due to rising public concern over its safety. The USA officially, at the moment, is not producing Bayer's transgenic rice. The General Court of the EU has annulled the approvals of BASF's Amflora potato, and the company stopped growing it in Europe in 2012. Currently, transgenic potato is grown just in the USA in small areas. Another newly approved crop, an apple from a Canadian biotech company that does not brown even after it has been sliced, has been produced only in the USA since 2016.

6.2 GMO Contaminations

Apart from officially sanctioned transgenic lines, some unapproved GMOs also penetrate the food chain through contamination. According to the data obtained from the GM Contamination Register as of April 2017 the following was recorded throughout the world: 258 cases of food contamination (Table 6.2), 42 cases of feed contamination, 87 cases of seed contamination, 11 cases of volunteer weeds, 4 cases of wild relatives, 4 cases of native landraces, 5 cases of neighbouring crops, 22 cases of feral crops, and 13 uncategorized cases (such as illegal trials). GeneWatch UK and Greenpeace recorded cases of food contamination in 57 countries. Ecuador, Guatemala, Nicaragua, Colombia, and Peru in the first half of the 2000s were exposed to incidents with transgenic maize and soybean through food aid. Failure of postharvest segregation was a route of contamination. European countries most often detected unauthorized rice and rice products from China and the USA, flax in frozen bakery products from Canada, and papaya from Thailand and Hawaii. Most unusual contaminations have occurred in the USA. As a consequence of poor record keeping and animal tracking, transgenic experimental pigs entered the food chain. A laboratory technician at the University of Florida stole three dead experimental GM pigs and had them turned into sausages, which were then eaten by at least nine people at a funeral (GeneWatch UK and Greenpeace, 2017).

It is important to note that the types of incident involving GMOs have changed over time. During the 2000s most incidents were connected with unapproved soybean and maize, while in recent years the majority of incidents have been with rice and rice products. This is important because no transgenic rice has ever been grown commercially (except in Iran for a short time). Regarding the US case 'the source of the contamination is believed to be field trials of herbicide-tolerant rice conducted between the mid-1990s and the early-2000s by Bayer Crop Science (or its precursor companies Aventis CropScience and AgrEvo). Bayer abandoned these trials in 2002. Despite two of their rice varieties, (LLRICE06 and LLRICE62) receiving deregulated status in 2002, none of Bayer's GM rice varieties have ever been placed on the market in the USA. The USDA official report into the incident identified

Table 6.2. Contaminations with GMOs around the world. (From GeneWatch UK and Greenpeace, 2017.)

Australia	Contamination of oilseed rape exports by unapproved transgenic variety; wheat exports bound for Columbia contaminated with transgenic maize.
Austria	Bt63 rice from China; unauthorized flax FP 967; KeFeng6 from China in rice noodles; LLRICE601; genetically modified papaya; linseed contamination with FP 967; Bt63 rice vermicelli; transgenic papaya from Thailand; LLRICE62 from the USA; Bt 63 rice in noodles from China; LLRICE601 rice.
Belgium	Bt176 maize in popcorn from Argentina; contaminated rice noodles from China; food supplement found to contain unauthorized recombinant human intrinsic factor; transgenic rice from the USA; unauthorized genetically modified papaya from Thailand; unauthorized flax (linseed) FP967; unauthorized rice from China.
Bolivia	Food aid contaminated by GMO ingredients (maize and soy from the USA's PL-480 aid programme); StarLink maize – maize intended for animal feed was found in US food aid.
Bulgaria	Unidentified genetically modified material found in soy protein from Brazil; unlabelled transgenic food products on sale (soy and maize in chocolate waffles).
Canada	StarLink maize – a US maize variety intended for animal feed was found in food products and grain elevators; Bayer's experimental LLRICE601 found in Canadian shops.
China	Illegal sale and growing of GM rice; unauthorized GMOs in a shipment of maize from the USA (several times); Heinz baby food containing illegal GM rice; LLRICE601 found in Beijing supermarket; quarantine unapproved transgenic maize from Argentina; unauthorized Bt63 rice found in baby food and other products.
Colombia	Food aid contaminated by transgenic ingredients (from the National Food and Nutrition Program imported from the USA).
Cyprus	Product Riceland Parboiled contained unauthorized transgenic rice LLRICE601; unauthorized Bt63 rice from China in a 100 tonne shipment of rice protein; unauthorized transgenic flax FP 967 (Triffid) in frozen bakery products imported from Canada via Germany (4 cases); KMD1 and KeFeng6 GM in rice noodles imported from China via the UK and Ireland.
Czech Republic	Unauthorized genetically modified papaya; unauthorized genetically modified rice vermicelli from China; LLRICE601 found in long grain rice; popcorn contaminated from transgenic maize grown in Hungary.
Denmark	Unauthorized genetically modified (Bt63) rice vermicelli from China; unauthorized Monsanto's GM maize GA21 found in a brand of Danish tortilla chips 'Kims Zapatas Tortilla Chips'; linseed FP967 cake animal feed imported from Belgium, in nuts and seeds from Switzerland (several cases).
Ecuador	Food aid contaminated by transgenic soy used in US food aid programmes 'Mi Papilla' and 'Mi Colada', for infants and breast-feeding mothers.
Egypt	StarLink and other transgenic contamination was found in maize imported from the USA and Argentina and sold in markets.
Finland	Illegal papaya in fruit smoothies (several cases); unauthorized genetically modified green papaya from Thailand; unauthorized genetically modified organic rice protein powder; unauthorized LL601 rice found in supermarket product Risofino Mexican Style rice meal marketed by Lidl supermarket; Rice LLRICE601 in Uncle Ben's dark rice; Bt63 rice found in rice vermicelli (several cases); linseed FP967 from Canada (several cases) in cereals and bakery products, in food supplements imported from Israel, in crushed linseed from the Russian Federation, via Sweden, in flour mix imported from Germany with raw material from Belgium.
France	Bt63 rice found in vermicelli from China; genetically modified rice from Thailand via Vietnam; Jellyfish gene lamb added to food chain (the lamb belonged to the INRA's animal research unit, UECA, which sells its unmodified animals to a local abattoir but has strictly no right to sell GM animals); contaminated long grain rice from the USA; unauthorized genetically modified rice in a batch of organic rice from the USA; unauthorized genetically modified frozen chocolate cake from China; unauthorized genetically modified green papaya from Thailand; GMOs in rice noodles from China; identified FP967 flax in bakery products, in wholegrain toasted bread; maize line MON88017 found in soy from the USA; unauthorized genetically modified (Pubi-Cry event) long basmati rice from Pakistan and India; unauthorized KeFeng6 in vermicelli from China.

Continued

Table 6.2. Continued.

Germany	Illegal transgenic flax imported from Canada; genetically modified red yeast extract imported from China; Transgenic papaya imported from Hawaii; LLRICE601 found; illegal insect resistant rice *Bt*63; unauthorized rice flour from China; rice crackers from China contain unauthorized rice varieties *Bt*63 and Keng6/KMD1; fresh papaya from Thailand; genetically modified MON88017 from Colombia; unauthorized genetically modified rice *Bt*63; unauthorized papaya from India; illegal transgenic rice imported from the Philippines; unauthorized transgenic basmati rice from Pakistan; unauthorized transgenic linseed from Canada; unauthorized LLRICE601 found on sale to the public; unlabelled GMO soy lecithin on sale imported from Brazil.
Ghana	Rice LL601 found in food supplies in six commercial brands of rice.
Greece	*Bt*63 rice from China; LLRICE601 at border; unauthorized transgenic rice from the USA; unauthorized genetically modified papaya from Thailand; transgenic DNA (CaMV35) in noodles; *Bt*63 rice found in vermicelli from China; unauthorized FP967 brown linseed in bakery products with raw ingredients from Canada; unauthorized DNA fragment identified in rice crackers; unauthorized maize found in 'Golden Griddle Syrup' from the USA; unauthorized MON88017 maize.
Guatemala	Food aid contaminated with StarLink maize; LLRICE601 found on supermarket shelves.
Hawaii	Transgenic papaya trees have contaminated both organic and conventional non-GM papaya on a wide scale.
Hong Kong	Transgenic papaya seedlings distributed to farmers by the Government.
Hungary	Canned meat products found containing transgenic soy in the supermarkets, Lidl and Tesco.
India	Transgenic contamination found in food in two popular products – Pringles Potato Chips (Procter and Gamble) and Isomil Baby food (Abbott Laboratories). Both products were manufactured in and imported from the USA.
Ireland	Unauthorized long grain rice from the USA; unauthorized *Bt*63 rice.
Italy	Chinese dumpling containing illegal transgenic rice; *Bt*63 rice from China; rice from the USA found to be contaminated by GMOs; unauthorized genetically modified popcorn from Argentina; contaminated popcorn from Argentina; illegal transgenic cherry, kiwi and olive trees found at Faculty of Agriculture of the University of Viterbo (field trial); unauthorized transgenic soy and wheat found in Japanese Eel Soup; Flax FP967.
Japan	Transgenic soy found in organic and conventional tofu products; transgenic Thai papayas; StarLink maize; noodles and rice flour from China found to be contaminated with *Bt*63 GM rice.
Korea	Organic soybean milk found with GMOs.
Kuwait	LLRice601 found in imported US rice.
Luxembourg	Unauthorized genetically modified FP967 flax (linseed); unauthorized genetic material in basmati long grain rice; unauthorized transgenic rice from USA.
Malta	Unauthorized long grain rice from the USA.
Mexico	Contaminated rice for sale LLRICE601.
Namibia	Three popular maize products contaminated with GMOs.
New Zealand	Rice product contaminated with unauthorized *Bt*63.
Nicaragua	US food aid contaminated with Monsanto's 'Roundup Ready' maize; LLRICE601.
Norway	*Bt*63 in wild rice mix; contaminated maize from Argentina; unauthorized transgenic rice from the USA; unauthorized genetically modified *Bt*63 rice; unauthorized papaya from Thailand; GM rice vermicelli from China.
Peru	Food aid contaminated by transgenic ingredients – in maize used in the 'Vaso de Leche' (Glass of Milk) programme at the 'La Libertad' district in Lima.
Philippines	Transgenic contamination of white corn (maize); baby food with GMO contamination (Gerber owned by Novartis, Isomil owned by Abbott Ros).
Poland	Unauthorized genetically modified (LL601 rice) rice fusilli from Vietnam; transgenic soybean contamination in food – a soy product sold by the Czech company Santé; unauthorized linseed FP967 imported from Canada; unauthorized GM rice from the US; *Bt*63 rice in noodles from China.

Continued

Table 6.2. Continued.

Romania	Insect resistant maize in popcorn; transgenic Flax FP967; transgenic potatoes grown in unauthorized trials; unauthorized trials with transgenic plums; unlabelled transgenic soy in food products.
Russia	Transgenic maize imported without a licence. The cargo carrier, Blue Zenith, arrived at St. Petersburg harbour on August 16, 1999 carrying 42,000 tonnes of US maize.
Saudi Arabia	Maize line Bt176, Bt11, T25, MON810, StarLink maize, transgenic soy in food product samples.
Sierra Leone	Rice LL601 found in food supplies.
Slovakia	Contaminated rice noodles from China; unauthorized genetically modified LLRICE601; unauthorized linseed FP967 imported from Russia.
Slovenia	Genetically modified rice in noodles from China, unauthorized Bt63 rice; unauthorized transgenic rice from the USA; unauthorized transgenic rice sticks from China; unauthorized genetically modified FP967 flax; illegal transgenic rice from China.
South Africa	Unlabelled baby food products containing genetically modified maize; unlabelled transgenic soy found in bread products.
South Korea	StarLink maize.
Spain	Organic maize contaminated by GMOs; Bt63 in rice sticks from China; unauthorized transgenic maize MIR604; GMO contamination of maize growing in farmers' fields; unauthorized maize MON88017 identified.
Sweden	LLRICE601 found in long grain rice; unapproved GM rice Bt63 contamination; unauthorized Bt63 rice found; unauthorized GM rice from the USA; Bt63 rice in vermicelli from Hong Kong; unauthorized genetically modified flax from Canada; Bt63 rice from China found in noodles; unauthorized KeFeng6 rice from China.
Switzerland	Transgenic rice from China; transgenic pollen in imported honey; illegal Bt63 rice found; Monsanto's maize found in products imported from Argentina; unauthorized flax FP967; unauthorized genetically modified green papaya from Thailand.
Taiwan	Transgenic papaya found in markets.
Thailand	Papaya trees contaminated.
The Netherlands	Contaminated papaya from the USA; unauthorized genetically modified rice from China; maize MON88017 and MIR604 found; unauthorized Bt63 rice found; unauthorized genetically modified maize, MIR604 found; unauthorized GM rice from the USA; unauthorized rice (KeFeng6) from China; honey found to contain transgenic oilseed rape pollen.
Turkey	Transgenic rice contamination from the USA; hazelnut spread found to contain GMOs.
United Arab Emirates	LLRice601 found in imported American rice products.
UK	Contamination of imported soy product; health and organic foods contaminated with GMOs; Bt63 rice from China; tortilla chips found to be contaminated with GMOs; unauthorized flax FP 967 from Canada; unauthorized genetically modified green papayas; unauthorized GMO rice from the USA.
USA	Soybeans destined for human consumption contaminated with stalks of ProdiGene's GM maize producing an animal vaccine; StarLink maize; experimental GM pigs enter the food chain; people eat GM sausage at funeral.
Zambia	GMO contaminated cornflakes.

the field trials as the source of contamination but was unable to decide whether gene flow (cross-pollination) or mechanical mixing was the mechanism responsible for the contamination. The incident had a major impact on US rice exports with US rice being pulled from shelves worldwide. Many countries including the EU, Japan, South Korea and the Philippines imposed a strict certification and testing regime on all rice imports, while Russia and Bulgaria imposed bans on US rice. By contrast, other rice exporting countries have seen an increase in trade. The contamination episode has also affected seed producers; an entire non-GM rice variety Clearfield 131 was banned by US regulators in early 2007 when it was found to be contaminated, costing producer BASF billions of dollars in losses.

Bayer has sought retrospective approval for the contaminating rice varieties. Approval for commercial growing of LL601 was granted in the USA in 2006 and approval for import was granted for LL62 in Canada in 2006 ... Seed companies in China that were found to have sold [genetically engineered] GE rice hybrid seed to farmers operate directly under the university developing GE rice and it has been reported that the key scientist even sat on the board of one of the seed companies. Following the first exposure of the illegal GE rice, more cases of contamination have been revealed involving almost all parts of the food chain. It was found in a wholesale market and unpackaged rice in supermarkets. In 2006 it was found as well in baby food sold in Beijing, Guangzhou and Hong Kong. In late 2006, the GE rice Bt63 was found for the first time outside the People's Republic of China in Europe: 10 cases of GE rice contaminated products were reported by European governments (Austria, France, the UK and Germany), and other cases were found by Greenpeace and Friends of the Earth. The Chinese government took several measures to try to stop the contamination, which included punishing seed companies, confiscating GE seed, destroying GE rice grown in the field and tightening control over the food chain. Despite these efforts and similarly to the StarLink corn (maize) incident in the past, the GE rice has still not been removed from the food chain. In 2007, it was found in 10 imports to Europe (Cyprus, Germany, Greece, Italy, and Sweden). The European Commission has now adopted emergency measures (commencing 15 April 2008) to require compulsory certification for the imports of Chinese rice products that could contain the unauthorized GMO Bt63. Bt63 was also found in a number of products imported to Japan, which had been testing for Bt63 since September 2006 but did not find contaminated products until January 2007' (GeneWatch UK and Greenpeace, 2017).

Recipient countries treat GMO incidents differently. For example, 'in April 2013, the Turkish authorities made several arrests of individuals working for companies importing rice into the country from the US. The arrests followed the seizure of 21–23,000 tonnes of rice believed to be contaminated with illegal GMOs' (GeneWatch UK and Greenpeace, 2017). Some other countries don't react at all. On the basis of Table 6.2 one may incorrectly conclude that most of the contaminations occurred in the EU. This is not the case; most probably a large number of incidents are recorded thanks to proper border control and thanks to one of the highest food safety standards in the world. 'A key tool to ensure the cross-border follow of information and to swiftly react when risks to public health are detected in the food chain is the Rapid Alert System for Food and Feed (RASFF)' (European Commission, 2017). Because of that system, the EU can react quickly, can detain products at the border and recall them from consumers.

6.3 Examples of Processed Final Products with GMO Ingredients

Knowing which transgenic crops are produced worldwide, knowing the producing countries and how GMOs entered the food chain, we can assume which products contain transgenic ingredients. First of all, a country that imports feedstuffs (soybean and corn) from GMO producing countries feeds its livestock with transgenic food. This means that their citizens eat meat and meat products, as well as milk and milk products originating from livestock fed on GMOs. By analysing the FAO detailed trade matrix we can gain an insight into countries that most likely import transgenic crops. The six most important soybean exporting countries – Brazil, the USA, Argentina, Paraguay, Canada, and Uruguay are transgenic producing countries. Considering that, it can be concluded that all their trade partners most probably import transgenic soybean. Of course, this issue cannot be considered narrowly. For instance, Turkey has allowed seven transgenic soybean events to be used as feedstuffs (GAIN, 2016a) and imports significant quantities of soybean from the main GMO producers (Table 6.3). In addition, in 2013 about 17% of Turkey's soybean import was from Ukraine, a non-GMO producing country. Similarly, Egypt also imported

Table 6.3. Trade matrix for soybean, 2013. (Adapted from FAOSTAT (http://www.fao.org/faostat/en).)

Exporter country	Main importing countries
Brazil	Bangladesh, China, Finland, France, Germany, Iran, Israel, Italy, Japan, Malaysia, Mexico, Netherlands, Norway, Portugal, Korea, Romania, Saudi Arabia, Singapore, Spain, Thailand, Turkey, United Arab Emirates
USA	Bangladesh, Barbados, China, Colombia, Costa Rica, Cuba, Egypt, France, Germany, Indonesia, Israel, Italy, Japan, Malaysia, Mexico, Morocco, Netherlands, Panama, Peru, Philippines, Republic of Korea, Russia, Saudi Arabia, Spain, Switzerland, Thailand, Turkey, UK, Venezuela, Vietnam
Argentina	Bangladesh, Bolivia, China, Colombia, Costa Rica, Cuba, Egypt, Indonesia, Iran, Italy, Spain, Syrian Arab Republic, Thailand, Tunisia, Turkey, United Arab Emirates, UK, Venezuela, Vietnam
Paraguay	Bangladesh, Costa Rica, France, Germany, Greece, Indonesia, Israel, Italy, Japan, Malaysia, Netherlands, Panama, Peru, Portugal, Republic of Korea, Russia, Saudi Arabia, Spain, Switzerland, Tunisia, Turkey, United Arab Emirates, UK, Vietnam
Canada	Bangladesh, Belgium, China, Egypt, France, Germany, Indonesia, Ireland, Israel, Italy, Japan, Malaysia, Netherlands, Norway, Republic of Korea, Singapore, Spain, Thailand, Turkey, United Arab Emirates, Vietnam
Uruguay	Bangladesh, China, Egypt, Germany, Greece, Indonesia, Italy, Japan, Netherlands, Russia, Tunisia, Turkey, Vietnam
Ukraine[a]	Bangladesh, China, Finland, France, Germany, Iran, Israel, Italy, Malaysia, Netherlands, Norway, Portugal, Republic of Korea, Romania, Saudi Arabia, Singapore, Spain, Thailand, Turkey, UK, Vietnam, Venezuela

[a]Non-GMO.

about 15% of its soybean needs from Ukraine, and Italy about 27% in the same year. This means that the aforementioned countries' livestock is fed with both non-GMO and GMO soy. However, it is not possible to separate the finished products in relation to feed. In fact, when someone buys meat it is not possible to know whether the animals were fed with GMOs or not. When this finding is transferred to everyday life issues, the answer to the question: 'Does famous Spanish or Italian prosciutto come from animals fed with GMOs?' is: 'Yes, it probably does!'

The situation with maize is less extreme because, in addition to large exporters of GMO maize (the USA, Argentina, Brazil), there are large exporters of non-GMO maize (Ukraine, Russia, Serbia) (Table 6.4). Therefore, much more choice is left to the importing countries. As can be seen from Table 6.4, Serbia exports maize mainly to neighbouring countries, including Montenegro. Montenegro also imports some quantity of non-GMO soybean from Serbia. Considering the non-GMO feed that Montenegro has at its disposal and its protected geographical brand – the prosciutto of Njegusi – the first answer to an everyday

question 'Is cured ham from Njegusi made from pigs fed with non-GMO feed?' will be 'Yes, it probably is!' However, by examining FAO statistics it can be seen that Montenegro imported 19,499 tonnes of pig and pork meat from the Netherlands, Germany, Spain, Belgium, and Austria in the last year with available data (2013). In the same year, indigenous pork production was 2,511 tonnes. This raises a serious suspicion that the protected brand, the prosciutto of Njegusi, is produced largely from imported meat. Thus, the answer to the question 'Is cured ham from Njegusi made from pigs fed with non-GMO feed?' is 'No, it probably isn't!'

Examples of the prosciutto from Spain, Italy, and Montenegro given in this text are just an illustration of how transgenic animal feed has broken almost all barriers. Although a superficial review of transgenic technology issues indicates that consumers in non-GMO producing countries consume food free from GMOs, a deeper consideration suggests something completely different. Just as it is generally agreed that the risks associated with each new GMO require an integrated and stepwise case-by-case approach (FAO, 1999),

Table 6.4. Trade matrix for maize, 2013. (Adapted from FAOSTAT (http://www.fao.org/faostat/en).)

Exporter country	Main importing countries
USA	Canada, China, Colombia, Costa Rica, Cuba, Dominican Republic, Guatemala, Honduras, Jamaica, Japan, Mexico, Panama, Peru, Korea, Saudi Arabia, Venezuela
Brazil	Algeria, Angola, China, Colombia, Costa Rica, Cuba, Dominican Republic, Egypt, Guatemala, Indonesia, Iran, Malaysia, Morocco, Netherlands, Portugal, Korea, Saudi Arabia, Spain, Tunisia, Vietnam, Yemen
Argentina	Algeria, Chile, Colombia, China, Cuba, Dominican Republic, Egypt, Guatemala, Indonesia, Japan, Jordan, Malaysia, Morocco, Peru, Korea, Tunisia, Yemen
Ukraine[a]	Belarus, Belgium, Egypt, France, Germany, Hungary, Ireland, Israel, Italy, Japan, Libya, Lebanon, Netherlands, Portugal, Korea, Spain, Syria, Tunisia, Turkey, UK
Russia[a]	Azerbaijan, Cyprus, Germany, Greece, Iran, Italy, Libya, Lebanon, Korea, Turkey
Serbia[a]	Albania, Austria, Bosnia and Herzegovina, Hungary, Croatia, Germany, Italy, Montenegro, Romania, FRY Macedonia

[a]Non-GMO.

we must add that any particularly sensitive consumers who want to avoid GMOs in their diet must check what they eat on a case-by-case basis. This is especially true for processed foods. In the case of prosciutto it means that a consumer will only be sure to consume a product that comes from animals that are not fed with GMO if the pigs are produced on their own farm, or if they buy finished products with a non-GMO label that guarantees both the production and processing. Almost as a rule, it is understood that such products have to come from small-scale production.

Further examples of final processed products that may contain GMO ingredients given in Table 6.5 confirm our assumption. To be more specific, this includes almost all processed food such as margarine, cookies, vegetable stocks in cartons, breakfast cereals, noodles, soups, sausages, beverages, etc. The table presents only a partial list of products that may contain the most widespread GMOs, soybean and maize. In addition, many processed products may contain transgenic sugar beet and canola (fried snacks, chocolate, mayonnaise).

If we add vitamins and other supplements to the list, the problem becomes even more complex. For example, 'vitamins E (tocopherols) and C are the most common vitamins raising GMO concerns, since E is derived from soy and C from corn [maize]... Another area of concern for non-GMO folks is probiotics (because substrates are often based on corn [maize])... In addition to the excipient

challenges (high risk precursor raw material) there are also GM risks relating to microencapsulation with gelatine and corn [maize] starch' (Daniells, 2013).

6.4 Who Owns Transgenic Foods?

In addition to the use of GMOs in the production of biofuels and other non-food use that was discussed in the previous chapters, the issue of the massive presence of transgenic ingredients in processed food leads directly to the essence of the diffusion of transgenic technology. Merged agriculture and industry during the second food regime (1950s–1970s) spawned a new manufactured diet based on 'fats and sweeteners, supplemented with starches, thickeners, proteins, and synthesized flavors' (Friedmann, 2009, p. 131). US consumers quickly accepted manufactured foods and changed their consumption pattern. Food manufacturing was carried out simultaneously with the spread of supermarkets through which these products were marketed. Changes in America's consumption pattern, embodied in the acceptance of processed food, opened space for the development of giants not only in manufacturing but also in retail industry. America's changes were reflected in the rest of the world, first in Western Europe as international trade caused convergence in food purchasing patterns (Connor, 1994), such that the USA was a precursor of changes in consumption patterns in

Table 6.5. Examples of final products that may contain GMOs. (From Health Canada (2016), Yadav and Supriya (2014) and GAIN (2013). Note: These lists are not complete.)

Food and products that contain or often contain soy	Bean sprouts; breadcrumbs; cereals and crackers; breaded foods; hydrolysed plant protein (HPP); hydrolysed soy protein (HSP) and hydrolysed vegetable protein (HVP); imitation dairy food; infant formula, follow-up formula, nutrition supplements for toddlers and children; meal replacements; meat products with fillers, for example, burgers and prepared ground meat products; Mexican foods, for example, chilli, taco fillings and tamales; miso; nutrition supplements; sauces, for example, soy, shoyu, tamari, teriyaki, Worcestershire; simulated fish and meat products, for example, surimi, imitation bacon bits, vegetarian burgers; stews, for example, in gravies; tempeh; vegetarian dishes; baked goods and baking mixes; beverage mixes, for example, hot chocolate and lemonade; canned tuna and minced hams, for example, seasoned or mixed with other ingredients for flavour; chewing gum; cooking spray, margarine, vegetable shortening and vegetable oil; dressings, gravies and marinades; frozen desserts; lecithin; milled maize; meat products with fillers, for example, pre-prepared hamburger patties, hotdogs and cold cuts; seafood-based products and fish; seasoning and spices; snack foods, for example, soy nuts; soups, broths, soup mixes and stocks; soy pasta; spreads, dips, mayonnaise and peanut butter; thickening agents; mono-diglycerides; monosodium glutamate (MSG) (may contain hydrolysed protein)
Non-food sources of soy	Cosmetics and soaps; craft materials; glycerine; milk substitutes for young animals; pet food; vitamins
Food and products that contain or often contain corn (maize)	Beer (corn (maize) starch, dry milled corn (maize) grits, corn (maize) syrup, dextrose); cake mix and bakery products (corn (maize) starch, dextrose); candies (corn (maize) syrup, corn (maize) starch); carbonated beverages such as Coke (high-fructose corn (maize) syrup HFCS); cookies (corn (maize) starch, corn (maize) flour, dextrose); corn (maize) flaxes; corn (maize) bread and muffins; granola dips (dextrose as a sweetener); instant coffee, tea, soup (maltodextrins); Mars bar and Twix bar (corn (maize) syrup); snack food-corn chips and Doritos (corn (maize) meal and flour); whiskey (corn (maize) major carbohydrate source); yogurt (corn (maize) syrup as a sweetener), processed seafood (corn (maize) oil); ice-cream, chocolate, cakes, frozen foods (corn (maize) starch); candy, cooked beans, jelly, condiments, processed fish (starch syrup); potato chips (hydrolysed protein)
Non-food sources of corn (maize)	Cosmetics (corn cob); paint and varnish; pharmaceuticals such as aspirin (dextrose); antibiotics (corn (maize) syrup, dextrose, corn (maize) starch); rubber (corn (maize) starch)

Western Europe. Market liberalization has given further stimulus to dietary changes worldwide, it 'pushes towards the reduction of trade barriers in the developing world, and the increasing penetration of international corporations into the commerce in each country' (Popkin, 2002, p. 206). Traditional diets with a limited range of staples are being substituted by a diet more composed of livestock products (meat, milk and eggs), vegetable oils, and sugar (Rayner et al., 2006).

'People in developing countries currently consume on average one-third of the meat and one-quarter of the milk products per capita compared to the richer North, but this is changing rapidly. The amount of meat consumed in developing countries over the past

[few years] has grown three times as much as it did in the developed countries. The Livestock Revolution is primarily driven by demand. Poor people everywhere are eating more animal products as their incomes rise above poverty level and as they become urbanized' (Delgado, 2003, p. 3907S). 'Principal vegetable oils include soybean, sunflower, rapeseed, palm, and groundnut oil. With the exception of groundnut oil, global availability of each approximately tripled between 1961 and 1990. While the broader macroeconomic shifts that affected this increase in edible vegetable fat intake, i.e. edible vegetable fat prices, supply, and consumption is unique because it affected rich and poor countries equally, the net impact is relatively much greater on low-income countries' (Popkin and Gordon-Larsen, 2004, p. S3). 'Urbanization has correlated highly in the developing world with access to processed foods higher in sugar... as income per capita and the proportion of the population residing in urban areas increased, so did sugar intake... we can clearly see the pronounced shift in the world's diet toward increased consumption of caloric sweetener and away from higher-fiber foods. Thus, we are increasingly consuming foods that provide energy but few other nutrients' (Popkin and Nielsen, 2003, pp. 1329–1331).

Dietary changes with respect to animal food sources, vegetable oil and sweeteners have touched almost every part of the world. A fairly good example is Serbia, an upper middle-income country in transition. Serbia is a country that is to a high extent self-sufficient in food production, a country with a strong culinary tradition and desire to preserve traditional food products. In a pre-transitional period, the typical diet consisted of a limited range of food items, but had a feature of high-income diets because of a significant share of meat and milk products. After trade liberalization in the 2000s, and supermarket penetration that brought diversification of food items (Lovre and Brankov, 2015), a growing trend has been observed in the consumption of processed products such as processed meat, fruit juices, chocolate and biscuits, and edible oils (Brankov et al., 2017). More importantly, in Serbia as in the rest of the world nutritional transition occurred in parallel with an increase of non-communicable diseases (NCDs) and general deterioration of public health. All in all, 'the most important factor now, when considering food, nutrition and public health, is not nutrients, and is not foods so much as what is done to foodstuffs and the nutrients originally contained in them, before they are purchased and consumed. That is to say, the big issue is food processing – or, to be more precise, the nature, extent and purpose of processing, and what happens to food and to us as a result of processing. Specifically, the public health issue is ultra-processing' (Monteiro, 2010, p. 238).

Going back, a GMO connection with the above can be easily found. The most widely used transgenic crops – soybean and maize – are an integral part of many processed food items. Simply put, the transgenic industry includes itself in a processed food diet and nutritional transition by covering all three changes that have taken place in the diet: livestock products, edible oil, and sugar. The Livestock Revolution and increased meat and milk demand gave wings to the industry of GMO feedstuff (soybean and maize); a huge increase in intake of vegetable oils stimulated production of transgenic soybean, maize, and canola; while the sweetening of the world's diet encouraged production of transgenic maize, soybean, canola, and sugar beet. As more and more industrial applications use certain transgenic crops, so more and more of them will be produced. In such a situation, everything is working in favour of multinationals. The most important factor – patent protection – gives them huge power over the global food system.

Out of 439 events that have got approval for direct food use or as additives, just 2.7%, or 12 events, were developed by universities, institutes or research institutions (Table 6.6). China's public institutions have developed five GMO events. Beijing University developed sweet pepper and tomato resistant to Cucumber mosaic cucumovirus (CMV); Huazhong Agricultural University developed rice with lepidopteran insect resistance and tomato with singular trait delayed ripening/senescence; and the Institute of Microbiology developed CAS tomato with modified product quality. US research institutions developed three GMO events: two varieties of papaya resistant to Papaya ringspot virus (PRSV) (University of Florida, and Cornell University and University of Hawaii) and a plum (resistant to Plum pox virus – PPV, USDA Agricultural Research Service). The Russian Academy of Science (Centre Bioengineering) protected two potato varieties resistant to Coleopteran insects with patents. The Agricultural Biotechnology Research Institute of Iran created a rice resistant to Lepidopteran insects. Canada's University of Saskatchewan developed one GMO event – a flax tolerant to sulfonylurea herbicide.

Every other GMO or the remaining 97.3% of events were developed by private companies. The top six companies according to their share in the total number of developed GMO events are: Monsanto, Syngenta, Bayer

Table 6.6. Owners of GMO events with approval for direct food use or as additives (Adapted from ISAAA, 2005.)

Developer	Total events	Crop	No. of events	Commercial trait
Agricultural Biotechnology Research Institute (Iran)	1	Rice	1	Singular IR
Agritope Inc. (USA)	3	Melon	2	Singular MPQ
		Tomato	1	
BASF	8	Argentine canola	5	Singular MPQ
		Potato	2	Singular MPQ
		Soybean	1	Singular HT
Bayer CropScience (including fully and partly owned companies)	49	Argentine canola	25	Singular HT, stacked HT, stacked HT+PCS
		Cotton	10	Singular HT, stacked HT, stacked IR+HT
		Maize	4	Singular PCS, singular HT
		Rice	3	Singular HT
		Soybean	6	Singular HT, stacked HT
		Sugar beet	1	Singular HT
Bayer CropScience and MS Technologies LLC	1	Soybean	1	Stacked HT
Bayer CropScience and Syngenta	1	Soybean	1	Stacked HT
Beijing University	2	Sweet pepper	1	Singular DR
		Tomato	1	Singular DR
Bejo Zaden BV (Netherlands)	3	Chicory	3	Stacked HT+PCS
Centre Bioengineering, Russian Academy of Science	2	Potato	2	Singular IR
Cornell University and University of Hawaii	1	Papaya	1	Singular DR
DNA Plant Technology Corporation (USA)	1	Tomato	1	Singular MPQ
Dow AgroSciences LLC	33	Cotton	6	Singular IR, stacked IR, stacked HT+IR
		Maize	22	Singular HT, stacked HT, stacked HT+IR, stacked HT+ASR
		Soybean	5	Stacked HT, stacked HT+IR
Dow AgroSciences LLC and DuPont (Pioneer Hi-Breed International Inc.)	9	Maize	9	Stacked HT+IR
DuPont (Pioneer Hi-Breed International Inc.)	40	Argentine canola	3	Singular HT, stacked HT+PCS
		Cotton	1	Singular HT
		Maize	32	Singular PCS, stacked HT, stacked IR, stacked HT+IR, stacked HT+PCS
		Soybean	4	Singular MPQ, stacked HT, stacked HT+MPQ

Continued

Table 6.6. Continued.

Developer	Total events	Crop	No. of events	Commercial trait
EMBRAPA (Brazil)	1	Bean	1	Singular DR
FuturaGene Group	1	Eucalyptus	1	Singular AGY
Genective S.A.	1	Maize	1	Singular HT
Huazhong Agricultural University (China)	2	Rice	1	Singular IR
		Tomato	1	Singular MPQ
Institute of Microbiology, CAS (China)	1	Tomato	1	Singular MPQ
J.R. Simplot Co.	14	Potato	14	Stacked MPQ, stacked DR+MPQ
Maharashtra Hybrid Seed Company (MAHYCO)	1	Eggplant (Aubergine)	1	Singular IR
Monsanto and BASF	1	Maize	1	Stacked HT+IR
Monsanto Company (including fully and partly owned companies)	127	Alfalfa	1	Stacked HT+MPQ
		Argentine canola	6	Singular MPQ, singular HT, stacked HT+PCS
		Cotton	26	Singular HT, singular IR, stacked IR, stacked HT+IR
		Maize	46	Singular HT, singular IR, stacked HT, stacked IR, stacked HT+IR
		Potato	28	Singular IR, stacked IR+DR, stacked HT+IR+DR
		Soybean	15	Singular HT, singular IR, stacked HT+IR, stacked HT+MPQ
		Sugar beet	1	Singular HT
		Tomato	3	Singular IR, singular MPQ
		Wheat	1	Singular HT
Monsanto Company and BASF	1	Maize	1	Singular ASR
Monsanto Company and Dow AgroScience LLC	12	Cotton	2	Stacked HT+IR
		Maize	10	Stacked HT+IR
Monsanto Company and DuPont (Pioneer Hi-Breed International Inc.)	1	Maize	1	Singular IR
Monsanto Company and Forage Genetics International	4	Alfalfa (Lucerne)	4	Singular HT, singular MPQ
Monsanto Company and Scott Seeds	1	Creeping bentgrass	1	Singular HT
Novartis Seed and Monsanto Company	1	Sugar beet	1	Singular HT
Oktanagan Specialty Fruits Incorporated	3	Apple	3	Singular MPQ
PT Perkebunan Nusantara XI (Persero)	3	Sugarcane	3	Singular ASR
Renessen LLC (Netherlands)	1	Maize	1	Singular MPQ
Renessen LLC (Netherlands) and Monsanto Company	1	Maize	1	Stacked IR+MPQ

Continued

Table 6.6. Continued.

Developer	Total events	Crop	No. of events	Commercial trait
Seminis Vegetable Seeds (Canada) and Monsanto Company (Asgrow)	2	Squash	2	Stacked DR
Stine Seed Farm, Inc. (USA)	1	Maize	1	Singular HT
Syngenta	96	Cotton	2	Singular IR
		Maize	94	Singular MPQ, singular IR, stacked IR, stacked HT, stacked IR+HT, stacked HT+MPQ, stacked IR+MPQ, stacked HT+IR+MPQ
Syngenta and Monsanto Company	1	Maize	1	Stacked HT+IR
USDA Agricultural Research Service	1	Plum	1	Singular DR
University of Florida	1	Papaya	1	Singular DR
University of Saskatchewan	1	Flax	1	Singular HT
Verdeca	1	Soybean	1	Singular ASR
Zeneca Plant Science and Petoseed Company	3	Tomato	3	Singular MPQ

AGY – altered growth/yield; ASR – abiotic stress tolerance; DR – disease resistance; HT – herbicide tolerance; IR – insect resistance; MPQ – modified product quality; PCS – pollination control system.

CropScience, Dow Agrosciences LLC, DuPont (Pioneer Hi-Breed International Inc.), and J.R. Simplot Co. By itself, Monsanto developed 28.9% of all GMO varieties approved for food use. Through cooperation with other companies Monsanto was involved in the development of 34.6% of the approved varieties. Syngenta alone developed 21.8% of transgenic varieties, Bayer CropScience 11.2%, DuPont 9.1%, and J.R. Simplot Co. 3.2%.[i] Quite differently from universities and institutes that developed simple GMOs with just one single modified feature mostly directed to disease resistance, multinationals created more complex GMOs. The most widely used transgenic traits are herbicide tolerance and insect resistance. This is understandable because multinationals that have patent protection over transgenic seed are at the same time herbicide producers. Thanks to such a tied sale of seeds and herbicides, multinationals can profit on two bases. In recent years, various combinations of these two properties have been used in the creation of stacked plants or plants in which two or more genes of interest are modified into a single plant. Recent research has shown the existence of a statistically significant correlation between stacked crops and the global market value of biotech crops, and this suggests the likelihood of 'stacked traits obtained through more complex transformations potentially leading to further enrichment of multinational companies' (Brankov et al., 2016, p. 14).

Bearing in mind projections about meal patterns and food choices, industry investments will be even more cost-effective in the future. 'More forward looking towards 2050, world population is expected to reach 9.3 billion, with a growth rate further slowing down. About 70% of the world's population will be urban, compared to 53% today. GDP and per capita income are assumed to increase strongly (2.5 and 1.8-fold resp.). Global demand for agricultural products in its turn is expected to grow at 1.1% per year, down from 2.2% per year in the past four decades. This means global production in 2050 should be 60% higher compared to 2007. Main relative increases in per capita consumption are expected for milk, meat and vegetable oils, while cereals remain stable, both in developed and developing countries' (European Commission, 2015:10). Simply put, a higher share of urban population in the total population and an increase in GDP will certainly increase demand for processed food. The more processed food is sold, the more multinational corporations will profit. Thus, in addition to the conclusion of the previous chapters that transgenic technology can be interpreted as the logical continuation of intensification and industrialization of agricultural production rather than environmentally friendly technology, it should be stressed that transgenic technology should also be interpreted as a loop of the food processing and retail industries. Based on the previous discussion it can be estimated that a small number of people in the world can claim to have never eaten foods containing certain GMOs. Those most likely to not have been in contact with GMOs are members of closed societies, such as members of tribes or residents of poor rural areas that fully produce food for their own needs. Whether or not GMO foods are harmful to health, or whether DNA may survive intact in the digestive tract or not, is not our field of expertise; we will therefore not engage in this segment of the global debate. What should be emphasized is that it is impossible for non-experts and laypeople to know which kinds of processed food contain GMO ingredients. Very often the ingredients are present in such a small amount that it is not subject to labelling even in those countries and regions where labelling is mandatory, such as the EU. However, ultra-processed food is unhealthy, just as Monteiro put it (2010, p. 260): 'Almost all types of ultra-processed products, including those advertised as *light, premium, supplemented, fortified,* or healthy in other ways, are intrinsically unhealthy.'

What solutions are there for avoiding GMOs in food? On the individual level, sensitive consumers should produce all food themselves (which is generally impossible) or buy raw food from some kind of 'green' system, and then prepare meals at home. Certainly, in conditions of urbanization and the growth of income this cannot be fulfilled. So,

as always when it comes to food production and public health a solution is the establishment of effective institutions. 'The swamping of food systems by ultra-processed products can be controlled and prevented only by statutory regulation' (Monteiro, 2010, p. 262). This can be done through labelling of ultra-processed food or additional taxation thereof that will lead to decreasing consumption. Because GMOs broke the majority of barriers by entering the food system either via feedstuffs or processed food, seeds are the last line of defence. States that wish to preserve food sovereignty should establish a proper system of food control. The swamping of seeds by GMOs should be prevented first at state borders. China uses the most sensitive PCR tests in order to detect GMO ingredients in shipments and refuses each container containing more than 0.1% (or less sometimes) of unapproved GMOs. Unlike China, India is flexible in relation to import shipments that may contain unapproved transgenic ingredients mostly 'due to lack of testing facilities at the ports of entry/exit', so 'there has not been any known instance of interception of import consignments containing unapproved events' (GAIN, 2016b). India is not unique, even countries in Europe face similar challenges. Montenegro, an upper middle-income country and a candidate for membership to the EU, harmonized its GMO Law according to WTO rules, but still has no laboratory for GMOs testing.

After preventing intentional or accidental GMO contaminations, the next step could be a vertical integration through production and processing that may result in some kind of 'high-quality' food label. This kind of label, named 'Serbian Quality', was adopted by the government of Serbia in 2017. The label is intended for dairy products, meat, fruits, vegetables, grains, oilseeds, grapes, honey, etc. This system requires complete traceability of production and processing. Labelled products must be produced from the basic raw material originating from the territory of the Republic of Serbia, meaning, inter alia, that it does not contain GMO ingredients. Alternative sourcing for many ingredients is possible. What remains to be seen is how much political will is there to use those sources, how much production costs, and how strong is consumer pressure.

All in all, GMOs have become an integral part of the food system and it is unlikely they will disappear from it in any way. Thus, we must fully agree with Hicks' criticisms about the predominance of the health and safety concerns in relations to GMOs. We argue that:

> ...while the predominance of the health and safety framing may appear democratic, it is actually democratically illegitimate... the predominance of the health and safety framing in the broader public controversy can be explained in part as a strategic response to the fact that this framing is institutionalized in the US regulatory scheme for GM crops... On the one hand, the predominance of this framing appears to be supported by arguments for 'public reason', shared assumptions and conceptual frameworks for decision making in a pluralist society. However... the predominance of this framing leads to the marginalization of food regime concerns, and thus GM opponents who hold these concerns... Feeding GM corn [maize] to lab rats does not tell us, one way or the other, whether certain large agricultural biotechnology companies have too much power over small farmers.
>
> (Hicks, 2017.)

Focusing the debate on health and safety concerns conceals the following: the flabbiness of the state system and its inability to protect the national seed and food industry; usurpation of the food chain by the multinationals; obvious need for removing patent protection; the need for more accurate and reliable methods for the creation of GMOs; the necessity to focus transgenic research towards the needs of the poorest; and the lack of transparent sources of information about GMOs and consumers' right to an informed choice. Thus, in order to calm the global debate on transgenic technology and more importantly, in order to gain benefits for all members of society, social movements should move food regime concerns to the centre of attention.

Note

[i] In this calculation we did not take into account any mergers or takeovers of companies.

References

Brankov, T., Lovre, K., Popovic, B. and Bozovic, V. (2016) Gene revolution in agriculture: 20 years of controversy. In: Jamal, F. (ed.) *Genetic Engineering: An Insight into the Strategies and Applications.* InTech, Rijeka, Croatia, pp.1–22.

Brankov, T., Zec, S., Gafare, C.E., Gregori D. and Lovre, K. (2017) Long term interaction between dietary patterns and disease incidence: Evidence from Serbia. *Archivos Latinoamericanos de Nutrition* 67, Suppl. 1.

CBAN (2015) *Where in the World are GM Crops and Foods?* Report 1, Canadian Biotechnology Network Action, Ontario, Canada. Available at: https://gmoinquiry.ca/wp-content/uploads/2015/03/where-in-the-world-gm-crops-foods.pdf (accessed 4 June 2018).

Center for Food Safety (2017) About Genetically Engineered Food. Available at: http://www.centerforfoodsafety.org/ISSUES/311/GE-FOODS/ABOUT-GE-FOODS# (accessed 4 June 2018).

Connor, J.M. (1994) North America as a precursor of changes in Western European food-purchasing patterns. *European Review of Agricultural Economics* 21, 155–173.

Daniells, S. (2013) Going non-GMO in dietary supplements: The supply community is not there with us yet, say manufacturers. Available at: http://www.nutraingredients-usa.com/Markets/Going-non-GMO-in-dietary-supplements-The-supply-community-is-not-there-with-us-yet-say-manufacturers (accessed 4 June 2018).

Delgado, C.L. (2003) Rising consumption of meat and milk in developing countries has created a new food revolution. *The Journal of Nutrition* 133, 3907S–3910S.

European Commission (2015) World food consumption patterns – trends and drivers. *EU Agricultural Markets Briefs* 6, June 2015. Available at: http://ec.europa.eu/agriculture/sites/agriculture/files/markets-and-prices/market-briefs/pdf/06_en.pdf (accessed 4 June 2018).

European Commission (2017) RASFF – Food and Feed Safety Alerts. Available at: https://ec.europa.eu/food/safety/rasff_en (accessed 4 June 2018).

FAO (1999) GMOs and Human Health. Report of The 23rd Session of the Codex Alimentarius Commission, Rome, 28 June – 3 July. Available at: http://www.fao.org/docrep/003/X9602E/x9602e06.htm (accessed 4 June 2018).

FAO/WHO (1996) Joint FAO/WHO Expert Consultation on Biotechnology and Food Safety. FAO, Rome, Italy.

Friedmann, H. (2009) Feeding the empire: The pathologies of globalized agriculture. *Socialist Register* 41, 124–142.

GAIN (2013) *Japan Agricultural Biotechnology Annual*. Available at: https://gain.fas.usda.gov/Recent%20GAIN%20Publications/Agricultural%20Biotechnology%20Annual_Tokyo_Japan_8-27-2013.pdf (accessed 4 June 2018).

GAIN (2016a) *Turkey Agricultural Biotechnology Annual*. Available at: https://gain.fas.usda.gov/Recent%20GAIN%20Publications/Agricultural%20Biotechnology%20Annual_Ankara_Turkey_12-2-2016.pdf (accessed 4 June 2018).

GAIN (2016b) *India Agriculture Biotechnology Annual*. Available at: https://gain.fas.usda.gov/Recent%20GAIN%20Publications/Agricultural%20Biotechnology%20Annual_New%20Delhi_India_12-12-2016.pdf (accessed 4 June 2018).

GeneWatch UK and Greenpeace (2017) GM Contamination Register. Available at: http://www.gmcontaminationregister.org/index.php?content=re_detail&gw_id=90&reg=cou.1&inc=0&con=1&cof=0&year=0&handle2_page (accessed 4 June 2018).

Health Canada (2016) Soy: A priority food allergen 2016. Available at: http://www.hc-sc.gc.ca/fn-an/alt_formats/pdf/pubs/securit/2016-soy-soja/soy-soja-eng.pdf (accessed 4 June 2018).

Hicks, D.J. (2017) Genetically Modified Crops, Inclusion, and Democracy. *Perspectives on Science* 25, 488–520. Available at: http://philsci-archive.pitt.edu/12584/1/Young.pdf (accessed 4 June 2018).

ISAAA (2005) ISAAA Brief 34–2005: Executive Summary. Available at: https://www.isaaa.org/resources/publications/briefs/34/executivesummary/default.html (accessed 4 June 2018).

Lovre, K. and Brankov, T. (2015) The supermarket revolution in the Balkan countries: The case of Serbia. *Agroeconomia Croatica* 5, 1–10.

Monteiro, C. (2010) The big issue is ultra-processing [Commentary]. *World Nutrition* 1, 237–269.

Popkin, B.M. (2002) The shift in stages of the nutrition transition in the developing world differs from past experiences! *Public Health Nutrition* 5, 205–214.

Popkin, B.M. and Gordon-Larsen, P. (2004) The nutrition transition: worldwide obesity dynamics and their determinants. *International Journal of Obesity* 28, S2–S9.

Popkin, B.M. and Nielsen, S.J. (2003) The sweetening of the world's diet. *Obesity* 11, 1325–1332.

Rayner, G., Hawkes, C., Lang, T. and Bello, W. (2006) Trade liberalization and the diet transition: A public health response. *Health Promotion International* 21, 67–74.

Yadav, V.K. and Supriya, P.S. (2014) Value addition in maize. In: Chaudhary, D.P., Paul, D., Kumar, S. and Singh, S. (eds) *Maize: Nutrition Dynamics and Novel Uses*. Springer, New Delhi, India, pp. 141–152.

Concluding Remarks

From Quintus Septimius Florens Tertullianus (3rd century CE), through Thomas Robert Malthus (18th century), to Paul Erlich (20th century), numerous authors have made their grim predictions of an overpopulated planet and mass hunger, believing in the impossibility of achieving food security for the growing population. However, science, technological advance and innovation have since negated these pessimistic predictions. During the past decades, the food production rate has managed to surpass the population growth rate. The Green Revolution made a great breakthrough in agricultural production by combining high-yield grain cultivars, artificial fertilizers, pesticides, and irrigation. Selected genetic traits have increased the yield, yield stability, and wide-scale adaptability of certain varieties. However, the problems of achieving food security have become relevant again, because of the alarming predicted population growth rate, and the consequent need to increase global food production by about 70% by 2050. One of the solutions offered to solve global food security has been transgenic technology. But, after 20 years of implementation and after occupying about 13% (i.e. c.180 million hectares) of arable land, the scientific debate on the consequences of GMOs has not slowed down.

In order to participate in the global debate, we analysed some of the changes in the world food system following the commercialization of GMOs. We have found that, in the periods before and after the introduction of GMOs into the market, there were no important differences in grain production per capita, but statistically significant differences in oilseed production. Essentially, this means that the multinationals have directed their research and investments toward the production of those crops that will bring greater profit. The needs of the poorest, whose diet is based on grains, have been completely ignored. The research shows the influence of transgenic technology on deepening the gap between the rich and the poor, measured by agriculture value added per worker: the gap has increased over time between the developed and developing GMO-producer countries; the gap has also increased over time between high-income countries and low-income countries, as well as between high-income and middle-income countries.

Although production of oilseed feedstuffs significantly increased after the commercialization of transgenic crops, prices of meat and animal products also increased, as elaborated in Chapter 3. Twenty years after the cultivation of GMOs, the prices of the most important transgenic crops, soybean, and maize, were higher than in 1995. The rise of maize prices and biofuel production, the decline in the stock/use ratio, and the increase of areas under HT and *Bt* maize in the USA, all changed at about the same time. Regardless of whether the cause is transgenic technology or not, there can be no doubt that the USA had significantly increased maize production in the previous period. But, unfortunately this escalation was invested in biofuels production, without any contribution to saving the world from hunger. In any case, transgenic technology, primarily through seed prices, has contributed to rising food prices. Seed costs have increased on a per kilogram produced basis, on a percentage of operating cost basis, and on a percentage of revenue basis. Increases in seed costs per kilogram of soybeans and maize produced, indicate that yield increase did not keep pace with seed cost increase. Thus, it seems reasonable to conclude that the significant increases in maize and soybean seed prices mostly occurred owing to the introduction of transgenic varieties into production. Comparisons between the seed prices of conventional and transgenic maize and soybean varieties lead to a similar conclusion. Average producer prices of non-transgenic maize in the last 20 years in Ukraine (the world's largest exporter of non-transgenic maize) have been lower than the average producer prices of transgenic maize in the three largest exporters: the USA, Brazil, and Argentina. Also, Ukrainian export maize prices have been lower than average world export prices since 2001, and since 2009 they have been continuously significantly lower than Argentinean and Brazilian prices. Although less competitive in soybean than in maize production, since 2008 Ukrainian soybean export prices were lower than Argentinian, Brazilian, and American prices. The example of Ukraine shows that all too often, the story about a lower market price for transgenic soy and maize is purely and simply a myth. A completely different dimension to the problem is the dominance of GMO soy and maize on the world market. Our calculation reveals that just slightly above 2% of soybeans and 20% of maize come from non-transgenic producing countries. Certainly, the participation

of non-GMO soybeans and especially maize in the global market is bigger because transgenic crop producing countries also produce their counterpart – conventional crops. However, GMO dominance over the market is undeniable.

The research reported in this book shows that transgenic technology can be interpreted rather as a logical continuation of the intensification and industrialization of agricultural production than as an environmentally friendly technology. To support this interpretation: GMO producing countries used two to four times more pesticides (measured as active ingredient use of pesticides in permanent arable land) than Ukraine, a non-GMO producing country. The biggest difference between Ukraine and GMO producing countries is related to herbicide consumption, while the smallest difference is observed in fungicide and bactericide consumption. Knowing that almost all commercialized transgenic crops are designed to be tolerant of certain herbicides or resistant to insecticides, the smaller differences in fungicide and bactericide consumption among GMO and non-GMO countries is quite understandable. However, the technology does not impede the trend of a slowing down in the drop of pesticide and fertilizer efficiency. Therefore, if we want to preserve our planet from further disintegration, serious changes in the way GMOs are created should be undertaken. Another solution for obtaining ecologically sustainable food production is a diffusion of some other 'greener' technology that would make use of the accumulated knowledge of the structure and behaviour of cultivated plants (target selection), ecological processes, disease dynamics, and soil and microbial processes.

The outcomes of this study suggest that national GMO politics should be analysed as an integral part of overall agricultural politics. Countries aspiring to achieve self-sufficiency are adopting strong regulatory oversights, while largely export-oriented countries adopt a weak regulatory approach to GMOs. States are trying to protect their own markets from GMOs where small farms cover a large part of agricultural output. If the predominant class of farms is large farms, nations are more open to the new technology.

Taking into account the rapid penetration of transgenic food and feed into US markets, we must consider the following facts: (i) the bulk of US production is concentrated in a small number of commercial farms (5% of farms participating, with 53% in the value of the total production) that are highly specialized (about half produce just one single commodity); (ii) there is a favourable regulatory climate – the FDA considers most GM crops as *substantially equivalent* to non-GM crops and labelling is only mandatory if a particular GM food product is no longer substantially equivalent to the corresponding conventional food in terms of composition, nutrition or safety; (iii) patent protection for new seed varieties has attracted large chemical, oil, and processing corporations (such as Dow, Dupont, and Monsanto) into acquiring many of the independent seed companies and to fund substantial R&D; (iv) the country relies heavily on export markets to sustain prices and revenues, since the productivity of agriculture is growing faster than domestic demand; and (v) public support for all kinds of transgenic technology has remained at rather high levels over time, mostly owing to an individualistic culture and the weak influence of NGOs. Contrary to the USA, the EU has possibly the most stringent GMO regulations in the world. It prescribes GMO safety, GMO thresholds, GMO labelling, GMO detection, and coexistence. Member states could provisionally prohibit or restrict the use of a GMO on their territory on the basis of environmental or agricultural policy objectives, town and country planning, land use, socio-economic impacts, coexistence and public policy. The EU fixes a threshold for the adventitious, or accidental, presence of GM material in non-GM food or feed sources of 0.9%, and this only applies to GMOs that have an EU authorization. Above this threshold, all foods should be labelled as follows: '*This product contains genetically modified organisms [or the names of the organisms].*' GMO cultivation in the EU is very limited (only four countries planting a single crop), but as a protein deficient area the EU imports large quantities of feeds. Such an EU policy is guided by: (i) the desire of the community to maintain a self-sufficiency in food production which has been gained by a highly expensive and protectionist CAP system; (ii) the necessity of importing animal feed; and (iii) the export of its standards and models for *precautionary regulation* abroad.

The USA and the EU transgenic policy choices limit the options of other countries. Countries economically dependent on them have simply applied regulatory alignment. Other countries have adopted transgenic neoregulation, the strictness of which lies in between the two dominant models. A good example for illustrating the above statements are the BRICS countries. Russia has adopted one of the world's strongest laws related to GMOs (*de jure* prohibition of transgenic crops cultivation), but the country tolerates transgenic feedstuffs imports. The reasons are the following: (i) in order to ensure self-sufficiency in the food supply, Russia has significantly increased its production – the production of wheat has increased by more than twice, sunflower seeds production by almost three times, while the production of soybeans and maize has increased ten times (2000–2015); (ii) Russia is still not self-sufficient in feedstuffs, because the increase in the area under feed crops, primarily soybeans, has not kept up with the enlargement of livestock and meat production; and (iii) the attitude of the public – Russians have an extremely negative attitude towards GMO cultivation, while they are totally indifferent to transgenic feed ingredients. Brazil's position on GMO issues is quite similar that of the USA. The biggest differences lie in labelling: Brazil does not fully adhere to the concept of substantial equivalence, since products containing more than 1% of transgenic ingredients must be labelled whatever the circumstances. Both Brazil and the USA have the same aspiration towards the removal of trade barriers, guided by the same agricultural policy goal: to expand exports. As in the USA, in Brazil the public attitude has also helped the government to adopt weak legislation on GMOs. Citizens are not interested in GMOs – three-quarters have never heard of them. The logical consequence is that, lacking awareness, they cannot be an engine of GMO resistance. China has adopted a regulatory oversight of transgenic technology that is both weak and strong at the same time. It is strong in relation to foreign seed entrance (it produces only GM crops obtained from its own research), weak in relation to importation of transgenic cotton, and maize and soybeans used as feed and for processing. This attitude is in line with: (i) the agricultural policy goal – to achieve food security through self-sufficiency in grain production; (ii) the fact that the country is already self-sufficient in rice, maize, wheat, cotton, and soybean seeds; (iii) the fact that 63% of its population is rural – national crop production is in the hands of the 200 million small family farms (average 0.6 ha) that are still important in the dairy industry and swine production; (iv) the feeds shortage caused by the growing livestock sector; and (v) a strong public rejection, that is slowing down further diffusion of transgenic crops. India is an example of resistance to international pressure in order to protect its indigenous peoples. It has developed a sui generis system for protection of plant varieties. Collective actions have given rise to massive and well-organized social movements which take the shape of large street protests and attract international attention. This approach is understandable with regard to: (i) the domination of small family farms in the country's farms structure – the share of farms less than 0.5 ha is 47%; (ii) the failure of *Bt* cotton to live up to its yield-return expectation; and (iii) the existence of food insecurity and malnutrition on a large scale. The last BRICS country, South Africa, has accorded agriculture the main role in building a strong economy, and has been trying to build an efficient and internationally competitive agricultural sector. As a strategic partner of the USA on the African continent, it has applied the neoregulatory paradigm of substantial equivalence. However, similar objectives in agricultural policy do not mean the same success in relation to GMOs. The USA, Brazil, and South Africa, all with a neoliberal stance with regard to GMOs, have achieved different results in the implementation of the technology. Unlike Brazil and

the USA, the two exporting giants, South Africa has failed to increase export competitiveness and remained dependent on import of all transgenic crops commercially produced in the country. Thus, this country is good proof that a strong neoliberal stance in respect to GMOs is not a guarantee of success.

The last country we discussed was Serbia. Serbia has unique GMO policies compared with those of the previously discussed countries: the production and commercialization as well as importation of transgenic crops and products is strictly forbidden by the Law of 2009. Unlike all the countries mentioned so far, Serbia is not a member of the WTO. Thanks to a well-organized social movement, 80% of its cities and municipalities (135 out of 169) have declared themselves GMO-free. A huge campaign, 'Serbia without GMOs', has been underway for several years. The reasons for this strict approach can be found in the country's history. In the 1970s, Serbia was greatly oversupplied in both food and agricultural products – its degree of self-sufficiency in the 1970s was 122.24%, and for its province of Vojvodina, as high as 237.03%. But the civil war, the NATO airstrikes, the international economic sanctions, the refugee influx, and one of the world's biggest hyperinflations contributed to the Serbian economy of 2000 being half the size it was in the 1990s. After the democratic changes in 2000, the country faced many of the negative effects of the market liberalization, including inadequate economic access to food. Knowing that all these hardships would be more difficult to handle without self-sufficiency in food production, Serbian citizens are united on the issue of resistance to GMOs. Such an attitude creates problems for the political elite who, exposed to frequent elections, have postponed the decision on amending the rigorous GMO law for several years, despite pressure from the USA, the WTO and the EU.

Finally, at the end of this book we demonstrate that, while the public is deeply divided, with heated public discussions, between passionate advocates and passionate opponents of transgenic technology, GMOs have broken almost all the barriers. Through feedstuffs, processed foods, and contamination, they have irretrievably penetrated into the food system, although the public still predominantly discusses health and safety issues. Such debate obscures the essential problems and paralyzes organized actions on the ultimate front line: seed defence. *Seeds* are an essential basis for achieving *food sovereignty. Without seeds, no country has its own food and agriculture systems.* Patent protection of transgenic technology, its processes, methods and products, is the crowning glory of neoliberal activities. With important help from supranational organizations, the multinational corporations have usurped the production and processing sectors, and have increasingly penetrated into the commercial sector. A perfect strategy to exclusively develop a narrow range of crops with a very wide range of industrial uses, as well to interlock crop production with certain herbicides whose owners are the same companies that have developed the crops, has made those multinationals the main beneficiaries of the Gene Revolution. These companies own more than 97% of all GMOs approved for direct food use or as additives, while universities, institutes, and research institutions have developed less than 3% of such GMOs. As Bloomberg has noted (McLaughlin, 2016), mergers between agricultural companies in seed and transgenic technology businesses have helped the biggest players to sharply consolidate their control over markets; the four biggest companies had a market share of 54% in 2009, an increase of more than double from 21% in 1994. The dominance of such companies can be seen as a continuation of the process from the past. The Green Revolution contributed to the transformation of agriculture into agribusiness, erased the borders between agriculture and industry, made farmers dependent on agro-input corporations, raised areas under monocropping, caused negative effects on the environment, and spawned a new manufactured diet. The Gene Revolution has deepened this process, and contributed to the even greater dependence of farmers on corporations, and the further increase of areas under monocropping and agrochemical use. And it has spawned a new ultra-processed diet, as an extension of the manufactured diet.

All in all, sovereign states have two choices: to surrender to the mercy of multinationals or to defend their borders. A national food policy, properly conceived, adequately implemented and respectful of the population's well-being, can result in a reduction in socio-economic inequality. Responsible government would be open to a prevalence of small farms or a bi-modal farm structure, and should not see large-scale farms only as the means to achieve food security. Small-scale farms can also be highly productive. Finally, national food policy, implemented with respect to the nation's well-being, should acknowledge the links between farming, diet, public health and the environment. As such, an adequate system of food controls should be an imperative.

Our results do have several implications for future research. First, our dataset is useful for other more in-depth analyses of GM food systems by region or by country, with more insights into regional or national issues. Second, those interested might find it useful to align policy, as described in Chapter 2, with their own country's needs. Third, sensitive consumers might find ways to avoid the GMOs present in food, after referring to the data provided in Chapter 6. Motivating further research on GM food systems and their economic impacts is a major objective of this book. We hope that we have achieved this objective.

Glossary

A

Accession to WTO: becoming a member of the WTO, signing up to its agreements. New members have to negotiate terms, bilaterally with individual WTO members, and multilaterally, in order to convert the results of the bilateral negotiations so that they apply to all WTO members, and as regards the required legislation and institutional reforms that are need to meet WTO obligations.

Aggregate Measurement of Support (AMS): annual level of support (subsidies) expressed in monetary terms, provided for an agricultural product in favour of the producers of the basic agricultural product (i.e. product specific), and non-product specific support provided in favour of agricultural producers in general. As per the WTO provision, AMS is a trade distorting subsidy. It includes (i) the sum total of subsidies on inputs like fertilizer, water, credit, power, etc.; and (ii) market price support, measured by calculating the difference between the domestic administered market price and the external reference price (world price) multiplied by the quantity of production eligible for obtaining the applied administered price. If domestic prices are lower than the world reference price, then (ii) is negative, and if this negative component is higher than input subsidies, then AMS turns out to be negative.

The reduction commitments of 28 Members (counting the EC as one) had non-exempt domestic support during the base period, and hence reduction commitments, specified in their schedules.

Agreement on Agriculture: an international treaty of the WTO. It was negotiated during the Uruguay Round of the GATT and entered into force with the establishment of the WTO on January 1, 1995.

Agricultural Adjustment Administration: major New Deal programme to restore agricultural prosperity by curtailing farm production, reducing export surpluses and raising prices. The Agricultural Adjustment Act (May 1933) was an omnibus farm-relief bill embodying the schemes of the major national farm organizations. Hoover's Federal Farm Board had tried to end the long-standing agricultural depression by raising prices without limiting production. Roosevelt's Agricultural Adjustment Act of 1933 was designed to correct the imbalance. Farmers who agreed to limit production would receive 'parity' payments to balance prices between farm and non-farm products, based on pre-war income levels. Farmers benefited also from numerous other measures, such as the Farm Credit Act of 1933, which refinanced a fifth of all farm mortgages in a period of 18 months, and the creation in 1935 of the Rural Electrification Administration (REA), which did more to bring farmers into the 20th century than any other single act. Thanks to the REA, nine out of ten farms were electrified by 1950, compared to one out of ten in 1935. These additional measures were made all the more important by the limited success of the Agricultural Adjustment Act. Production did fall as intended, aided by the severe drought of 1933–36, and prices rose in consequence; but many, perhaps a majority, of farmers did not prosper as a result. The Agricultural Adjustment Act was of more value to big operators than to small family farmers, who often could not meet their expenses if they restricted their output and therefore could not qualify for parity payments. The farm corporation, however, was able to slash its labour costs by cutting acreage, and could cut costs further by using government subsidies to purchase machinery. Thus, even before the Supreme Court invalidated the Agricultural Adjustment Act in 1936, support for it had diminished.

Agricultural capital stock: the total value of a producer's holdings of a defined set of fixed assets. Fixed assets consist of tangible or intangible assets that are used repeatedly or continuously in other processes of production over periods of 1 year or longer. The physical assets included are land development, livestock, machinery and equipment, plantation crops (trees, vines and shrubs yielding repeated products) and structures for livestock.

Agricultural reform: can refer either, narrowly, to government-initiated or government-backed redistribution of agricultural land or, more broadly, to an overall redirection of the agrarian system of the country, which often includes land reform measures. The World Bank evaluates agrarian reform using five dimensions: (i) stocks and market liberalization; (ii) land reform (including the development of land markets); (iii) agro-processing and input supply channels; (iv) urban finance; and (v) market institutions.

Agricultural support: the annual monetary value of gross transfers to agriculture from consumers and taxpayers arising from government policies that support agriculture, regardless of their objectives and economic impacts. This indicator includes the total support estimate (TSE), measured as a percentage of GDP, the producer support estimate (PSE), measured as a percentage of gross farm receipts, the consumer support estimate (CSE), measured as a percentage of agricultural consumption, and the general services support estimate (GSSE), measured as a percentage of total support. Agricultural support is also expressed in monetary terms, in million US dollars and million euros.

Agricultural transformation: the process by which individual farms shift from highly diversified, subsistence-oriented production towards more specialized production oriented towards the market or other systems of exchange (e.g. long-term contracts). The process involves a greater reliance on input and output delivery systems and increased integration of agriculture with other sectors of the domestic and international economies.

Agriculture, value added (% of GDP): agriculture corresponds to International Standard Industrial Classification (ISIC) divisions 1–5, and includes forestry, hunting, and fishing, as well as cultivation of crops and livestock production. Value added is the net output of a sector after adding up all outputs and subtracting intermediate inputs. It is calculated without making deductions for depreciation of fabricated assets or depletion and degradation of natural resources.

Agriculture value added per worker: a measure of agricultural productivity. Value added in agriculture measures the output of the agricultural sector less the value of intermediate inputs.

Agrobacterium tumefaciens: A bacterium that causes crown gall disease in some plants. The bacterium characteristically infects a wound, and incorporates a segment of Ti plasmid DNA into the host genome. This DNA causes the host cell to grow into a tumour-like structure that synthesizes specific opines that only the pathogen can metabolize. This DNA-transfer mechanism is exploited in the genetic engineering of plants.

Aid in kind: flow of goods and services with no payment in money or debt instruments in exchange. In some cases, 'commodity aid' goods (such as grain) are subsequently sold and the receipts are used in the budget or, more commonly through a special fund, for public expenditure.

Allele: a variant form of a gene. In a diploid cell there are two alleles of every gene (one inherited from each parent, although they can be identical). Within a population there may be many alleles of a gene. Alleles are symbolized with a capital letter to denote dominance, and lower case for recessive. In heterozygotes with co-dominant alleles, both are expressed.

Alternative trade system: a trading system that is not regulated as an exchange, but is a venue for matching the buy and sell orders of its subscribers. Alternative trading systems are gaining popularity around the world and account for much of the liquidity found in publicly traded issues. An alternative trading organization (ATO) is usually a NGO or mission-driven business aligned with the Fair Trade Movement, aiming to contribute to the alleviation of poverty in developing regions of the world by establishing a system of trade that allows marginalized producers in developing regions to gain access to developed markets.

Amber box: supports considered to distort trade and therefore subject to reduction commitments.

Antibiotic resistance: the ability of a microorganism to disable an antibiotic or prevent its transport into the cell.

Anthrax: an infectious, often fatal disease of cattle, sheep and other mammals, caused by *Bacillus anthracis*, transmitted to humans by contaminated wool, raw meat, or other animal products.

Apartheid: policy that governed relations between South Africa's white minority and non-white majority, and sanctioned racial segregation and political and economic discrimination against non-whites.

Applied tariffs: duties that are actually charged on imports. These can be below the bound rates.

Arable land: the land under temporary agricultural crops (multiple-cropped areas are counted only once), temporary meadows for mowing or pasture, land under kitchen and market gardens and land temporarily fallow (less than 5 years).

Artificial insemination: the delivery of semen into the uterus of the female animal usually by injection with a syringe-like apparatus for the purpose of achieving fertilization and sexual reproduction.

Artificial sweeteners: substances that are used in place of sweeteners with sugar (sucrose) or sugar alcohols. They may also be called sugar substitutes, non-nutritive sweeteners, and non-caloric sweeteners.

Average dietary energy supply adequacy: an indicator of food availability. It expresses on a percentage basis to what extent the daily food calorie availability satisfies the specific nutritional needs of the population of each country, in conformity with the national health authorities' recommendations.

Average protein supply: an indicator of food availability. It represents the available protein quantity (g/capita/day), calculated as a 3-year average. It also represents a food quality indicator.

Average supply of protein of animal origin: an indicator of food availability. It is an indicator identical with the average protein supply, but it refers only to the daily animal protein availability (g/capita/day).

Average value of food production: an indicator of food availability that represents the food production value per capita in each country, expressed in international dollars.

B

***Bacillus thuringiensis* (Bt):** a Gram-positive, soil-dwelling bacterium, commonly used as a biological pesticide. *B. thuringiensis* also occurs naturally in the gut of caterpillars of various types of moths and butterflies, as well as on leaf surfaces, aquatic environments, animal faeces, insect-rich environments, and flour mills and grain-storage facilities. *Bt* is largely used in agriculture, especially organic farming. *Bt* is also used in urban aerial spraying programmes, and in transgenic crops. Since 1996, plants have been modified with short sequences of genes from *Bt* to express the crystal protein *Bt* makes. With this method, plants themselves can produce the proteins and protect themselves from insects without any external *Bt* and/or synthetic pesticide sprays. *Bt* transgenic crops are protected specifically against European corn borer, southwestern corn borer, tobacco budworm, cotton bollworm, pink bollworm, and the Colorado potato beetle. Other benefits attributed to using *Bt* include: (i) reduced environmental impacts from pesticides; (ii) increased opportunity for beneficial insects; and (iii) reduced pesticide exposure to farm workers and non-target organisms. However, potential risks to using *Bt* also exist, such as: (i) invasiveness (genetic modifications can potentially change the organism to become invasive); and (ii) resistance to *Bt*; cross-contamination of genes.

Bactericide: a substance able to destroy bacteria.

Bacteriophage: a virus that infects bacteria. Altered forms are used as cloning vectors.

Biochemical engineering: the branch of science concerned with the use of biochemical processes and techniques for industrial purposes.

Bioenergy: renewable energy made available from materials derived from biological sources.

Biofuels: fuels produced directly or indirectly from organic material, including plant material and animal waste.

Biopesticide: a compound that kills organisms by virtue of specific biological effects rather than as a broader chemical poison.

Bioreactor: a tank in which cells, cell extracts or enzymes carry out a biological reaction. Often refers to a fermentation vessel for cells or micro-organisms.

Biotechnology: any technological application that uses biological systems, living organisms, or derivatives thereof, to make or modify products or processes for specific use; or, interpreted in a narrow sense, a range of different molecular technologies such as gene manipulation and gene transfer, DNA typing and cloning of plants and animals.

Bioterrorism: terrorism using biological agents.

Blue Box: permitted supports linked to production, but subject to production limits and therefore minimally trade-distorting. Direct payments under production limiting programmes are exempt from commitments if such payments are made on fixed areas and yield or a fixed number of livestock. Such payments also fit into this category if they are made on 85% or less of production in a defined base period. While the Green Box covers decoupled payments, in the case of the Blue Box measures, production is still required in order to receive the payments, but the actual payments do not relate directly to the current quantity of that production.

Bound tariffs: specific commitments made by individual WTO member governments. The bound tariff is the maximum most favoured nation (MFN) tariff level for a given commodity line. When countries join the WTO or when WTO members negotiate tariff levels with each other during trade rounds, they make agreements about bound tariff rates, rather than actually applied rates.

Breeding: making deliberate crosses or matings of plants or animals, so the offspring will have particular desired characteristics derived from one or both of the parents.

C

CAGR: an abbreviation for compound annual growth rate. CAGR is the rate of increase in the value of a quantity compounded over several years.

Cairns Group: in full, the Cairns Group of Fair Trading Nations, a coalition of agricultural countries advocating market-oriented reforms in the international agricultural trading system. It was established in 1986 during the early phases of the Uruguay Round of the GATT negotiations. It was against this background of rising protectionism, and the corruption of international agricultural trade, that the Cairns Group was formed. The original members are Argentina, Australia, Brazil, Canada, Chile, Colombia, Fiji, Hungary, Indonesia, Malaysia,

New Zealand, the Philippines, Thailand, and Uruguay.

CAP: an abbreviation for the Common Agricultural Policy. CAP is the European Union's agricultural policy.

Carbohydrate: a linear or branched polymer (e.g. starch, cellulose, etc.) composed of covalently linked monosaccharides, including cellulose, pectin and starch. Synonym: polysaccharide.

Cartagena Protocol: an internationally agreed protocol set up to protect biological diversity from the potential risks posed by the release of genetically modified organisms. Synonym: Biosafety Protocol.

CBD: an abbreviation for Convention on Biological Diversity, the international treaty governing the conservation and use of biological resources around the world.

CCC: abbreviation for Commodity Credit Corporation, a crop loan and storage programme, established to make price-supporting loans and purchases of specific commodities under the New Deal.

Cell: the fundamental level of structural organization in complex organisms.

Cell culture: the *in vitro* growth of cells isolated from multicellular organisms.

Cell fusion: formation *in vitro* of a single hybrid cell from the coalescence of two cells of different species origin. In the hybrid cell, the donor nuclei may remain separate, or may fuse, but during subsequent cell divisions, a single spindle is formed so that each daughter cell has a single nucleus containing complete or partial sets of chromosomes from each parental line. Synonym: cell hybridization.

Cell therapy: treatments in which stem cells are induced to differentiate into the specific cell type required to repair damaged or depleted adult cell populations or tissues.

Cereal import dependency ratio (%): an indicator of food supply stability. It tells how much of the available domestic food supply of cereals has been imported and how much comes from the country's own production. It is computed as [(cereal imports – cereal exports) / (cereal production + cereal import – cereal export) *100]. Given this formula the indicator assumes only values ≤ 100. Negative values indicate that the country is a net exporter of cereals.

CGIAR: an abbreviation for Consultative Group on International Agricultural Research. It is a global partnership that unites organizations engaged in research for a food secure future.

Chemical engineering: a branch of chemistry which deals with the manufacturing of finished products using raw materials by applying chemical procedures like nanotechnology, fuel cells etc., so as to change their chemical or physical composition, structure or energy content.

China Grain Reserves Corporation (Sinograin): a state-owned enterprise which plays a key role in carrying out state initiatives to ensure China's food security and economic growth. Founded in 2000 with a registered capital of CNY16.68 billion, Sinograin has grown into one of the largest and most wide-ranging grain storage and transportation corporations in China and is a leader in the research and use of innovative technology and equipment.

Chloroplast: specialized plastid that contains chlorophyll. Chloroplasts have their own DNA; these genes are inherited only through the female parent, and are independent of nuclear genes.

Chromosome: in eukaryotic cells, these are the nuclear bodies containing most of the genes largely responsible for the differentiation and activity of the cell.

CIF: cost-insurance-freight. CIF-trade values include the transaction value of the goods, the value of services performed to deliver goods to the border of the exporting country, and the value of the services performed to deliver the goods from the border of the exporting country to the border of the importing country. Import values are mostly reported as CIF.

Clone: (i) a group of cells or individuals that are genetically identical as a result of asexual reproduction, breeding of completely inbred organisms, or forming of genetically identical organisms by nuclear transplantation; (ii) a group of plants genetically identical, in which all are derived from one selected individual by vegetative propagation; or (iii) verb: to clone – to insert a DNA segment into a vector or host chromosome.

Cloning: the synthesis of multiple copies of a chosen DNA sequence using a bacterial cell or another organism as a host. The gene of interest is inserted into a vector, and the resulting recombinant DNA molecule is amplified in an appropriate host cell.

Codex Alimentarius Commission: UN commission that drafts food and hygiene standards and publishes them in the Codex Alimentarius. These non-binding (voluntary) standards become enforceable when accepted as national standards by the member countries. Jointly sponsored in 1962 by two UN bodies (FAO and WHO), it now has over 145 countries as its members.

Cold War: the open yet restricted rivalry that developed after World War II between the USA and the Soviet Union and their respective allies.

Coleopteran: any member of the insect order Coleoptera, consisting of the beetles and weevils. It is the largest order of insects, representing about 40% of the known insect species. Among the over 360,000 species of Coleoptera are many of the largest and most conspicuous insects, some of which also have brilliant metallic colours, showy patterns, or striking forms. Most of the beetles and weevils harmful to humans are phytophagous (plant feeders). Of primary importance are the leaf beetles (*Chrysomelidae*) and the weevils and their relatives (*Curculionoidea*).

Colonialism: the policy and practice of a power in extending control over weaker peoples or areas.

Comparative advantage: economic theory, first developed by 19th-century British economist David Ricardo, that attributed the cause and benefits of international trade to the differences in the relative opportunity costs (costs in terms of other goods given up) of producing the same commodities within countries. In Ricardo's theory, based on the labour theory of value (in effect, making labour the only factor of production), the fact that one country could produce everything more efficiently than another was not an argument against international trade.

Conservation: study of the loss of Earth's biological diversity and the ways this loss can be prevented. Biological diversity, or biodiversity, is the variety of life either in a particular place or on the entire Earth, including its ecosystems, species, populations, and genes. Conservation thus seeks to protect life's variety at all levels of biological organization.

Constant prices: 'real' prices, the data for each year, are calculated on the value of a particular base year. Constant series are used to measure the true growth of a series, i.e. adjusting for the effects of price inflation.

Corn rootworms: important insect pests of maize in the Midwest of the USA. Include: western corn rootworm (*Diabrotica virgifera virgifera* LeConte), northern corn rootworm (*Diabrotica barberi* Smith & Lawrence), and southern corn rootworm (*Diabrotica undecimpunctata howardi* Barber). Two species of rootworm that may cause severe damage to maize as both larvae and adults are the western and northern corn rootworms. Southern corn rootworm adults may damage maize leaves, however, because they cannot overwinter in most areas of the Midwest; but southern corn rootworm larvae do not present a major threat to maize in this region. Both corn rootworm larvae and adults may damage maize plants. Newly hatched larvae feed primarily on root hairs and outer root tissue. As larvae grow and their food requirements increase, they burrow into the roots to feed. Larval damage is usually most severe after the secondary root system is well established and brace roots are developing. Root tips will appear brown and are often tunnelled into and chewed back to the base of the plant. Larvae may be found tunnelling into larger roots and occasionally into the plant crown.

Corporate governance: rules and practices by which companies are governed or run. Corporate governance is important because it refers to the governance of what is arguably the most important institution of the capitalist economy.

CPI: an abbreviation for Consumer Price Index. A CPI measures changes over time in the general level of prices of goods and services that a reference population acquires, uses or pays for consumption.

CPVO: an abbreviation for Community Plant Variety Office, an EU agency managing the system of plant variety rights.

Cross: the mating of two individuals or populations.

Cross-breeding: mating between members of different populations (lines, breeds, races or species).

Crossing over: the process by which homologous chromosomes exchange material at meiosis through the breakage and reunion of non-sister chromatids.

Cross-fertilization (cross-pollination): transference of pollen from one plant to another to effect the latter's fertilization.

Cryopreservation: the preservation of germplasm resources in a dormant state by storage at ultra-low temperatures, often in liquid nitrogen. Currently it is applied to storage of plant seeds and pollen, microorganisms, animal sperm and tissue culture cell lines.

CSE: acronym for the Consumer Support Estimate. CSE transfers from consumers of agricultural commodities are measured at the farm-gate level. If negative, the CSE measures the burden (implicit tax) on consumers through Market Price Support (higher prices), that more than offsets consumer subsidies that lower prices for consumers.

Cucumber mosaic cucumovirus (CMV): a pathogenic plant virus in the family *Bromoviridae*. This virus has a worldwide distribution and a very wide host range. Some of the most important fruits and vegetables affected by CMV are peppers, bananas, tomatoes and cucurbits.

Culture: a population of plant or animal cells or microorganisms grown under controlled conditions.

Current prices: 'nominal' prices for each year in the value of the currency for that particular year. Current series are influenced by the effect of price inflation.

Cytology: the study of the structure and function of cells.

D

Decoupling: the removal of the link between the receipt of a direct payment and the production of a specific product.

Deficiency payment: an output subsidy in which the rate per unit of output of a commodity is the difference between the administered price and the market price.

De minimis: a Latin expression meaning 'with regard to minimal matters', 'of minimal importance', and sometimes referring the minimal amounts of domestic support that are allowed even though they distort trade – at up to 5% of the value of production for developed countries, and 10% for developing.

Depleted uranium: radioactive metal that is primarily used in the production of ammunition and armour plating. Depleted uranium rounds penetrate conventional armour easily and ignite on impact, usually causing extensive damage to the target.

Depth of the food deficit: an indicator of access to food. It indicates how many calories would be needed to lift the undernourished from their status, everything else being constant.

Deregulation: a subset of regulatory reform and referring to complete or partial elimination of regulation in a sector, to improve economic performance.

Developing and developed countries: There is no established convention for the designation of 'developed' and 'developing' countries or areas in the UN system. In common practice, Japan in Asia, Canada and the USA in North America, Australia and New Zealand in Oceania, and Europe are to be considered 'developed' regions or areas. In international trade statistics, the Southern African Customs Union is also treated as developed region and Israel as a developed country; countries emerging from the former Yugoslavia are treated as developing countries; and the countries of Eastern Europe and former USSR countries in Europe are not included under either developed or developing regions. In the 2016 edition of its World Development Indicators, the World Bank no longer distinguishes between 'developed' and 'developing' countries in the presentation of its data. Each year on July 1, the World Bank revises its analytical classification of the world's economies based on estimates of gross national income (GNI) per capita for the previous year. The updated GNI per capita estimates are also used as input to the World Bank's operational classification of economies that determines lending eligibility.

Dietary energy supply: food available for human consumption, expressed in kilocalories per person per day (kcal/person/day).

Dietary energy supply adequacy: dietary energy supply as a percentage of the average dietary energy requirement.

Direct payments: payments granted to farmers in order to support their incomes and to remunerate them for their production of public goods. Direct payments were established by the 1992 reform of the CAP. Prior to this reform, the CAP supported prices, i.e. the prices at which farmers sold their products in the market (such support is therefore not paid directly to farmers). The 1992 reform reduced the level of price support. To prevent a corresponding fall in the incomes of farmers, direct payments were introduced.

Disease resistance: the genetically determined ability to prevent the reproduction of a pathogen, thereby preserving health. Some resistances operate by pathogen exclusion, some by preventing pathogen spread, and some by tolerating pathogen toxins.

DNA: abbreviation for deoxyribonucleic acid – a long chain polymer of deoxyribonucleotides. DNA constitutes the genetic material of most known organisms and organelles, and usually is in the form of a double helix, although some viral genomes consist of a single strand of DNA, and others of a single- or double-stranded RNA.

DNA fingerprinting: the derivation of unique patterns of DNA fragments obtained using a number of marker techniques; historically these were RFLPs, but latterly they are generally polymerase chain reaction (PCR) based.

Doha Round: the latest round of trade negotiations among the WTO membership. Its aim is to achieve major reform of the international trading system through the introduction of lower trade barriers and revised trade rules. The work programme covers about 20 areas of trade. The Round is also known semi-officially as the Doha Development Agenda, since a fundamental objective is to improve the trading prospects of developing countries. The Round was officially launched at the WTO's Fourth Ministerial Conference in Doha, Qatar, in November 2001.

Dolly: the first mammal (a sheep) to be created (via nuclear transfer) by the cloning of an adult cell (from the mammary tissue of a ewe).

Domestication: the process of breeding for the purpose of furthering desirable characteristics in plants and animals.

Domestic food price index: an indicator of access to food. It is an indicator of relative prices that is calculated on the basis of the International Comparison Program of the World Bank and of the consumer price index developed by the FAO. The reference period is 2003–2006. Practically, this index reveals the relative level of domestic food prices compared with those from the USA. For the USA, this index is equal to 1.

Domestic food price volatility: an indicator of food stability. The domestic food price volatility index measures the variability in the relative price of food in a country.

Domestic supply: production + imports – exports + changes in stocks (decrease or increase) = supply for domestic utilization.

Domestic support: any domestic subsidy or other measure which acts to maintain producer prices at levels above those prevailing in international trade: direct payments to producers, including deficiency payments, and input and marketing cost reduction measures available only for agricultural production.

Double helix: describes the coiling of the two strands of the double-stranded DNA molecule, resembling a spiral staircase in which the base pairs form the steps and the sugar-phosphate backbones form the rails on each side. One strand runs 3'-5', while the complementary one runs 5'-3'.

Dominant: of alleles, one whose effect with respect to a particular trait is the same in heterozygotes as in homozygotes. The opposite is termed 'recessive'.

Drosophila melanogaster: the fruit fly, used for many years as a model for eukaryotic genetics. Of the nearly 300 disease-causing genes in the human genome, more than half have an analogous gene in the *Drosophila* genome.

E

EAP: an abbreviation for the total Economically Active Population. It includes all employed and unemployed persons.

EAP in agriculture: the number of people engaged in or seeking work in agriculture, hunting, fishing or forestry.

Early adopter: an individual or business who uses a new product or technology before others.

Effective demand: the actual demand for particular goods or services that is supported by a capacity to purchase.

Elasticity: a measure of the responsiveness of one variable, such as demand or supply, to changes in another, such as price or income.

Embryo: an immature organism in the early stages of development. In mammals, it develops in the first months in the uterus. In plants, it is the structure that develops in the megagametophyte, as a result of the fertilization of an egg cell, or occasionally without fertilization. Somatic embryos can often be induced in *in vitro* plant cell cultures.

Embryo transfer: the transfer of mammalian embryos from an *in vivo* or *in vitro* environment to a suitable host to improve pregnancy or gestational outcome in a human or animal.

Embryology: the science dealing with the formation, development, structure and functional activities of embryos.

Endangered species: a plant or animal species in immediate danger of extinction because its population number has reached a critical level, or its habitat has been drastically reduced.

Engel's Law: the observation made by Ernst Engel that people tend to spend a smaller share of their budget on food as their income rises.

Enzyme: a protein which, even in very low concentrations, catalyses specific chemical reactions but is not used up in the reaction.

EPC: an abbreviation for European Patent Convention. Also known as the Convention on the Grant of European Patents of 5 October 1973, it is a multilateral treaty instituting the European Patent Organisation and providing an autonomous legal system according to which European patents are granted.

Epidemic: a disease that spreads rapidly among many people in a community at the same time.

Escherichia coli: a commensal bacterium inhabiting the colon of many animal species, including humans. *E. coli* is widely used as a model of cell biochemical function, and as a host for cloning DNA. In environmental studies, its presence is a key indicator of water pollution due to human sewage effluent.

Eugenics: the adjective eugenic traces its roots from the Greek word *eugenes*, meaning 'well-born.' The field of eugenics aimed in the 19th and 20th centuries to improve the characteristics of a race by promoting certain qualities in its offspring, which they then would pass on to future generations. But it also tried to prevent people with 'unfavourable' qualities from procreating. This philosophy has fallen out of favour greatly since World War II, especially among biologists. It is the study of methods of improving genetic qualities by selective breeding (especially as applied to human mating).

European corn borer: *Ostrinia nubilalis* (Hübner) is a pest of grain, particularly maize. The insect is native to Europe, originally infesting varieties of millet, including broom corn. The European corn borer was first reported in North America in 1917 in Massachusetts, but was probably introduced from Europe several years earlier. This is a very serious pest of both sweet corn and grain corn (maize), and before the availability of modern insecticides this insect caused very marked reductions in maize production.

European Federation of Biotechnology: Europe's non-profit federation of national biotechnology associations, learned societies, universities, scientific institutes, biotech companies and individual biotechnologists working to promote biotechnology throughout Europe and beyond, established in 1978.

Export quota: a restriction on exports to protect local business from shortages, maintain international business, and create a restraint agreement.

Extensive stage of capitalism: early or concurrent stage of capitalism, with plenty of room for the extension of capitalist relations of production, which is to say, of wage labour and therefore of commodity production.

Extrachromosomal: in eukaryotes, non-nuclear DNA, present in cytoplasm organelles such as mitochondria and chloroplasts; in prokaryotes, non-chromosomal DNA, i.e. plasmids.

F

F_1: abbreviation for filial generation 1; the initial hybrid generation resulting from a cross between two parents.

F_2: the second hybrid generation, produced either by intercrossing two F_1 individuals, or by self-fertilizing an F_1 individual.

Fair trade: a social movement whose stated goal is to help producers in developing countries achieve better trading conditions and to promote sustainable farming; a trading partnership, based on dialogue, transparency and respect, that seeks greater equity in international trade. It contributes to sustainable development by offering better trading conditions to, and securing the rights of, marginalized producers and workers – especially in the South.

FAO: abbreviation for the Food and Agriculture Organization of the UN, an agency that leads international efforts to defeat hunger.

Fermentation: the anaerobic breakdown of complex organic substances, especially carbohydrates, by microorganisms, yielding energy; often misused to describe large-scale aerobic cell culture in specialized vessels (fermenters, bioreactors) for secondary product synthesis.

Fertilization: the union of two gametes from opposite sexes to form a zygote. Typically, each gamete contains a haploid set of chromosomes. Hence the zygotic nucleus contains a diploid set of chromosomes. Several categories can be distinguished: (i) self-fertilization (selfing): fusion of male and female gametes from the same individual; (ii) cross-fertilization (crossing): fusion of male and female gametes from different individuals; or (iii) double fertilization: restricted to flowering plants, in which the fusion of one male gamete with the ovum occurs at about the same time as the second male gamete nucleus fuses with the female polar nuclei (or secondary nucleus) to form the endosperm.

Fertilizer use efficiency: yield per unit of input. Fertilizers are considered efficient when maximum yield is obtained with the minimum possible amount of fertilizer application.

FOB: an abbreviation for Free-On-Board. FOB-trade values include the transaction value of the goods and the value of services performed to deliver goods to the border of the exporting country. Export values are mostly reported as FOB.

Food access: a household's ability to acquire adequate amounts of food regularly through a combination of production, purchases, barter, borrowing, food assistance or gifts.

Food additive: the term refers to 'any substance the intended use of which results or may reasonably be expected to result – directly or indirectly – in its becoming a component or otherwise affecting the characteristics of any food.' This definition includes any substance used in the production, processing, treatment, packaging, transportation, or storage of food. The Food and Drug Administration (FDA) maintains a list of over 3000 ingredients in its database, 'Everything Added to Food in the US', many of which we use at home every day (e.g. sugar, baking soda, salt, vanilla, yeast, spices and colours). **Direct** food additives are those that are added to a food for a specific purpose. For example, xanthan gum – used in salad dressings, chocolate milk, bakery fillings, puddings and other foods to add texture – is a direct additive. Most direct additives are identified on the ingredient label of foods. **Indirect** food additives are those that become part of the food in trace amounts owing to its packaging, storage or other handling. For instance, minute amounts of packaging substances may find their way into foods during storage. See more: Food ingredients.

Food aid: the international sourcing of concessional resources in the form or for the provision of food; the provision of food or cash to purchase food in times of emergency or to provide longer-term solutions in areas where food shortages exist.

Food availability: the amount of food that is present in a country or area through all forms of domestic production, imports, food stocks and food aid.

Food import bills: the annual value of food imported under Standard International Trade Classification (SITC) sections 0 + 22 + 4 (0: Food and live animals; 22: Oil-seeds and oleaginous fruits; 4: Animal and vegetable oils, fats and waxes), as expressed in current US dollars.

Food ingredients: the types of common food ingredients are: preservatives; sweeteners; colour additives; flavours and spices; flavour enhancers; fat replacers (and components of formulations used to replace fats); nutrients; emulsifiers; stabilizers and thickeners; binders; texturizers; pH control agents and acidulants; leavening agents; anti-caking agents; humectants; yeast nutrients; dough strengtheners and conditioners; firming agents; enzyme preparations; and gases. *Preservatives* prevent food spoilage from bacteria, moulds, fungi, or yeast (antimicrobials); slow down or prevent changes in colour, flavour or texture, and delay rancidity (antioxidants); maintain freshness (ascorbic acid, citric acid, sodium benzoate, calcium propionate, sodium erythorbate, sodium nitrite, calcium sorbate, potassium sorbate, BHA, BHT, EDTA, tocopherols – vitamin E). Examples of use: fruit sauces and jellies, beverages, baked goods, cured meats, oils and margarines, cereals, dressings, snack foods, and fruits and vegetables. *Sweeteners* add sweetness with or without the extra calories (sucrose i.e. sugar, glucose, fructose, sorbitol, mannitol, corn (maize) syrup, high fructose corn (maize) syrup, saccharin, aspartame, sucralose, acesulfame potassium i.e. acesulfame-K, neotame). They are used in beverages, baked goods, confections, table-top sugar, substitutes, and many other processed foods. *Colour additives* offset colour loss due to exposure to light, air, temperature extremes, moisture, and storage conditions; correct natural variations in colour; enhance colours that occur naturally; provide colour to colourless and 'fun' foods (FD&C Blue Nos. 1 and 2, FD&C Green No. 3, FD&C Red Nos. 3 and 40, FD&C Yellow Nos. 5 and 6, Orange B, Citrus Red No. 2, annatto extract, beta-carotene, grape skin extract, cochineal extract or carmine, paprika oleoresin, caramel colour, fruit and vegetable juices, saffron). Colour additives are used in many processed foods (candies, snack foods margarine, cheese, soft drinks, jams/jellies, gelatines, pudding and pie fillings). *Flavours and spices* add specific flavours (natural and synthetic). They are used in pudding and pie fillings, gelatine dessert mixes, cake mixes, salad dressings, candies, soft drinks, ice cream, BBQ sauce. *Flavour enhancers* enhance flavours already present in foods, without providing their own separate flavour (the names found on product labels are usually: monosodium glutamate (MSG), hydrolysed soy protein, autolyzed yeast extract, disodium guanylate or inosinate). They are used in many processed foods. *Fat Replacers (and components of formulations used to replace fats)* provide expected texture and a creamy 'mouth-feel' in reduced-fat foods (the names found on product labels are usually: olestra, cellulose gel, carrageenan, polydextrose, modified food starch, microparticulated egg white protein, guar gum, xanthan gum, whey protein concentrate). They are used in baked goods, dressings, frozen desserts, confections, cake and dessert mixes, dairy products. *Nutrients* replace vitamins and minerals lost in processing (enrichment), adding nutrients that may be lacking in the diet (fortification) (the names found on product labels are usually: thiamine hydrochloride, riboflavin (vitamin B_2), niacin, niacinamide, folate or folic acid, beta carotene, potassium iodide, iron or ferrous sulfate, alpha tocopherols, ascorbic acid, vitamin D, amino acids i.e. L-tryptophan, L-lysine, L-leucine, L-methionine). They are used in flour, breads, cereals, rice, macaroni, margarine, salt, milk, fruit beverages, energy bars, and instant breakfast drinks. *Emulsifiers* allow smooth mixing of ingredients, preventing separation; keep emulsified products stable, reduce stickiness, control crystallization, keep ingredients dispersed, and help products dissolve more easily (names found on product labels are usually: soy lecithin, mono- and diglycerides, egg yolks, polysorbates, sorbitan monostearate). They are used in salad dressings, peanut butter, chocolate, margarine, and frozen desserts. *Stabilizers and thickeners, binders, texturizers* produce uniform texture, improve 'mouth-feel' (names found on product labels are usually gelatine, pectin, guar gum, carrageenan, xanthan gum, whey). They are used in frozen desserts, dairy products, cakes, pudding and gelatine mixes, dressings, jams and jellies, and sauces. *pH control agents and acidulants* control acidity and alkalinity and prevent spoilage (names usually found on product labels: lactic acid, citric acid, ammonium hydroxide, sodium carbonate). They are used in beverages, frozen desserts, chocolate, low-acid canned foods, baking powder. *Leavening agents* promote rising of baked goods (baking soda, monocalcium phosphate, calcium carbonate). They are used in breads and other baked goods. *Anti-caking agents* keep powdered foods free-flowing, prevent moisture absorption (calcium silicate, iron ammonium citrate, silicon dioxide). They are used in salt, baking powder, and confectioner's sugar. Humectants retain moisture (glycerine, sorbitol). They are used in shredded coconut, marshmallows, soft candies, and confections. *Yeast nutrients* promote growth of yeast (calcium sulfate, ammonium phosphate). They are used in breads and other baked goods. *Dough strengtheners and conditioners* produce more stable dough (names usually found on labels: ammonium sulfate, azodicarbonamide, L-cysteine). They are used in breads and other baked goods. *Firming agents* maintain crispness and firmness (calcium chloride, calcium lactate). They are used in processed fruits and vegetables. *Enzyme preparations* modify proteins, polysaccharides, and fats (enzymes, lactase, papain, rennet, and chymosin). They are used in cheese, dairy products, and meat. *Gases* serve as propellant, aerate, or create carbonation (carbon dioxide, nitrous oxide). They are used in oil cooking spray, whipped cream, and carbonated beverages.

Food insecurity: a situation that exists when people lack secure access to sufficient amounts of safe and nutritious food for normal growth and development and an active and healthy life.

Food miles: the distance food travels from the farm where it is produced to the plate of the final consumer.

Food regime: a relatively bounded historical period in which complementary expectations govern the behaviour of all social actors, such as farmers, firms, and workers engaged in all aspects of food growing, manufacturing, services, distribution and sales, as well as government agencies, citizens and consumers.

Food safety: a term referring to the extent to which food is safe to eat. The term is sometimes confused with food security, which refers to the extent to which food is available, i.e. whether it is physically available and can be bought at a price that people can afford.

Food security: a situation that exists when all people, at all times, have physical, social, and economic access to sufficient, safe, and nutritious food that meets their dietary needs and food preferences for an active and healthy life.

Food self-sufficiency: the extent to which a country can satisfy its food needs from its own domestic production. The self-sufficiency ratio (SSR) is defined as: SSR = production × 100/ (production + imports – exports).

Food sovereignty: the right of peoples to healthy and culturally appropriate food produced through ecologically sound and sustainable methods, and their right to define their own food and agriculture systems.

Food stability: a dimension of food security. Even if one's food intake is adequate today, one is still considered to be food insecure if one has inadequate access to food on a periodic basis, risking a deterioration of one's nutritional status.

Food system: a system that embraces all the elements (environment, people, inputs, processes, infrastructure, institutions, markets and trade) and activities that relate to the production, processing, distribution and marketing, preparation and consumption of food, and the outcomes of these activities, including socio-economic and environmental outcomes.

Food utilization: the selection and intake of food and the absorption of nutrients. Food utilization depends on adequate diet, clean water, sanitation and health care.

Forensics: scientific tests or techniques used in connection with the detection of crime.

Free trade: a policy also called 'laissez-faire', by which a government does not discriminate against imports or interfere with exports by applying tariffs (to imports) or subsidies (to exports).

Fungicide: a specific type of pesticide that controls fungal disease by specifically inhibiting or killing the fungus causing the disease.

G

Gamete: a mature reproductive cell which is capable of fusing with a cell of similar origin but of opposite sex to form a zygote from which a new organism can develop. Gametes normally have a haploid chromosome content. In animals, a gamete is a sperm or egg; in plants, it is pollen, a spermatic nucleus, or an ovum.

GATT: an abbreviation for the General Agreement on Tariffs and Trade, a set of multilateral trade agreements aimed at the abolition of quotas and the reduction of tariff duties among the contracting nations.

GDP PPP: is the Gross Domestic Product (GDP) converted to international dollars using PPP rates. An international dollar has the same purchasing power over the GDP as the US dollar has in the US.

Gelatine: a glutinous, proteinaceous gelling and solidifying agent. It is used to gel or solidify nutrient solutions for tissue culture, and as a food additive.

Gene: the unit of heredity transmitted from generation to generation during sexual or asexual reproduction. More generally, the term is used in relation to the transmission and inheritance of particular identifiable traits. The simplest gene consists of a segment of nucleic acid that encodes an individual protein or RNA.

Gene expression: the process by which a gene produces mRNA and protein, and hence exerts its effect on the phenotype of an organism.

Gene insertion: the incorporation of one or more copies of a gene into a chromosome.

Gene stacking: the process of combining two or more genes of interest into a single plant.

Gene therapy: the proposed treatment of an inherited disease by the transformation of an affected individual with a wild-type copy of the defective gene causing the disorder. In germ-line (or heritable) gene therapy, reproductive cells are transformed; in somatic-cell (or non-inheritable) gene therapy, cells other than reproductive ones are modified.

Genetic diversity: the heritable variation within and among populations which is created, enhanced or maintained by evolutionary or selective forces.

Genetic engineering: modifying a genotype, and hence a phenotype, by transgenesis.

Genetically modified food: food that contains more than a certain legal minimum content of raw material obtained from genetically modified organisms.

Genetically modified organism (GMO): an organism (i.e. plant, animal or microorganism) in which the genetic material (DNA) has been altered in a way that does not occur naturally by mating and/or natural recombination.

Genetics: the science of heredity.

Genome: (i) the entire complement of genetic material (genes plus non-coding sequences) present in each cell of an organism, virus or organelle; or (ii) the complete set of chromosomes (hence of genes) inherited as a unit from one parent.

Genomic imprinting: the differential expression of a single gene according to its parental origin.

Genomics: the research strategy that uses molecular characterization and cloning of whole genomes to understand the structure, function and evolution of genes and to answer fundamental biological questions.

Genotype: (i) the genetic constitution of an organism; (ii) the allelic constitution at a particular locus, e.g. Aa or aa; or (iii) the sum effect of all loci that contribute to the expression of a trait.

Germination: (i) the initial stages in the growth of a seed to form a seedling; or (ii) the growth of spores (fungal or algal) and pollen grains.

GFSI: an abbreviation for the Global Food Safety Initiative, an industry-driven initiative providing thought, leadership and guidance on the food safety management systems necessary for safety along the supply chain.

GHI: an abbreviation for the Global Hunger Index published by IFPRI. The GHI regards hunger as multidimensional and uses three indicators: the FAO estimate of the proportion of the population with insufficient access to food; WHO's estimate of the proportion of underweight children under the age of 5; and UNICEF's figures on mortality among children under the age of 5. An average of the three percentage rates is taken and countries are then classified in the index as serious, alarming or extremely alarming.

GIEWS: an abbreviation for the FAO's Global Information and Early Warning System. GIEWS continuously monitors food supply and demand and other key indicators for assessing the overall food security situation in all countries of the world.

Gold standard: the monetary system in which the standard unit of currency is a fixed quantity of gold, or kept at the value of a fixed quantity of gold.

Grain Stabilization Corporation: the institution created by the Farm Board in 1930 to bolster sagging prices by buying surpluses.

Great Depression: worldwide economic downturn that began in 1929 and lasted until about 1939. It was the longest and most severe depression ever experienced by the industrialized Western world, sparking fundamental changes in economic institutions, macroeconomic policy, and economic theory.

Greenpeace: an international organization dedicated to preserving endangered species of animals, preventing environmental abuses, and heightening environmental awareness through direct confrontations with polluting corporations and governmental authorities.

Green Box: support measures that are exempt from reduction commitments and, indeed, can even be increased without any financial limitation under the WTO. The Green Box covers many government service programmes, including general services provided by governments, public stockholding programmes for food security purposes and domestic food aid – as long as the general criteria and some other measure-specific criteria are met by each measure concerned. The Green Box also provides for the use of direct payments to producers which are not linked to production decisions, i.e. although the farmer receives a payment from the government, this payment does not influence the type or volume of agricultural production ('decoupling'). The conditions preclude any linkage between the amount of such payments and production, prices or factors of production in any year after a fixed base period. In addition, no production shall be required in order to receive such payments.

Green Revolution: refers to a set of research and development of technology transfer initiatives that increased agricultural production worldwide, particularly in the developing world, beginning most markedly in the late 1960s. The initiatives resulted in the adoption of new technologies, including 'new, high-yielding varieties of cereals, especially dwarf wheats and rices, in association with chemical fertilizers and agro-chemicals, and with controlled water-supply (usually involving irrigation) and new methods of cultivation, including mechanization. All of these together were seen as a package of practices to supersede traditional technology and to be adopted as a whole.' The initiatives were led by Norman Borlaug, the 'Father of the Green Revolution', who received the Nobel Peace Prize in 1970.

GSSE: an abbreviation for the General Services Support Estimate. GSSE transfers are linked to measures creating enabling conditions for the primary agricultural sector through development of private or public services, institutions and infrastructure. GSSE includes policies where primary agriculture is the main beneficiary, but does not include any payments to individual producers. GSSE transfers do not directly alter producer receipts or costs or consumption expenditure.

H

Habitats: the natural environment where an organism, population or community lives, including those biotic and abiotic factors affecting it.

Hunter-gatherer society: a social system having the simplest mode of production. Most hunter-gatherer societies depend primarily on gathering existing food, while meat is an occasional source of nutrition. They rarely generate a surplus, since they have no means of storing what they cannot consume in the near future. They also must move from place to place, making it impractical to accumulate possessions. Production is communal and cooperative, and the distribution of food is based on sharing.

Harvested area: the area from which a crop is gathered.

Helix: a structure with a spiral shape. The normal state of double-stranded DNA is in the form of a double helix.

Heredity: resemblance among individuals related by descent; transmission of traits from parents to offspring.

Herbicides: chemical substances used to control unwanted plants. Selective herbicides control specific weed species, while leaving the desired crop relatively unharmed, while non-selective herbicides (sometimes called 'total weedkillers' in commercial products) can be used to clear waste ground, industrial and construction sites, railways and railway embankments, as they kill all plant material with which they come into contact.

Herbicide tolerance (HT): weed control technology widely used in creation of HT crops. HT crops have the ability to tolerate the broad-spectrum herbicides – in particular glyphosate and glufosinate. Glyphosate herbicide kills plants by blocking the enzyme 5-enolpyruvylshikimate-3-phosphate synthase (EPSPS), an enzyme involved in the biosynthesis of aromatic amino acids, vitamins and many secondary plant metabolites. There are several ways by which crops can be modified to be glyphosate-tolerant. One strategy is to incorporate a soil bacterium gene that produces a glyphosate-tolerant form of EPSPS. Another way is to incorporate a different soil bacterium gene that produces a glyphosate degrading enzyme. Glufosinate herbicides contain the active ingredient phosphinothricin, which kills plants by blocking the enzyme responsible for nitrogen metabolism and for detoxifying ammonia, a by-product of plant metabolism. Crops modified to tolerate glufosinate contain a bacterial gene that produces an enzyme that detoxifies phosphinothricin and prevents it from doing damage. Other methods by which crops are genetically modified to survive exposure to herbicides include: (i) producing a new protein that detoxifies the herbicide; (ii) modifying the herbicide's target protein so that it will not be affected by the herbicide; or (iii) producing physical or physiological barriers preventing the entry of the herbicide into the plant. The first two approaches are the most common ways scientists develop herbicide tolerant crops. Potential benefits of using HT technology include simplifying weed control to the use of a single herbicide and with a more flexible timing than that required for conventional herbicides. In glyphosate-resistant crops, optimal weed control often requires sequential applications with glyphosate, and the timing relative to weed emergence is important. When glyphosate is sprayed two to three times annually at high rates, it imposes a high selection pressure on the weed flora. In 5–8 years, this may cause shifts in weed composition towards species that naturally tolerate glyphosate and other herbicides may be needed to control these weeds. Furthermore, it may become difficult to control volunteer crops in subsequent years. If farmers grow glyphosate-resistant varieties of both maize and soybean in a soybean-maize rotation, then glyphosate cannot control the volunteer maize, which can present a serious weed problem for soybean. Gene-flow from crops to other crops or related species is another route to the development of resistant weed populations in the field. Once the resistance gene is present in crop volunteers or related weed species, then it is expected that the same weed control practices (consistent sprayings with herbicides having the same mode of action), which cause herbicide resistance to occur in naturally tolerant/resistant weed biotypes, will lead to a rapid build-up of herbicide resistant weeds and volunteers. Increased herbicide use is considered a risk in some parts of the world, although the effects on human health or the environment are seldom specified in detail; but the derived effects from pesticide use, such as groundwater pollution and pesticide residues in vegetables, for example, have caused public concern. Biodiversity within the field may be influenced if the herbicide, to which the HT crops are resistant, is used at a higher level of efficacy than before, in order to achieve an improved weed control.

Herbicide resistance: the ability of a plant to remain unaffected by the application of an herbicide.

Herbivores: an animal that feeds on plant substances.

High-fructose corn (maize) syrup: a corn (maize) syrup to which enzymes have been added to change some of the glucose to fructose, making the product sweeter than regular corn (maize) syrup.

High-income economies: those with a GNI per capita of US$12,476 or more.

Homo erectus: an extinct species of the human lineage, formerly known as *Pithecanthropus erectus*, having an upright stature and a well-evolved postcranial skeleton, but with a smallish brain, low forehead, and protruding face.

Horizontal integration: the merger of companies at the same stage of production in the same or different industries.

Human Genome Project: an extensive international research effort to determine the sequence in which human DNA is arranged.

Hunger: synonymous with chronic undernourishment.

Household plot: a legally defined farm type in all former socialist countries in the CIS and CEE. It is a small plot of land (typically less than 0.5 ha) attached to a rural residence.

HVFs: an abbreviation for high value foods. It includes meats, dairy products, fish, edible horticultural products, spices, oilseeds, animal and vegetable oils, and animal feedstuffs.

Hybrid: the offspring of two genetically unlike parents.

Hybrid seed: (i) seed produced by crossing genetically dissimilar parents; or (ii) in plant breeding, used colloquially for seed produced by specific crosses of selected pure lines, such that the F_1 crop is genetically uniform and displays hybrid vigour. As the F_1 plants are heterozygous with respect to many genes, the crop does not breed true, and so new seed must be purchased each season.

Hybrid selection: the process of choosing individuals possessing desired characteristics from among a hybrid population.

Hybrid vigour: the extent to which a hybrid individual outperforms both its parents with respect to one or many traits. The genetic basis of hybrid vigour is not well understood, but the phenomenon is widespread, particularly in inbreeding plant species. Synonym: heterosis.

Hybridization: (i) the process of forming a hybrid by cross-pollination of plants or by mating animals of different types; (ii) the production of offspring, normally from sexual reproduction, but also asexually by the fusion of protoplasts or by transformation; or (iii) the pairing of two DNA strands, often from different sources, by hydrogen bonding between complementary nucleotides.

I

IEFR: an abbreviation for the International Emergency Food Reserve. An international standby arrangement established in 1975 in order to respond quickly to emergencies wherever they occur.

IFAD: an acronym for International Funds for Agricultural Development. This specialized agency of the UN was established as an international financial institution in 1977, in one of the major outcomes of the 1974 World Food Conference, in order to finance agricultural development projects primarily for food production in the developing countries.

IFPRI: an abbreviation for the International Food Policy Research Institute that provides research-based policy solutions to sustainably reduce poverty and end hunger and malnutrition in developing countries. Established in 1975, IFPRI currently is active in over 50 countries. It is a research centre of the CGIAR, a worldwide partnership engaged in agricultural research for development.

IMF: an acronym for the International Monetary Fund, an international organization of 184 member countries. It was established to promote international monetary cooperation, exchange stability, and orderly exchange arrangements; to foster economic growth and high levels of employment; and to provide temporary financial assistance to countries to help ease balance of payments adjustments.

Import dependency ratio (IDR): it is defined as: [IDR = imports × 100/ (production + imports − exports)]. IDR refers to the extent of dependency on importation in relation to domestic consumption. However, there is a caveat to be kept in mind: these ratios hold only if imports are mainly used for domestic utilization and are not re-exported.

Import quota: the maximum quantity of a good that a country's importers may import at zero or reduced duty.

Import tariff: a tax or duty imposed on imported goods.

In vitro: outside the organism, or in an artificial environment; applied for example to cells, tissues or organs cultured in glass or plastic containers.

Inbreeding: matings between individuals that have one or more ancestors in common, the extreme condition being self-fertilization, which occurs naturally in many plants and some primitive animals. Synonym: endogamy.

Inbreeding depression: the reduction in vigour over generations of inbreeding. This affects species which are normally outbreeding and highly heterozygous.

Income distribution: national income divided among groups of individuals, households, social classes, or factors of production,

to compute an average for purposes of comparison.

Independent assortment: the random distribution during meiosis of alleles (within different genes) to the gametes, which is the case when the genes in question are located on different chromosomes or are unlinked on the same chromosome.

Indigenous communities, peoples and nations: those which, having a historical continuity with the pre-invasion and pre-colonial societies that developed on their territories, consider themselves distinct from other sectors of the societies now prevailing on those territories, or parts of them. They form at present the non-dominant sectors of society and are determined to preserve, develop and transmit to future generations their ancestral territories, and their ethnic identity, as the basis of their continued existence as peoples, in accordance with their own cultural patterns, social institutions and legal systems.

Inheritance: the transmission of genes and phenotypes from generation to generation.

Inputs: resources used in the production process, e.g. seeds, pesticides, feed, materials, labour, and machinery, measured in physical or financial terms.

Insectide: a substance used to kill insects.

Insulin: a peptide hormone secreted by the Langerhans islets of the pancreas that regulates the level of sugar in the blood.

Intensive agriculture: agricultural practices that produce high output per unit area, usually by intensive use of manure, agrochemicals, mechanization and so on.

Intensification (agricultural): an increase in agricultural production per unit of inputs (which may be labour, land, time, fertilizer, seed, feed or cash).

Intensive stage of capitalism: second stage of capitalism: when the extensive stage becomes exhausted, the expansion of commodity production is reduced to the increase in productivity of labour, or to the intensification of production.

Inter-American Coffee Agreement: the precursor to the International Coffee Agreement. The International Coffee Agreement is an international commodity agreement between coffee producing countries and consuming countries. First signed in 1962, it is aimed at maintaining exporting countries' quotas and keeping coffee prices high and stable in the market, mainly using export quotas to steer the price. The International Coffee Organization (ICO), the controlling body of the agreement, represents all major coffee producing countries and most consuming countries.

International Office of Epizootics (OIE): an organization formed through the international agreement signed on January 25th 1924 to fight animal diseases at global level. In May 2003 the Office became the World Organisation for Animal Health, but kept its historical acronym OIE.

International Plant Protection Convention (IPPC): a 1951 multilateral treaty overseen by the FAO that aims to secure coordinated, effective action to prevent and control the introduction and spread of pests among plants and plant products.

International Union of Pure and Applied Chemistry (IUPAC): the international body that represents chemistry and related sciences and technologies.

Intervention purchase: act of purchasing a commodity once its market price drops below a set administered price (the intervention price), so as to raise its market price to at least the level of the intervention price.

IPR: the acronym for Intellectual Property Right. An exclusive right that is possessed by a person or a company to use their own plans, ideas, or other intangible assets without the worry of competition, at least for a specific period of time.

ITO: the acronym for International Trade Organization. It was the proposed name for an international institution for the regulation of trade. Led by the USA in collaboration with its allies, the efforts to form this organization between 1945 to 1948, with the successful passing of the Havana Charter, eventually failed, owing to lack of approval by the US Congress. Until the creation of the WTO in 1994, international trade was managed through the GATT.

J

Joint venture: a partnership or alliance among two or more businesses or organizations based on shared expertise or resources to achieve a particular goal. A joint venture is distinct from other forms of partnerships among organizations, such as mergers or simple contractual arrangements. Partners in a joint venture maintain separate legal identities but are bound by agreements about how to share the equity, liability, and profits of their partnership.

K

Karyokinesis: the division of a cell nucleus.

Kilocalorie (kcal): a unit of measurement of energy. One kilocalorie equals 1,000 calories. In the International System of Units (SI), the universal unit of energy is the joule (J). One kilocalorie = 4.184 kilojoules (kJ).

L

La Via Campesina: an international movement of poor peasants and small farmers from the global South and North. Initiated by Central, South, and North American peasant and farmers' movements and European farmer's groups, Via Campesina was formally launched in 1993 in Mons, Belgium. Existing transnational networks of activists located in peasant movements and non-governmental funding agencies in the North facilitated the earlier contacts between key national peasant movements, most of which had emerged already in the 1980s. By 2008, Via Campesina represented more than 150 (sub)national rural social movement organizations from 56 countries in Latin America and the Caribbean, North America, Western Europe, Asia, and Africa.

Land grabs: the seizing of land, often unfairly, illegally, or deceptively, by a nation, state, or organization. It is the acquisition of valuable or strategic territory for much less than its actual worth. The term was defined in the Tirana Declaration (2011) by the

International Land Coalition, consisting of 116 organizations ranging from community groups to the World Bank.

Land reform: the redistribution of private or public lands.

Large-scale farming: industrial farming. Unlike small-scale farms, large-scale farms utilize various industrial methods to maximize production.

Least-developed countries (LDCs): low-income countries confronting severe structural impediments to sustainable development. There are currently 48 countries on the list of LDCs.

Lepidopteran: any of more than 155,000 species of butterflies, moths, and skippers. This order of insects is second in size only to Coleoptera, the beetles. Many hundreds of Lepidoptera injure plants useful to humans, including some of the most important sources of food, fabrics, fodder, and timber.

Livestock Revolution: the term coined in an influential 1999 IFPRI publication. The basic tenet of the Livestock Revolution paradigm is that the combination of population growth, rising per capita incomes, and progressive urbanization are creating an unprecedented growth in demand for food of animal origin in developing countries, giving rise to major opportunities and threats for mankind.

Low income country: for the current 2018 fiscal year, low-income economies are defined as those with a GNI per capita, calculated using the World Bank Atlas method, of US$1005 or less in 2016.

Lower middle-income economies: are those countries with a GNI per capita between US$1006 and US$3955.

M

Macronutrients: the proteins, carbohydrates and fats that are available to be used for energy. They are measured in grams.

Malnutrition: an abnormal physiological condition caused by inadequate, unbalanced, or excessive consumption of macronutrients and/or micronutrients. Malnutrition includes undernutrition and overnutrition as well as micronutrient deficiencies.

Micronutrients: vitamins, minerals, and certain other substances that are required by the body in small amounts. They are measured in milligrams or micrograms.

Mammalian: any warm-blooded vertebrate having a skin more or less covered with hair; the young are born alive, except for the small subclass of monotremes, and nourished with milk.

Market intervention: the act, usually by a government or central bank, of buying or selling in a market in order to influence the price.

Marshall Plan: formally called the European Recovery Program, (April 1948–December 1951), it was US-sponsored programme designed to rehabilitate the economies of 17 Western and Southern European countries in order to create stable conditions in which democratic institutions could survive. The USA feared that the poverty, unemployment, and dislocation of the post-World War II period were reinforcing the appeal of communist parties to voters in western Europe. On June 5, 1947, in an address at Harvard University, Secretary of State George C. Marshall advanced the idea of a European self-help programme to be financed by the USA. On the basis of a unified plan for Western European economic reconstruction presented by a committee representing 16 countries, the US Congress authorized the establishment of the European Recovery Program, which was signed into law by President Harry S. Truman on April 3, 1948. Aid was originally offered to almost all the European countries, including those under military occupation by the Soviet Union. The Soviets early on withdrew from participation in the plan, however, and were soon followed by the other eastern European nations under their influence. This left the following countries to participate in the plan: Austria, Belgium, Denmark, France, Greece, Iceland, Ireland, Italy, Luxembourg, the Netherlands, Norway, Portugal, Sweden, Switzerland, Turkey, the UK, and West Germany.

Marker: an identifiable DNA sequence that is inherited in Mendelian fashion, and which facilitates the study of the inheritance of a trait or a linked gene.

Market demand: the aggregate of the demands of all potential customers (market participants) for a specific product over a specific period in a specific market.

Market liberalization: the loosening of government controls. It refers to reductions in restrictions on international trade and capital.

Marker-assisted selection: the use of DNA markers to improve response to selection in a population. The markers will be closely linked to one or more target loci, which may often be quantitative trait loci.

Marketing margin: the difference between prices at different levels of the marketing chain, such as between the price paid by a consumer and that received by a farmer. Margins can be calculated all along the marketing chain. Each margin reflects the value added at that level of the chain.

Mass production: application of the principles of specialization, division of labour, and standardization of parts to the manufacture of goods. Such manufacturing processes attain high rates of output at low unit costs, with lower costs expected as volume rises.

MDGs: an abbreviation for the Millennium Development Goals. The MDGs are the world's time-bound and quantified targets for addressing extreme poverty in its many dimensions – income poverty, hunger, disease, lack of adequate shelter, and exclusion – while promoting gender equality, education, and environmental sustainability. They are also the basic human rights of each person on the planet to health, education, shelter and security.

Membrane cell: the lipid bilayer and associated proteins and other molecules that surrounding the protoplast, within the cell wall. Synonyms: plasmalemma; plasma membrane.

Mendel's Law: the two laws summarizing Gregor Mendel's theory of inheritance. The Law of Segregation states that each hereditary characteristic is controlled by two 'factors' (now called alleles), which segregate and pass into separate germ cells. The Law of Independent Assortment states that pairs of 'factors' segregate independently of each other when germ cells are formed.

Mercantilism: an economic system developing during the decay of feudalism to unify and increase the power and especially the monetary wealth of a nation by a strict governmental regulation of the entire national economy, usually through policies designed to secure an accumulation of bullion, a favourable balance of trade, the development of agriculture and manufacturing, and the establishment of foreign trading monopolies.

Metabolic disease: a disease caused by some defect in the chemical reactions of the cells of the body.

Metabolite: a low-molecular-weight biological compound that is usually synthesized enzymatically.

Metabolomics: the large-scale study of small molecules, commonly known as metabolites, within cells, biofluids, tissues or organisms.

MFN: an abbreviation for Most Favoured Nation. Treating other people equally under the WTO agreements, countries cannot normally discriminate between their trading partners.

Microbiology: the branch of biology dealing with the structure, function, uses, and modes of existence of microscopic organisms.

Minimum dietary energy requirement: in a specified age/sex category, the minimum amount of dietary energy per person that is considered adequate to meet the energy needs at a minimum acceptable body mass index (BMI) of an individual engaged in low physical activity. It is expressed as kilocalories per person per day.

Minimum guaranteed prices: a form of market intervention by the government to insure agricultural producers against any sharp fall in farm *prices*.

Mitochondria: compartments or organelles in the cell that are the cell's main energy source and often are called the powerhouse of the cell. The mitochondria also contain their own DNA and therefore genes. Mitochondrial genes follow maternal inheritance.

Modern biotechnology: the application of: (i) *in vitro* nucleic acid techniques, including recombinant DNA and direct injection of nucleic acid into cells or organelles; or (ii) fusion of cells beyond the taxonomic family, that overcome natural physiological reproductive or recombination barriers, and that are not techniques used in traditional breeding and selection.

Molecular biology: the study of living processes at the molecular level.

Molecular genetics: the study of the expression, regulation and inheritance of genes at the level of DNA and its transcription products.

Molecule: the stable union of two or more atoms; some organic molecules contain very large numbers of atoms.

Monetary policy: measures employed by governments to influence economic activity, specifically by manipulating the supplies of money and credit and by altering rates of interest.

Monoclonal antibody: an antibody, produced by a hybridoma, directed against a single antigenic determinant of an antigen.

Monoculture: the repeated planting of the same crop in the same field year after year. The opposite is crop rotation, whereby a series of different crops are planted in the same field following a defined order (i.e. maize-cotton-sunn hemp, or maize-soyabeans). In monocultures, increases in crop-specific pests and diseases are often observed over time. Also, continuously growing the same crop will tend to exploit the same soil root zone, which can lead to a decrease in available nutrients for plant growth and a decrease in root development.

Monopoly: when the production of a good or service with no close substitutes is carried out by a single firm with the market power to decide the price of its output.

Morphology: shape, form, external structure or arrangement.

MPS: an acronym for Market Price Support. An indicator of the annual monetary value of gross transfers from consumers and taxpayers to agricultural producers arising from policy measures creating a gap between the domestic producer prices and reference prices of a specific agricultural commodity measured at the farm-gate level.

mRNA: an abbreviation for messenger RNA, the RNA molecule resulting from transcription of a protein-encoding gene, following any splicing. The information encoded in the mRNA molecule is translated into a gene product by the ribosomes.

Multifunctionality: a term generally used to indicate that agriculture can produce various non-commodity outputs in addition to food. Synonym: multifunctional agriculture.

Multinational corporation: an enterprise producing goods or delivering services in more than one country. Stated alternatively, a multinational enterprise has its management headquarters in one (or, rarely, more than one) country, the home country, while also operating in other countries, the host countries. Synonyms: multinational, multinational enterprise.

Mutation: any change in the genome with respect to a defined wild type. It can occur at the level of ploidy, karyotype, or nucleotide sequence. Most of the latter mutations are silent (i.e. cannot be associated with any change in phenotype), either because the DNA sequence affected is in the non-coding part of the genome, or because the specific change does not alter the function of a coding sequence.

N

NACs: an abbreviation for New Agricultural Countries, those middle-income Third World countries pursuing agro-industrialization and agro-export.

NATO: an abbreviation for the North Atlantic Treaty Organization, an alliance of 28 countries bordering the North Atlantic Ocean. It includes Canada, the USA, Turkey and most members of the EU.

Neocolonialism: the continuation of a colonial system in spite of the formal recognition of independence of a formerly colonized nation. After World War II, when the European colonial system collapsed, the international community recognized the right to self-determination of peoples living under colonial rule. This right was used to justify and promote the transition of former colonies to sovereign states, causing a rapid emergence of newly independent countries through the transfer of power from empire to nation-state. As a result, the total number of independent countries increased considerably from the late 1940s to the mid-1970s. Despite their newly recognized sovereignty, however, these decolonized countries became subject to de facto control by western powers.

Neoliberalism: ideology and policy model that emphasizes the value of free market competition.

New Deal: the domestic programme of the administration of US President Franklin D. Roosevelt between 1933 and 1939, which took action to bring about immediate economic relief as well as reforms in industry, agriculture, finance, waterpower, labour, and housing, vastly increasing the scope of the federal government's activities. Opposed to the traditional American political philosophy of laissez-faire, the New Deal generally embraced the concept of a government-regulated economy aimed at achieving a balance between conflicting economic interests.

New Development Bank: formerly referred to as the BRICS Development Bank; a multilateral development bank established by the BRICS states. Acronym: NDB.

New international division of labour: an outcome of globalization. The term was coined by theorists seeking to explain the spatial shift of manufacturing industries from advanced capitalist countries to developing countries, with an ongoing geographic reorganization of production. It is a spatial division of labour which occurs when the process of production is no longer confined to national economies. Under the 'old' international division of labour, until around 1970, underdeveloped areas were incorporated into the world economy principally as suppliers of minerals and agricultural commodities. However, as developing economies are merged into the world economy, more production takes place in these economies.

NICs: abbreviation for Newly Industrializing Countries, sometimes also called newly industrializing economies. It represents a group of states in the developing world that developed substantial manufacturing bases in the late 20th century. Definitions of this group vary, but they are generally taken to be states that have exhibited rapid and sustained increases in GDP, incomes, and industrial employment. Most are located in Asia, although several are also found in Latin America. Some definitions of NICs include South Africa and Turkey.

NGOs: acronym for Non-Governmental Organizations.

Nomad: a member of a people or tribe that has no permanent abode but moves about from place to place, usually seasonally and often following a traditional route or circuit according to the state of the pasturage or food supply.

Nominal prices: prices charged by providers of general government services such as health and education, and prices that are heavily subsidized through government funding or regulated by government policy. Such prices are not economically significant and therefore do not provide signals of market-driven inflation.

Non-communicable disease (NCD): a medical condition or disease that is not caused by infectious agents (non-infectious or non-transmissible). The main types of NCDs are cardiovascular diseases (like heart attacks and stroke), cancers, chronic respiratory diseases (such as chronic obstructive pulmonary disease and asthma) and diabetes. NCDs disproportionately affect people in low- and middle-income countries where more than three quarters of global NCD deaths – 31 million – occur.

Non-product specific subsidy: a subsidy which refers to the total level of support given to the agricultural sector as a whole, i.e. subsidies on inputs such as fertilizers, electricity, irrigation, seeds, credit, etc. Usually, these non-product subsidies are given for all crops.

NPK fertilizers: three-component fertilizers providing nitrogen (N), phosphorus (P), and potassium (K). Nitrogen fertilizers are made from ammonia (NH_3), which is sometimes injected into the ground directly. All phosphate fertilizers are obtained by extraction from minerals containing the anion PO_4^{3-}. Potash is a mixture of potassium minerals used to make potassium fertilizers. These three main macronutrients enhance the growth of plants, specifically: leaf growth (N); development of roots, flowers, seeds, fruit (P); and strong stem growth, movement of water in plants, promotion of flowering and fruiting (K).

Nuclein: the term used by Friedrich Miescher to describe the nuclear material he discovered in 1869, which today is known as DNA.

Nucleic acid: a macromolecule consisting of polymerized nucleotides. Two forms are found, DNA and RNA. Nucleic acids may be linear or circularized, and single- or double-stranded.

Nucleus: the dense protoplasmic membrane-bound region of a eukaryotic cell that contains the chromosomes, separated from the cytoplasm by a membrane; present in all eukaryotic cells except mature sieve-tube elements and red blood cells.

Nutrient: any substance that can be metabolized by an organism to give energy and build tissue.

O

OECD: abbreviation for the Organization for Economic Cooperation and Development, an intergovernmental economic organization with 35 member countries, founded in 1960 to stimulate economic progress and world trade. Most OECD members are high-income economies with a very high Human Development Index (HDI), and are regarded as developed countries.

Offspring: new individuals resulting from sexual or asexual reproduction. Synonym: progeny.

Oil-bearing crops (permanent only): perennial plants whose seeds (kapok), fruits, mesocarp (olives), or nuts (coconuts) are used mainly for extraction of culinary or industrial oils and fats. Consequently, dessert or table nuts, such as walnuts, are excluded because, although they are high in oil content, they are not used mainly for extraction of oil.

Oil-bearing crops (temporary only): annual plants whose seeds are used mainly for extraction of culinary and industrial oils,

but excluding essential oils. Temporary oil-bearing crops are usually called oilseeds.

OPEC: the acronym for the Organization of the Petroleum Exporting Countries, a permanent, intergovernmental organization, created at the Baghdad Conference on September 10–14, 1960, by Iran, Iraq, Kuwait, Saudi Arabia, and Venezuela. OPEC's objectives are to co-ordinate and unify petroleum policies among Member Countries, in order to secure fair and stable prices for petroleum producers; to provide an efficient, economic and regular supply of petroleum to consuming nations; and to secure a fair return on capital to those investing in the industry.

Organelle: a membrane-bounded specialized region within a cell, such as the mitochondrion or dictyosome, which carries out a specialized function in the life of a cell.

Organic farming: an overall system of farm management and food production that combines best environmental practices, a high level of biodiversity, the preservation of natural resources, the application of high animal welfare standards and production methods in line with the preference of certain consumers for products produced using natural substances and processes.

P

Paleobiology: the branch of paleontology dealing with fossil life forms, especially with reference to their origin, structure, evolution, etc.

Paleontology: the scientific study of life in the geologic past, especially through the study of animal and plant fossils.

Pandemic: an epidemic occurring worldwide, or over a very wide area, crossing international boundaries and usually affecting a large number of people.

Papaya ringspot virus (PRSV): an aphid-transmitted plant virus belonging to the genus *Potyvirus*, family *Potyviridae*. Papaya exhibits yellowing, leaf distortion, and severe mosaic. Oily or water-soaked spots and streaks appear on the trunk and petioles. The fruit will exhibit bumps and the classic 'ringspot.' A severe isolate of PRSV has also been shown to cause tissue necrosis. In Hawaii, PRSV has had dramatic effects: between 1992 and 1997 nearly all fields in the Puna region had been affected.

Patent: the exclusive right granted by a government to an inventor to manufacture, use, or sell an invention for a certain number of years.

Pathogen: a disease-causing organism (generally microbial: bacteria, fungi, viruses; but can be extended to other organisms: e.g. nematodes, etc.). Synonym: infectious agent.

Pathogen-free: uncontaminated with a pathogen.

Per capita food production variability: an indicator of food stability. This is an expression of the net food production variability per capita expressed in international dollars.

Per capita food supply variability: an indicator of food stability. It is expressed in calories/capita/day.

Permanent arable land: the total of 'arable land' and 'land under permanent crops'. For arable land, see above; land under permanent crops is the land cultivated with crops that occupy the land for long periods and need not be replanted after each harvest.

Pest management: a means to reduce pest numbers to an acceptable threshold. An acceptable threshold, in most cases, refers to an economically justifiable threshold where application of pest control measures reduces pest numbers to a level below which additional applications would not be profitable (i.e., where additional costs of control exceed additional benefits). Pest eradication (i.e., complete removal) is usually not a viable option.

Pesticide: a toxic chemical product that kills harmful organisms (e.g. insecticides, fungicides, herbicides, rodenticides, bactericides).

Pesticide use efficiency: yield per unit input. Pesticides are considered as being efficient when maximum yield is obtained with minimum possible amount of pesticide application.

Phenotype: the visible appearance of an individual (with respect to one or more traits) which reflects the reaction of a given genotype with a given environment.

Plant pests: any insect, mite, nematode, other invertebrate animal, disease, noxious weed, plant or animal parasite in any stage of development which is injurious to plants and plant products.

Plasmid: a circular self-replicating non-chromosomal DNA molecule found in many bacteria, capable of transfer between bacterial cells of the same species, and occasionally of different species. Antibiotic resistant genes are frequently located on plasmids. Plasmids are particularly important as vectors for genetic engineering.

Plum pox virus (PPV): a cause of sharka, the most devastating viral disease of stone fruit of the genus *Prunus*.

Pluralistic society: a diverse society, where the people in it believe all kinds of different things and tolerate each other's beliefs even when they don't match their own. Pluralism serves as a model of democracy, where different groups can voice their opinions and ideas.

Plutonomy: an economy whose growth and health are heavily influenced by the activities of a wealthy minority. The US economy was described as a plutonomy between 2003 and 2005, as growth continued despite high interest rates and a high national debt.

Polymerase: an enzyme that synthesizes long chains or polymers of nucleic acids.

Polymerase chain reaction (PCR): a technique used to reproduce (amplify) selected sections of DNA or RNA for analysis. Previously, amplification of DNA involved cloning the segments of interest into vectors for expression in bacteria, and took weeks. But now, with PCR conducted in test tubes, it takes only a few hours.

Population: a defined group of interbreeding organisms.

Population density: number of cells or individuals per unit.

Poverty headcount ratio: the percentage of the population living below the defined poverty lines.

PPP: an abbreviation for Power Purchasing Parity. The concept that compares different countries' currencies through a 'basket of goods' market approach. Two currencies are in equilibrium or at par when a market basket of goods (taking into account the exchange rate) is priced the same in both countries.

Precautionary principle: the concept that states if an action or policy has a suspected risk of causing severe harm to the public domain (affecting general health or the environment globally), the action should not be taken in the absence of scientific near-certainty about its safety. Under these conditions, the burden of proof about absence of harm falls on those proposing an action, not those opposing it. The principle is intended to deal with uncertainty and risk in cases where the absence of evidence and the incompleteness of scientific knowledge carries profound implications, and where the presence of the risk of 'a black swan' threatens unforeseen and unforeseeable events of extreme consequence.

Price elasticity in demand: the percentage change in quantity demanded divided by the percentage change in price. Since the demand curve is normally downward sloping, the price elasticity of demand is usually a negative number. However, the negative sign is often omitted.

Price index: an average of the proportionate changes in the prices of a specified set of goods and services between two periods of time.

Price volatility: also known as market volatility, it refers to the fluctuation of prices of agricultural goods on the market.

Primary commodities: food and live animals, beverages and tobacco excluding manufactured goods; crude materials, inedibles, excluding fuels, synthetic fibres, waste, and scrap; mineral fuels, lubricants, and related materials, excluding petroleum products; animal and vegetable oils, fats, and waxes.

Primary crops: crops which come directly from the land without having undergone any real processing, apart from cleaning.

Primary metabolite: a kind of metabolite that is directly involved in normal growth, development, and reproduction. It usually performs a physiological function in the organism (such as certain amino acids).

Producer's price: the amount receivable by the producer from the purchaser for a unit of a good or service produced as output minus any VAT, or similar deductible tax, invoiced to the purchaser; it excludes any transport charges invoiced separately by the producer.

Product-specific subsidies: subsidies which refer to the total level of support provided for each individual agricultural commodity.

Protectionism: the policy of protecting domestic industries against foreign competition by means of tariffs, subsidies, import quotas, or other restrictions or handicaps placed on the imports of foreign competitors.

Protein: a macromolecule composed of one or more polypeptides, each comprising a chain of amino acids linked by peptide bonds.

Proteomics: an approach that seeks to identify and characterize complete sets of proteins, and protein–protein interactions, in a given species.

Protoplast fusion: the induced or spontaneous coalescence of two or more protoplasts of the same or different species origin. Where fused protoplasts can be regenerated into whole plants, the opportunity exists for the creation of novel genomic combinations.

PSE: an abbreviation for the Producer Support Estimate. PSE transfers to agricultural producers are measured at farm-gate level and comprise market price support, budgetary payments and the cost of revenue foregone.

Purchasing power: the financial ability to buy goods and services. It depends on income and prices.

Q

Quota: in international trade, a government-imposed limit on the quantity, or in exceptional cases the value, of goods or services that may be exported or imported over a specified period of time.

R

Radiation genetics: a subdiscipline of genetics that studies radiation effects on the components and processes of biological inheritance.

Recessive: used to describe an allele whose effect with respect to a particular trait is not evident in heterozygotes. Opposite: dominant.

Recombinant: a term used in both classical and molecular genetics. (i) in classical genetics: An organism or cell that is the result of meiotic recombination; (ii) in molecular genetics: A hybrid molecule made up of DNA obtained from different organisms. Typically used as an adjective, e.g. recombinant DNA.

Recombinant DNA: the result of combining DNA fragments from different sources.

Recombinant DNA technology: a set of techniques for manipulating DNA, including: (i) the identification and cloning of genes; (ii) the study of the expression of cloned genes; and (iii) the production of large quantities of gene product.

Regulationism: the theory that discusses historical changes in the political economy through two central concepts, regime of accumulation and mode of regulation. The concept of the regime of accumulation allows theorists to analyse the ways of production; and circulation, consumption, and distribution organize and expand capital in a way that permits the stabilization and regulation of the economy over time.

Replication: the *in vivo* synthesis of double-stranded DNA by copying from a single-stranded template.

Resistance: the ability to withstand abiotic (high temperature, drought, etc.) or biotic (disease) stress, or a toxic substance; often in the context of genetic determination of resistance.

Restriction enzyme: a class of enzymes that cut DNA after recognizing a specific sequence.

Retail revolution: rapid rise of supermarkets around the globe, increasing spread of retail chains throughout the world, especially into developing and transition economies.

Retinol: the predominant circulating form of vitamin A in the blood.

Ribosome: the sub-cellular structure that contains both RNA and protein molecules and is the site for the translation of mRNA into protein. Ribosomes comprise large and small sub-units.

RNA: abbreviation for ribonucleic acid, an organic acid polymer composed of adenosine, guanosine, cytidine and uridine ribonucleotides. It is the genetic material of some viruses, but more generally it is the molecule, derived from DNA by transcription, that either carries information (messenger RNA), provides sub-cellular structure (ribosomal RNA), transports amino acids (transfer RNA), or facilitates the biochemical modification of itself or other RNA molecules.

Rural population: refers to people living in rural areas as defined by national statistical offices. It is calculated as the difference between total population and urban population. The most used variable for defining 'rural' is population density: a territory is rural if population density is below 150 inhabitants per square kilometre.

S

Sanitary: of, or relating to, health or the conditions affecting health, especially with reference to cleanliness, precautions against disease, etc.

Sanitary and phytosanitary measures: measures dealing with food safety and animal and plant health.

Secondary metabolite: product of secondary metabolism.

Sedentary: abiding in one place; not migratory.

Seed dispersal: the movement or transport of seeds away from the parent plant.

Seed dormancy: an innate seed property that defines the environmental conditions in which the seed is able to germinate. Dormancy is a mechanism to prevent germination during unsuitable ecological conditions, when the probability of seedling survival is low.

Segregation: for genes, the separation of allele pairs from one another and their resulting assortment into different cells at meiosis; for chromosomes, the separation and re-assortment of the two homologues in anaphase of the first meiotic division; for individuals, the occurrence of different genotypes and/or phenotypes among offspring, resulting from chromosome or allele separation in their heterozygous parents.

Selection: (i) differential survival and reproduction of phenotypes; or (ii) a system for either isolating or identifying specific genotypes in a mixed population.

Semi-sedentary: sedentary during part of the year and nomadic otherwise.

Settler-states: societies founded by migrant groups who assume a superordinate position vis-à-vis the native inhabitants and build self-sustaining states that are de jure or de facto independent from their mother countries and organized around the settlers' political domination over the indigenous population.

Share of food expenditures of the poor: an indicator of access to food expressed as the share of the population from the lowest quintile (population with the lowest incomes) in total expenditures.

Singular trait: characteristic or attribute of an organism that is expressed by gene (one gene of interest is inserted into a single plant).

Social movements: sustained, organized forms of collective action that strive to bring about social or political change. Typically originating in civil society, they seek to change the policies and actions of the state, the practices of corporations, or the organization of society generally. Like all social relations, social movements are spatially constituted. Social movements mobilize and operate in a spatial context, are themselves spatially constituted by actors both internal and external to social movement organizations, and seek to alter sociospatial power relations.

Somatic: referring to cell types, structures and processes other than those associated with the germ line.

Somatic hybridization: (i) a sexual fusion of protoplasts from somatic cells of genetically different parents; or (ii) hybridization by induced fusion of cells (protoplasts) from two contrasting genotypes for production of hybrids or cybrids which contain various mixtures of nuclear and/or cytoplasmic genomes, respectively.

Species: a class of individuals capable of interbreeding, but which is reproductively isolated from other such groups having many characteristics in common. It is a somewhat arbitrary and sometimes blurred classification, but still quite useful in many situations.

SPS: an abbreviation for Agreement on the Application of Sanitary and Phytosanitary Measures, which entered into force with the establishment of the WTO on 1 January 1995. It concerns the application of food safety and animal and plant health regulations.

Stockpiling: the storage of something in order to have it available in the future if the need for it increases.

Stacked trait: the combined traits resulting from the gene-stacking process. An example of a stack is a plant transformed with two or more genes that code for Bt proteins having different modes of action. It is a plant expressing both insect resistant and herbicide tolerant genes derived from two parent plants.

StarLink corn: a plant modified with a gene that encodes the Bt protein Cry9c, which was a severe test of US regulatory agencies. The US Environmental Protection Agency had restricted its use to animal feed due to concern about its potential allergenicity. However, StarLink maize was later found throughout the human food supply, resulting in food recalls by the FDA and significant disruptions of the supply.

Structural Adjustment Programme: a World Bank instrument prevalent in the 1980s that focused on correcting the major macroeconomic distortions hindering development.

Stunting: low height for age; reflecting a past episode or episodes of sustained undernutrition.

Sugar crops: those crops cultivated primarily for the manufacture of sugar, secondarily for the production of alcohol (food and non-food) and ethanol. There are two main sugar crops: sugar beets and sugar cane.

Subsidy: a direct or indirect payment, economic concession, or privilege granted by a government to a private firm, household, or other governmental unit0+ in order to promote a public objective.

Subsidy boxes: a WTO term for subsidies, these 'boxes' are identified by the colours of traffic lights: green (permitted), amber (slowing down, i.e. being reduced), red (forbidden). The Agriculture Agreement, notable for its complexity, has no red box; however, although domestic support exceeding the reduction commitment levels in the amber box is prohibited; and there is a blue box for subsidies that are tied to programmes that limit production.

Substantial equivalence: the concept that the safety of a new food, particularly one that has been transgenetically produced, may be assessed by comparing it with a similar traditional food that has proven safe in normal use over time. It implies that the GMO does not inherently introduce additional risk, and therefore, transgenic products may be assessed in the same way as conventionally bred products. The term 'substantial equivalence' was formulated as a food safety policy by the OECD, and first described in their 1993 report, 'Safety Evaluation of Foods Derived by Modern Biotechnology: Concepts and Principles.' The term substantial equivalence and the underlying approach were borrowed from the US Food and Drug Administration's (FDA) definition of a class of new medical devices that do not differ materially from their predecessors and thus do not raise new regulatory concerns.

Subspecies: population(s) of organisms sharing certain characteristics that are not present in other populations of the same species.

Supply and demand: the relationship between the quantity of a commodity that producers wish to sell at various prices and the quantity that consumers wish to buy.

Supranational organization: an international organization in which member states transcend national boundaries or interests to share in the decision making and vote on issues pertaining to the wider grouping. The EU and the WTO are both supranationals. In the EU, each member votes on policies that will affect each member nation. The benefits of this construct are the synergies derived from social and economic policies,

along with a stronger presence on the international stage.

Systematics: the science of the classification of organisms; the arrangement of organisms into systematic groups, such as species, genus, family, and order. Synonym: taxonomy.

Systems biology: the computational and mathematical modelling of complex biological systems based on the understanding that the whole is greater than the sum of the parts.

T

Target income: the profit that is expected to be attained for a designated accounting period.

Target price: a price that is not guaranteed, but rather, serves as a policy guideline. In the EU, a target price is fixed annually by the Council of Ministers for products of standard quality. In Switzerland, a non-binding target price is set annually for milk to provide market guidance.

Tariff: the term has two meanings. Firstly, it means the list, book or database of charges that are imposed by a government on goods when these are imported or exported. Secondly it means the charge itself. In its second meaning, the term 'tariff' is synonymous with the term 'customs duty'.

Technology: the application of scientific knowledge to practical purposes in any field. It includes methods, techniques, and instrumentation.

Technical barriers to trade: technical regulations, minimum standards and certification systems for health, safety and environmental protection, and to enhance the availability of information about products, which may result in the erection of technical barriers to trade.

Third World: former political designation originally used (1963) to describe those states not part of the First World, i.e. the capitalist, economically developed states led by the USA, or the Second World, the communist states led by the Soviet Union. When the term was introduced, the Third World principally consisted of the developing world, the former colonies of Africa, Asia, and Latin America. With the end of the Cold War and the increased economic competitiveness of some developing countries, the term lost its analytic clarity.

Tissue: a group of cells of similar structure which sometimes performs a special function.

Tissue culture: the *in vitro* culture of cells, tissues or organs in a nutrient medium under sterile conditions.

Totipotency: the ability of a cell or tissue to be induced to regenerate into a complete organism.

Traceability: the ability for any food, feed, food-producing animal or substance that will be used for consumption to be tracked, through all stages of production, processing and distribution. Traceability should mean that movements can be traced one step backwards and one step forward at any point in the supply chain.

Trade distortive measures: any measures that cause distortion of trade, if prices are higher or lower than normal, and if the quantities produced, bought, and sold are also higher or lower than normal, i.e. than the levels that would usually exist in a competitive market. For example, import barriers and domestic subsidies can make crops more expensive in a country's internal market. The higher prices can encourage overproduction. If the surplus is to be sold on world markets, where prices are lower, then export subsidies are needed. As a result, the subsidizing countries can be producing and exporting considerably more than they normally would.

Traditional breeding: modification of plants and animals through selective breeding. Practices used in traditional plant breeding can include aspects of biotechnology such as tissue culture and mutation breeding.

Trait: one of the many characteristics that define an organism. The phenotype is a description of one or more traits. Synonym: character.

Transcription: synthesis of RNA from a DNA template via RNA polymerase.

Transcriptomics: transcriptomics is the study of the transcriptome – the complete set of RNA transcripts that are produced by the genome, under specific circumstances or in a specific cell, using high-throughput methods, such as microarray analysis. Comparison of transcriptomes allows the identification of genes that are differentially expressed in distinct cell populations, or in response to different treatments.

Transduction: (i) genetic: the transfer by means of a viral vector of a DNA sequence from one cell to another; or (ii) signal: any process that helps to produce biological responses to events in the environment (e.g. transduction of hormone binding into cellular events by hormone receptors).

Transgene: an isolated gene sequence used to transform an organism. Often, but not always, the transgene has been derived from a different species than that of the recipient.

Transgenesis: the introduction of a gene or genes into animal or plant cells, which leads to the transmission of the input gene (transgene) to successive generations.

Transgenic foods: foods derived from organisms whose genetic material (DNA) has been modified in a way that does not occur naturally, e.g. through the introduction of a gene from a different organism.

Transgenic plant: an individual plant in which a transgene has been integrated into its genome. In transgenic eukaryotes, the transgene must be transmitted through meiosis to allow its inheritance by the offspring.

Trangenic organism: a plant, animal, or other organism with different traits from the parent organism, resulting from the use of recombinant DNA techniques to insert genetic material from another organism.

Transition (economy): an economy which is changing from a centrally planned economy to a market economy. Transition economies undergo a set of structural transformations intended to develop market-based institutions.

Transnational corporation: any corporation that is registered and operates in more than one country at a time.

Transposon: a DNA element that can move from one location in the genome to another. Synonym: transposable genetic element.

TRIPS: acronym for the Agreement on Trade-Related Aspects of Intellectual Property Rights; an international legal agreement between the member nations of the WTO, which sets down minimum standards for the regulation by national governments of many forms of intellectual property as applied to nationals of other WTO member nations.

TSE: acronym for the Total Support Estimate. TSE transfers represent the total support granted to the agricultural sector, and consist of the PSE, CSE, and GSSE.

U

Ultra-processed food products: food products obtained through processes such as salting, sugaring, baking, frying, deep frying, curing, smoking, pickling, canning, and also, frequently, the use of preservatives and cosmetic additives, the addition of synthetic vitamins and minerals, and sophisticated types of packaging. These industrial processes are all designed to create durable, accessible, convenient, attractive ready-to-eat or ready-to-heat products. Many of them are 'fast' foods or convenience foods. They are formulated to reduce microbial deterioration ('long shelf-life'), to be transportable for long distances, to be extremely palatable ('high organoleptic quality') and often to be habit-forming. Typically they are designed to be consumed anywhere – in fast-food establishments, at home in place of dishes and meals prepared from scratch, while watching television, at desks or elsewhere at work, in the street, and while driving. Ultra-processed products can be sub-divided into: (i) ready-to-eat snacks or products likely to be consumed as snacks or desserts; and (ii) pre-prepared ready-to-heat products created to replace home-prepared dishes and meals. Their processing is usually undertaken by food manufacturers, or else by caterers (such as burger outlets), or food retailers (such as bakeries), for sale to consumers.

UN: acronym for the United Nations, an international organization founded in 1945 following the devastation of World War II, with one central mission: the maintenance of international peace and security.

Undernourishment: a state, lasting for at least 1 year, of inability to acquire enough food, defined as a level of food intake insufficient to meet dietary energy requirements.

Undernutrition: the outcome of undernourishment and/or poor absorption and/or poor biological use of nutrients consumed and/or as a result of repeated infectious disease. It includes being underweight for one's age, too short for one's age (stunted), dangerously thin for one's height (wasted), and deficient in vitamins and minerals (micronutrient malnutrition).

Underweight: low weight for age in children, and a BMI of less than 18.5 in adults, reflecting a current condition resulting from inadequate food intake, past episodes of undernutrition or poor health conditions.

Upper middle-income countries: those economies with a GNI per capita of between US$4,036 and US$12,475.

UPOV: an abbreviation for the International Union for the Protection of Plant Varieties. This system of plant variety protection came into being with the adoption of the International Convention for the Protection of New Varieties of Plants at a Diplomatic Conference in Paris on December 2, 1961. This was the point at which there was recognition at an international level of the intellectual property rights of plant breeders, with respect to the varieties bred. The UPOV Convention came into force on August 10, 1968, having been ratified by the UK, the Netherlands and West Germany. The UPOV Convention was revised on November 10, 1972, on October 23, 1978, and on March 19, 1991, in order to reflect technological developments in plant breeding and the experience acquired with the application of the UPOV Convention.

Uruguay Round: 8th round of multilateral trade negotiations conducted within the framework of the GATT, spanning from 1986 to 1994, and embracing 123 countries as 'contracting parties'. The Uruguay Round led to the creation of the WTO, with the

GATT remaining as an integral part of the WTO agreements.

USDA: an abbreviation for the US Department of Agriculture, an executive division of the US federal government in charge of programmes and policies relating to the farming industry and the use of national forests and grasslands.

US National Science Federation: an independent US government agency responsible for promoting science and engineering through research programmes and education projects.

Utilized agricultural area: the total area taken up by arable land, permanent pasture and meadow, land used for permanent crops, and kitchen gardens.

V

Value-added products: (i) a product resulting from a change in its original physical state or form (such as wheat milled for flour or strawberries made into jam); (ii) a product produced in a manner that enhances its value, as demonstrated through a business plan (such as organically produced products); or (iii) an agricultural commodity or product physically segregated in a manner that results in the enhancement of the value of that commodity or product. As a result of the change in the physical state or the manner in which the agricultural commodity or product is produced and segregated, the customer base for the commodity or product is expanded, and a greater portion of revenue derived from the marketing, processing or physical segregation is made available to the producer of the commodity or product.

Value of food imports over total merchandise exports: an indicator of food stability. Provides a measure for vulnerability and captures the adequacy of foreign exchange reserves to pay for food imports, which has implications for national food security, depending on production and trade patterns.

Variety: (i) a naturally occurring subdivision of a species, with distinct morphological characters; or (ii) a defined strain of a crop plant, selected on the basis of phenotypic (sometimes genotypic) homogeneity.

Vector: (i) an organism, usually an insect that carries and transmits pathogens; or (ii) a small DNA molecule (plasmid, virus, bacteriophage, artificial or cut DNA molecule) that can be used to deliver DNA into a cell. Vectors must be capable of being replicated and contain cloning sites for the introduction of foreign DNA.

Vertical integration: a company's control over several or all of the production and/or distribution steps involved in the creation of its product or service. In microeconomics and management, vertical integration is an arrangement by which the supply chain of a company is owned by that company. Usually each member of the supply chain produces a different product or (market-specific) service, and the products combine to satisfy a common need.

Virgin soil: soil which has never been cultivated.

Virus: an infectious particle composed of a protein capsule and a nucleic acid core (DNA or RNA), which is dependent on a host organism for replication.

Volatility: the directionless variability of an economic variable, i.e. the dispersion of that variable within a given time horizon. For example, high (/low) price volatility is described by situations when prices fluctuate significantly (/little) over a short time period in either direction.

W

Wheat Agreement: the first international agreement related to wheat. It was signed by 46 nations in 1949, under the leadership of the USA, the world's biggest wheat exporter. A big world surplus was keeping wheat prices low, and it seemed both good international policy and smart business to set fixed prices for world wheat sales. Roughly, the agreement protected importing nations by giving them the right to buy fixed quotas of wheat at a ceiling price of US$1.80 a bushel. Exporters were protected by a floor of US$1.50 a bushel (later reduced to US$1.20).

WFC: an abbreviation for the World Food Council, a UN organization established by the UN General Assembly in December 1974 at the recommendation of the World Food Conference. The WFC's goal was to serve as the coordinating body for national ministries of agriculture to help reduce malnutrition and hunger. The WFC was officially suspended in 1993, one of very few (if not the only) UN organizations to have been suspended. The WFC's functions were absorbed by the FAO and the World Food Programme.

WHO: an abbreviation for World Health Organization. The WHO began when its constitution came into force on 7 April 1948; its primary role is to direct and coordinate international health within the UN system.

World Bank Group: a group made up of five organizations: the International Bank for Reconstruction and Development (IRBD); the International Development Association (IDA); the International Finance Corporation (IDC); the Multilateral Investment Guarantee Agency (MIGA); and the International Centre for Settlement of Investment Disputes (ICSID). Established in 1944 at a conference of world leaders in Bretton Woods, New Hampshire, the World Bank is a lending institution whose aim is to help integrate developing and transition economies within the global economy, and to reduce poverty by promoting economic growth. The Bank lends for policy reforms and development projects and policy advice, and offers technical assistance and non-lending services to its 181 member countries.

World Food Summit: a world meeting convened by the FAO in order to examine the progress made in eradicating hunger. The first food summit, the 'World Food Conference', took place in Rome in 1974.

World price: the reference (border) price is the import (CIF) or export (FOB) price of a commodity used for calculating the market price support price gap, measured at the farm-gate level. An implicit border price may be calculated as, for example, the unit value of imports or exports.

WTO: acronym for the World Trade Organization. It deals with the global rules of trade between nations. Its main function is to ensure that trade flows as smoothly, predictably and freely as possible. As of July 2016, the organization has 164 members.

World-systems theory: as one of the main models of how the global system of states and markets operates, world-systems theory, introduced by the sociologist Immanuel Wallerstein in the 1970s, provides an enormously influential perspective on the changing structure and dynamics of the world economy. In many respects, it retains a fundamentally Marxist version of the world, one that puts production, class, uneven development, and historical context at the centre. The focus of world-systems theory is on the entire world rather than on individual nation-states. This view maintains that one cannot study the internal dynamics of countries without also examining their external ones; thus, the boundary between foreign and domestic effectively disappears. From this perspective, all regions are interconnected; that is, they never exist in isolation.

Y

Yeast: a unicellular ascomycete fungus, commonly found as a contaminant in plant tissue culture.

Yield: the amount of an agricultural crop, such as a grain, fruit, or vegetable, produced in a season. It can be measured in pounds or bushels per acre, or kilograms or metric tonnes per ha.

Suggested Reading

Business Dictionary (2017) Business Dictionary. Available at: http://www.businessdictionary.com/ (accessed 4 June 2018).

Dictionary (2016) Available at: http://www.dictionary.com/ (accessed 4 June 2018)

EC (2017) Glossary of terms related to the Common Agricultural Policy. Available at: https://definedterm.com/a/document/11266 (accessed 10 June 2018).

Encyclopedia (2017) Available at: www.encyclopedia.com/ (accessed 4 June 2018).

Encyclopedia Britannica (2017) Available at: https://www.britannica.com (accessed 4 June 2018)

Encyclopedia of Geography (2017) Available at: http://sk.sagepub.com/reference/geography (accessed 4 June 2018).

Eurostat (2016) Statistics Explained – Thematic Glossaries. Available at: http://ec.europa.eu/eurostat/statistics-explained/index.php/Thematic_glossaries (accessed 4 June 2018).

FAO Statistics (2011) Crops Statistics: Concepts, Definitions and Classifications. Available at: http://www.fao.org/economic/the-statistics-division-ess/methodology/methodology-systems/crops-statistics-concepts-definitions-and-classifications/en/ (accessed 4 June 2018).

FAO (2014) What is Family Farming? Available at: http://www.fao.org/family-farming-2014/home/what-is-family-farming/en/ (accessed 4 June 2018).

FAO (2015) The State of Food Insecurity in the World. FAO, Rome, Italy.

FAO (2017) Benefits and Risks of the Use of Herbicide-resistant Crops. Available at: http://www.fao.org/docrep/006/Y5031E/y5031e0i.htm (accessed 4 June 2018).

FAO (2017) Agricultural Intensification. Available at: http://www.fao.org/docrep/007/j0902e/j0902e03.htm (accessed 4 June 2018).

FDA (2010) Overview of Food Ingredients, Additives & Colors. Available at: https://www.fda.gov/food/ingredientspackaginglabeling/foodadditivesingredients/ucm094211.htm (accessed 4 June 2018).

Free Dictionary by Farlex (2016) Medical Dictionary. Available at: http://medical-dictionary.thefreedictionary.com/ (accessed 4 June 2018).

Investopedia (2017) Dictionary of Terms. Available at: https://www.investopedia.com/dictionary/ (accessed 4 June 2018).

ISAAA (2017) Herbicide Tolerance Technology: Glyphosate and Glufosinate. Available at: http://isaaa.org/resources/publications/pocketk/10/default.asp (accessed 4 June 2018).

Law Dictionary (2017) Black's Law Dictionary – Free Online Legal Dictionary 2nd edn. Available at: http://thelawdictionary.org/ (accessed 4 June 2018).

McLaughlin, D. (2016) Bayer, Monsanto Face Global Review as Farmer Options Shrink. Bloomberg, September 14. Available at: https://www.bloomberg.com/news/articles/2016-09-14/bayer-monsanto-confront-global-review-as-farmer-options-shrink (accessed June 10, 2017).

Merriam-Webster (2017) Merriam-Webster Dictionary. Available at: https://www.merriam-webster.com/dictionary/ (accessed 4 June 2018).

Monteiro, C.A, Levy, R.B., Claro, R.M., de Castro, I.R.R. and Cannon, G. (2011) Increasing consumption of ultra-processed foods and likely impact on human health: evidence from Brazil. Public Health Nutrition 14, 5–13.

OECD (2017) Glossary of Statistical Terms. Available at: https://stats.oecd.org/glossary/index.htm (accessed 4 June 2013).

Pica-Ciamarra, U. and Otte, J. (2010) *The Livestock Revolution: Rhetoric and Reality*. Available at: http://www.fao.org/3/a-bp263e.pdf (accessed 4 June 2018).

Purdue University (2017) Corn Rootworms. Available at: https://extension.entm.purdue.edu/fieldcropsipm/insects/corn-rootworms.php (accessed 4 June 2018).

Sudan Agricultural Information System (2017) Food Security Monitoring System. http://fsis.sd/SD/EN/FoodSecurity/Indicators/DEAI/10/ (accessed 12 June 2018).

Taleb, N.N., Read, R., Douady, R., Norman, J. and Bar-Yam, Y. (2014) The Precautionary Principle (with Application to the Genetic Modification of Organisms). Available at: http://www.fooledbyrandomness.com/pp2.pdf (accessed 4 June 2018).

The Economist (2016) Economics A–Z terms. Available at: http://www.economist.com/economics-a-to-z/a (accessed 4 June 2018).

ThoughtCo (2016) Hunter Gatherers – People Who Live on the Land. Available at: https://www.thoughtco.com/hunter-gatherers-people-live-on-land-171258 (accessed 4 June 2018).

University of California San Diego (2017) *Bacillus thuringiensis*. Available at: http://www.bt.ucsd.edu/index.html (accessed 4 June 2018).

University of Florida (2017) Featured Creatures: European corn borer. Available at: http://entnemdept.ufl.edu/creatures/field/e_corn_borer.htm (accessed 4 June 2018).

USDA (2016) Glossary of Agricultural Terms. Available at: https://agclass.nal.usda.gov/glossary.shtml (accessed 4 June 2018).

WFP (2009) World Hunger Series. Available at: http://www.wfp.org/content/world-hunger-series-hunger-and-markets (accessed 5 June 2018).

Zaid, A., Hughes, H.G., Porceddu, E. and Nicholas, F. (2001) *Glossary of Biotechnology for Food and Agriculture – A Revised and Augmented Edition of the Glossary Of Biotechnology and Genetic Engineering*. FAO, Rome, Italy.

Index

CABI – who we are and what we do

This book is published by **CABI**, an international not-for-profit organisation that improves people's lives worldwide by providing information and applying scientific expertise to solve problems in agriculture and the environment.

CABI is also a global publisher producing key scientific publications, including world renowned databases, as well as compendia, books, ebooks and full text electronic resources. We publish content in a wide range of subject areas including: agriculture and crop science / animal and veterinary sciences / ecology and conservation / environmental science / horticulture and plant sciences / human health, food science and nutrition / international development / leisure and tourism.

The profits from CABI's publishing activities enable us to work with farming communities around the world, supporting them as they battle with poor soil, invasive species and pests and diseases, to improve their livelihoods and help provide food for an ever growing population.

CABI is an international intergovernmental organisation, and we gratefully acknowledge the core financial support from our member countries (and lead agencies) including:

Discover more

To read more about CABI's work, please visit: **www.cabi.org**

Browse our books at: **www.cabi.org/bookshop**,
or explore our online products at: **www.cabi.org/publishing-products**

Interested in writing for CABI? Find our author guidelines here:
www.cabi.org/publishing-products/information-for-authors/